MICROGRAPHIA

or Some Physiological Descriptions of
Minute Bodies
Made by Magnifying Glasses
with Observations and
Inquiries thereupon

by Robert Hooke
With a Preface by Dr. R. T. Gunther

DOVER PUBLICATIONS, INC.
Mineola, New York

DOVER PHOENIX EDITIONS

Bibliographical Note

This Dover edition, first published in 2003, is an unabridged republication of the 1961 Dover reprint of the first edition, published by the Royal Society in 1665, and the Index from the 1745 and 1780 editions. The publisher is grateful to the University of Pennsylvania Library for making its copy of the first edition available for this facsimile reproduction.

This edition also contains a Preface and a Supplementary Index by Dr. R. T. Gunther which first appeared in Volume XIII of *Early Science in Oxford*, Oxford, 1938, and which are reprinted by permission of Mr. A. E. Gunther, F.G.S. Dr. Gunther, a Fellow of Magdalen College, was the first Curator of the Museum of the History of Science in Oxford.

Library of Congress Cataloging-in-Publication Data

Hooke, Robert, 1635–1703.
    Micrographia, or, Some physiological descriptions of minute bodies made by magnifying glasses, with observations and inquiries thereupon / by Robert Hooke ; with a preface by R.T. Gunther.
        p. cm. – (Dover phoenix editions)
    Originally published: London : Printed by J. Martyn and J. Allestry, 1665. With index from the 1745 and 1780 editions.
    ISBN 0-486-49564-7
    1. Microscopy–Early works to 1800. 2. Magnifying glasses–Early works to 1800. 3. Natural history–Pre-Linnean works. I. Title: Micrographia. II. Title: Some physiological descriptions of minute bodies made by magnifying glasses, with observations and inquiries thereupon. III. Title. IV. Series.

QH271.H79 2003
570'.28'2–dc21

2003047261

Manufactured in the United States of America
Dover Publications, Inc., 31 East 2nd Street, Mineola, N.Y. 11501

PREFACE

PEPYS on the 20th of January 1665 went to his booksellers, 'and there took home Hook's book of microscopy, a most excellent piece, and of which I am very proud'. And well he might be. It was the masterpiece of the English father of Microscopy. It bore the imprimatur of Lord Brouncker, the President of the Royal Society, and was dedicated to Charles II.

To-day owners of copies of the first edition of the *Micrographia* are also very proud. The piece has stood the test of time.

A second edition was called for two years after the printing of the first, and thirty-three of the plates were issued with abbreviated letterpress in 1745 and in 1780 under the title of *Micrographia Restaurata*.

In December last the International Organization for Intellectual Co-operation at Prague strongly advocated a reissue of the book in facsimile, an idea that already in 1930 had commended itself to me when I edited *The Life and Work of Robert Hooke*.

The discovery of the great power of the microscope as an engine of research was due to the workings of that widespread spirit of inquiry that prevailed in Oxford and in London during the two decades preceding the foundation of the Old Ashmolean Museum. To this spirit of curiosity, of endeavour to find out all about everything, and without any ulterior motive of benefit to mankind, must be attributed the bringing to perfection of the microscope, the most valuable scientific instrument that the world has ever seen.

The applications of this invention have now become so numerous and so far reaching that it has become impossible for numbers of our fellow men to lead a civilized and healthy life in crowded cities or tropical countries without the service of those who are skilled in its use. Modern medicine is unthinkable without the subsidiary micrographic sciences of histology, pathology, protozoology, and bacteriology, all of which are dependent upon the new worlds that have been discoverable only by this aid to vision. To the Physicist and Astronomer it has afforded a ready means of accuracy of measurement ; to the Chemist and Mineralogist a means of observing the early beginnings and forms of growth of the subjects of their study. Engineers rely upon its revelations for indication of the strength and durability of their constructions whether of wood or metal. By its means many a criminal, who might otherwise have escaped, has been brought to justice.

To ROBERT HOOKE alone must be ascribed the honour of having caused the great capability of the microscope to be realized in England, and that within a few months of the exhibition of his new model microscope at a meeting of the Royal Society. To this end Hooke worked hard in spare hours snatched from the routine of his busy life as Curator of Experiments for the Royal Society.

The minutes of the Society indicate the order in which his observations were made and the dates of their official recognition.

Thus on March 25th, 1663, Mr. Hooke was solicited by the Royal Society to prosecute his microscopical observations in order to publish them. A week later he was charged to bring in at every meeting one microscopical observation at least.

To this charge he replied on April 8th by delighting the company with a scheme of the appearance of Common Moss under the microscope (p. 131).

On April 13th he showed what is now regarded as an epoch-making discovery, a microscopical scheme, representing the Pores of Cork, cut both transverse and perpendicular (p. 115). He also showed drawings of Kettering-stone (p. 93) appearing to be composed of hollow globules, each having three coats sticking to one another, and so making up one entire firm stone (p. 93).

Subsequent demonstrations were made on the following dates:

April 22nd. Leeches in Vinegar (p. 216). Bluish Mould on Leather (p. 125).
April 29th. A Mine of Diamonds in Flint (p. 82). Spider with Six Eyes (p. 200).
May 6th. Female and male Gnats (p. 195).
May 20th. Head of Ant. Fly like a Gnat. Point of a Needle (p. 1).
May 27th. Pores in Petrified Wood (p. 107). Male Gnat.
June 10th. Sage-leaves appearing not to have Cavities.
June 17th. Pores in Petrified Wood.

On June 24th Dr. Wilkins, Dr. Wren, and Mr. Hooke were appointed to join together for more

microscopical observations, and on July 6th Mr.
Hooke was ordered to show to King Charles II his
microscopical observations in a handsome book, to
be provided by him for that purpose. There is no
evidence that King Charles ever attended a meeting
of the Society at Gresham College, but he used to
invite men of science to Whitehall and to show
him their experiments, perhaps in his laboratory
there.

July 8th. Edge of Razor (p. 4). Fine Taffeta Ribbon
(p. 6). Millepede.
July 16th. Fine Lawn (p. 5). Gilt-edge of Venice Paper.
Tinea argentea (p. 195).
August 5th. Honey-comb Sea-weed (p. 140). Teeth of
a Snail (p. 180). Plant growing on Rose-leaves (p. 121).
August 17th. Insects in Rain-water.
September 2nd. Gnat Larva.
September 9th. Parts of Fly (p. 169).
September 30th. Silk from Virginia. Scales of a Sole's
Skin (p. 162). Tabby (p. 7). Beard of Wild Oat (p. 147).
October 7th. Common Fly. Moss with the Seed (p. 131).
October 19th. Wing of Fly (p. 172).
October 28th. Pismire (p. 203).
November 4th. Mite (pp. 205, 213). Sparks of a Flint
(p. 44). Hair of Man, Cat, Horse and some Bristles
(p. 156).
November 25th. Egg of Silkworm (p. 181).
December 9th. Hair of Deer (p. 158).
December 16th and 23rd. Hair of Indian Deer (p. 158).

1663–4

March 23rd. It was ordered that Mr. Hooke produce at
every meeting of the Society one of his microscopical dis-
courses, in order to their being printed by order of the
Society. And on June 22nd Lord Brouncker was desired to
peruse Mr. Hooke's microscopical observations.

July 6th. Common Flies.

September 21st. Ciron or Wheal-worm.

October 26th. Poison fangs of Viper.

November 2nd. Poison fangs of Viper.

On November 23rd the President was desired to sign a licence for the printing of Mr. Hooke's microscopical book, which according to a letter from Hooke to Boyle had been printed off before October 24th. The delay was due to the several members of the Society who read the sheets.

A review appeared in the *Philosophical Transaction* for April 1665. But the universally curious Mr. Pepys, as we have seen, was earlier in the field. He is known to have had a microscope of his own when at the Navy Office.

Doubtless there were many persons then, as there are now, who considered such observations trivial, even as Charles II 'mightily laughed' at those who spent their time 'only in weighing of ayre'. But Hooke did much else beside.

The book is a fundamental classical work in the development of several Sciences and presents many ingenious ideas. The author's views on Combustion were familiar to Mayow, who further elaborated them, and are frequently quoted. Heat he conceived to be a mode of motion, and Light 'a very short vibrative motion transverse to straight lines of propagation through a homogeneous medium'. The interference colours of their plates, the black spot in soap-bubbles, and the phenomenon of 'Newton's rings' were all known to him. The behaviour of liquids in capillary tubes, and the inventions of the microscope, the hygrometer, the wheel-barometer, and many others are noticed in this volume.

His ingenious anticipation of the possibility of producing artificial silk has often been quoted. He winds up by asserting that his hint 'may give some Ingenious inquisitive Person an occasion of making some trials, which if successfull, I have my aim, and I suppose he will have no occasion to be displeas'd'.

Few prophesies have come more true!

R. T. GUNTHER

THE MUSEUM OF THE HISTORY OF SCIENCE
OXFORD, 1938

By the Council of the Royal Society of *London* for Improving of Natural Knowledge.

Ordered, *That the Book written by* Robert Hooke, *M.A. Fellow of this Society, Entituled*, Micrographia, or fome Phyfiological Defcriptions of Minute Bodies, made by Magnifying Glaffes, with Obfervations and Inquiries thereupon, *Be printed by* John Martyn, *and* James Alleftry, *Printers to the faid Society.*

Novem. 23.
1664.

BROUNCKER. *P. R. S.*

MICROGRAPHIA:

OR SOME

Physiological Descriptions

OF

MINUTE BODIES

MADE BY

MAGNIFYING GLASSES.·

WITH

Observations and Inquiries thereupon.

By *R. HOOKE*, Fellow of the Royal Society.

Non possis oculo quantum contendere Linceus,
Non tamen idcirco contemnas Lippus inungi. Horat. Ep. lib. 1.

LONDON, Printed by *Jo. Martyn*, and *Ja. Allestry*, Printers to the
Royal Society, and are to be fold at their Shop at the *Bell* in
S. *Paul's* Church-yard. M DC LX V.

TO THE
KING.

SIR,

I Do here moſt humbly lay this *ſmall* Preſent at *Your Majeſties* Royal feet. And though it comes accompany'd with two *diſadvantages*, the *meanneſs* of the *Author*, and of the *Subject*; yet in both I am *incouraged* by the *greatneſs* of your *Mercy* and your *Knowledge*. By the *one* I am taught, that you can

A   *forgive*

forgive the moſt *preſumptuous Offendors:*
And by the *other,* that you will not e-
ſteem the leaſt work of *Nature,* or *Art,*
unworthy your *Obſervation.* Amidſt the
many *felicities* that have accompani'd
your Majeſties happy *Reſtauration* and
Government, it is none of the leaſt conſiderable, that *Philoſophy* and *Experimental
Learning* have *proſper'd* under your *Royal
Patronage.* And as the calm proſperity
of your Reign has given us the *leiſure*
to follow theſe *Studies* of *quiet* and *re-
tirement,* ſo it is juſt, that the *Fruits* of
them ſhould , by way of *acknowledge-
ment ,* be return'd to *your Majeſty.*
There are, Sir, ſeveral other of your
Subjects, of your *Royal Society,* now
buſie about *Nobler* matters : The *Improvement* of *Manufactures* and *Agricul-
ture,* the *Increaſe* of *Commerce ,* the *Advantage* of *Navigation:* In all which
they are *aſſiſted* by *your Majeſties Incouragement* and *Example.* Amidſt all thoſe

greater

greater Designs, I here presume to bring in that which is more *proportionable* to the *smalness* of my Abilities , and to offer some of the *least* of all *visible* things, to that *Mighty King*, that has established an *Empire* over the best of all *Invisible things* of this World, the *Minds* of Men.

Your Majesties most humble

and most obedient

Subject and Servant,

ROBERT HOOKE.

TO THE

ROYAL SOCIETY.

Fter my *Addreſs* to our *Great Founder* and *Patron*, I could not but think my ſelf oblig'd, in conſideration of thoſe *many Ingagements* you have laid upon me, to offer theſe my *poor Labours* to this MOST ILLU-
STRIOUS ASSEMBLY. YOU have been pleas'd formerly to accept of theſe rude *Draughts*. I have ſince added to them ſome *Deſcriptions*, and ſome *Conjectures* of my own. And therefore, together with YOUR *Acceptance*, I muſt alſo beg YOUR *pardon*. The Rules YOU have preſcrib d YOUR ſelves in YOUR Philoſophical Progreſs do ſeem the beſt that have ever yet been practis d. And particularly that of avoiding *Dogmatizing*, and the *eſpouſal* of any *Hypotheſis* not ſufficiently grounded and confirm'd by *Experiments.* This way ſeems the moſt excellent, and may preſerve both *Philoſophy* and *Natural Hiſtory* from its former *Corruptions*. In ſaying which, I may ſeem to condemn my own Courſe in this Treatiſe ; in which there may perhaps be ſome *Expreſſions*, which may ſeem more *poſitive* then YOUR Preſcriptions will permit : And though I deſire to have them underſtood only as *Conjectures* and *Quæries* (which YOUR Method does not altogether diſallow) yet if even in thoſe I have exceeded, 'tis fit that I ſhould declare, that it was not done by YOUR Directions. For it is moſt unreaſonable, that YOU ſhould undergo the *imputation* of the *faults* of my *Conjectures*, ſeeing YOU can receive ſo *ſmall advantage* of reputation by the *ſleight Obſervations* of

> *YOUR* moſt *humble and*
> moſt *faithful Servant*

ROBERT HOOKE.

THE

PREFACE.

I is the great prerogative of Mankind above other Creatures, that we are not only able to behold the works of Nature, or barely to suftein our lives by them, but we have also the power of confidering, comparing, altering, affifting, and improving them to various ufes. And as this is the peculiar priviledge of humane Nature in general, fo is it capable of being fo far advanced by the helps of Art, and Experience, as to make fome Men excel others in their Obfervations, and Deductions, almoft as much as they do Beafts. By the addition of fuch artificial Inftruments and methods, there may be, in fome manner, a reparation made for the mifchiefs, and imperfection, mankind has drawn upon it felf, by negligence, and intemperance, and a wilful and fuperftitious deferting the Prefcripts and Rules of Nature, whereby every man, both from a deriv'd corruption, innate and born with him, and from his breeding and converfe with men, is very fubject to flip into all forts of errors.

The only way which now remains for us to recover fome degree of thofe former perfections, feems to be, by rectifying the operations of the Senfe, the Memory, and Reafon, fince upon the evidence, the ftrength, the integrity, and the right correfpondence of all thefe, all the light, by which our actions are to be guided, is to be renewed, and all our command over things is to be eftablifht.

It is therefore moft worthy of our confideration, to recollect their fefeveral defects, that fo we may the better underftand how to fupply them, and by what affiftances we may inlarge their power, and fecure them in performing their particular duties.

As for the actions of our Senfes, we cannot but obferve them to be in

many

The PREFACE.

*many particulars much outdone by those of other Creatures, and when at best, to be far short of the perfection they seem capable of : And these infirmities of the Senses arise from a double cause, either from the dis-*proportion of the Object to the Organ, *whereby an infinite number of things can never enter into them, or else from* error in the Perception, *that many things, which come within their reach, are not received in a right manner.*

The like frailties are to be found in the Memory ; *we often let many things* slip away *from us, which deserve to be retain'd ; and of those which we treasure up, a great part is either* frivolous *or* false ; *and if good, and substantial, either in tract of time* obliterated, *or at best so* overwhelmed *and buried under more frothy notions, that when there is need of them, they are in vain sought for.*

The two main foundations being so deceivable, it is no wonder, that all the succeeding works which we build upon them, of arguing, conclu-ding, defining, judging, and all the other degrees of Reason, are lyable to the same imperfection, being, at best, either vain, or uncertain : So that the errors of the understanding *are answerable to the two other, being defective both in the quantity and goodness of its knowledge ; for the li-mits, to which our thoughts are confi'nd, are small in respect of the vast extent of* Nature it *self; some parts of it are* too large *to be comprehen-ded, and some* too little *to be perceived. And from thence it must fol-low, that not having a full sensation of the Object, we must be very lame and imperfect in our conceptions about it , and in all the propositions which we build upon it ; hence we often take the* shadow *of things for the* substance, *small* appearances *for good* similitudes, *similitudes for* definitions; *and even many of those, which we think to be the most solid definitions, are rather expressions of our own misguided apprehen-sions then of the true nature of the things themselves.*

The effects of these imperfections are manifested in different ways, ac-cording to the temper and disposition of the several minds of men, some they incline to gross ignorance *and stupidity, and others to a* pre-sumptuous imposing *on other mens Opinions, and a* confident dog-matizing *on matters, whereof there is no assurance to be given.*

Thus

The Preface.

Thus all the uncertainty, and mistakes of humane actions, proceed either from the narrowness and wandring of our Senses, *from the flipperiness or delusion of our* Memory, *from the confinement or rashness of our* Understanding, *so that 'tis no wonder, that our power over natural causes and effects is so slowly improv'd, seeing we are not only to contend with the obscurity and* difficulty *of the things whereon we work and think, but even the forces of our own minds conspire to betray us.*

These being the dangers in the process of humane Reason, *the remedies of them all can only proceed from the* real, *the* mechanical, *the* experimental Philosophy, *which has this advantage over the Philosophy of* discourse *and* disputation, *that whereas that chiefly aims at the subtilty of its* Deductions *and* Conclusions, *without much regard to the first ground-work, which ought to be well laid on the* Sense *and* Memory; *so this intends the right ordering of them all, and the making them serviceable to each other.*

The first thing to be undertaken in this weighty work, is a watchfulness *over the failings and an* inlargement of the dominion, *of the* Senses.

To which end it is requisite, first, That there should be a scrupulous *choice, and a* strict examination, *of the* reality, constancy, *and* certainty *of the* Particulars *that we admit: This is the first rise whereon truth is to begin, and here the most severe, and most impartial diligence, must be imployed ; the storing up of all, without any regard to* evidence *or* use, *will only tend to darkness and confusion.* We must *not therefore esteem the riches of our* Philosophical treasure *by the* number *only, but chiefly by the* weight; *the most vulgar* Instances *are not to be neglected, but above all, the most* instructive *are to be entertain'd ; the footsteps of Nature are to be trac'd, not only in her* ordinary course, *but when she seems to be put to her shifts, to make many* doublings *and* turnings, *and to use some kind of art in indeavouring to* avoid *our* discovery.

The next care to be taken, in respect of the Senses, *is a supplying of their infirmities with* Instruments, *and, as it were, the adding of* artificial Organs *to the* natural ; *this in one of them has been of late years*

The PREFACE.

accomplisht with prodigious benefit to all forts of useful knowledge, by the invention of Optical Glasses. By the means of Telescopes, *there is nothing fo far distant but may be represented to our view;* and by the *help of* Microscopes, *there is nothing fo small, as to escape our inquiry;* hence there is a new visible World discovered to the understanding. *By this means the Heavens are open'd, and a vast number of new Stars, and new Motions, and new Productions appear in them, to which all the antient Astronomers were utterly Strangers.* By this the Earth it self, *which lyes fo neer us, under our feet, shews quite a new thing to us, and in every* little particle *of its matter,* we now behold almost as great a *variety of* Creatures, *as we were able before to reckon up in the whole* Universe it self.

It seems not improbable, but that by these helps the subtilty of the *composition of Bodies, the structure of their parts, the various texture of their matter, the instruments and manner of their inward motions, and all the other possible appearances of things, may come to be more fully discovered;* all which the antient Peripateticks were content to *comprehend in two general and (unless further explain'd) useless words of Matter and* Form. *From whence there may arise many admirable advantages, towards the increase of the* Operative, *and the* Mechanick Knowledge, *to which this Age feems fo much inclined, because we may perhaps be inabled to discern all the secret workings of Nature, almost in the same manner as we do those that are the productions of Art, and are manag'd by Wheels, and Engines, and Springs, that were devised by humane Wit.*

In this kind I here present to the World *my imperfect Indeavours;* which though they shall prove *no other way considerable, yet, I hope, they may be in some measure useful to the main Design of a* reformation *in Philosophy, if it be only by shewing, that there is not fo much requir'd towards it, any strength of* Imagination, *or exactness of* Method, *or depth of* Contemplation *(though the addition of these, where they can be had, must needs produce a much more perfect composure) as a* sincere Hand, *and a* faithful Eye, *to examine, and to record, the things themselves as they appear.*

And

The PREFACE.

And I beg my Reader, to let me take the boldneſs to aſſure him, that in this preſent condition of knowledge, a man ſo qualified, as I have indeavoured to be, only with reſolution, and integrity, and plain intentions of imploying his Senſes *aright,may venture to compare the reality and the uſefulneſs of his ſervices, towards the true* Philoſophy, *with thoſe of other men, that are of much ſtronger,and more acute* ſpeculations,*that ſhall not make uſe of the ſame method by the* Senſes.

The truth is, the Science of Nature has been already too long made only a work of the Brain *and the* Fancy *: It is now high time that it ſhould return to the plainneſs and ſoundneſs of* Obſervations *on* material *and* obvious things. *It is ſaid of great Empires, That* the beſt way to preſerve them from decay, is to bring them back to the firſt Principles, and Arts, on which they did begin. *The ſame is undoubtedly true in* Philoſophy,*that by wandring far away into* inviſible Notions,*has almoſt quite deſtroy'd it ſelf,and it can never be recovered, or continued, but by returning into the ſame* ſenſible paths, *in which it did at firſt proceed.*

If therefore the Reader expects from me any infallible Deductions, or certainty of Axioms, *I am to ſay for my ſelf, that thoſe ſtronger Works of Wit and Imagination are above my weak Abilities ; or if they had not been ſo, I would not have made uſe of them in this preſent Subject before me : Whereever he finds that I have ventur'd at any ſmall Conjectures, at the cauſes of the things that I have obſerved, I beſeech him to look upon them only as* doubtful Problems,*and* uncertain gheſſes, *and not as unqueſtionable Concluſions, or matters of unconfutable Science ; I have produced nothing here, with intent to bind his underſtanding to an* implicit conſent ; *I am ſo far from that, that I deſire him, not abſolutely to rely upon theſe Obſervations of my eyes, if he finds them contradicted by the future Ocular Experiments of ſober and impartial Diſcoverers.*

As for my part, I have obtained my end, if theſe my ſmall Labours ſhall be thought fit to take up ſome place in the large ſtock of natural Obſervations, *which ſo many hands are buſie in providing. If I have contributed the* meaneſt foundations *whereon others may raiſe nobler*

The PREFACE.

Superstructures, *I am abundantly satisfied*; *and all my ambition is, that I may serve to the great Philosophers of this Age, as the makers and the grinders of my Glasses did to me*; *that I may prepare and furnish them with some* Materials, *which they may afterwards order and manage with better skill, and to far greater advantage.*

The next remedies in this universal cure of the Mind are to be applyed to the Memory, *and they are to consist of such Directions as may inform us, what things are best to be* stor'd up *for our purpose, and which is the best way of so* disposing *them, that they may not only be* kept in safety, *but ready and convenient,to be at any time* produc'd *for* use, *as occasion shall require. But I will not here prevent my self in what I may say in another Discourse, wherein I shall make an attempt to propose some Considerations of the manner of compiling a Natural and Artificial History, and of so ranging and registring its Particulars into Philosophical Tables, as may make them most useful for the raising of* Axioms *and* Theories.

The last indeed is the most hazardous *Enterprize, and yet the most* necessary; *and that is, to take such care that the* Judgment *and the* Reason of Man *(which is the third Faculty to be repair'd and improv'd) should receive such assistance, as to avoid the dangers to which it is by nature most subject. The Imperfections, which I have already mention'd, to which it is lyable, do either belong to the extent, or the goodness of its knowledge; and here the difficulty is the greater, least that which may be thought a remedy for the one should prove* destructive *to the other, least by seeking to inlarge our Knowledge, we should render it weak and uncertain; and least by being too scrupulous and exact about every Circumstance of it, we should confine and streighten it too much.*

In both these the middle wayes are to be taken, nothing is to be omitted, *and yet every thing to pass* a mature deliberation: *No* Intelligence *from Men of all Professions, and quarters of the World, to be* slighted, *and yet all to be so severely examin'd,that there remain no room for doubt or instability; much* rigour *in admitting, much* strictness *in comparing,and above all, much* slowness *in debating, and*

shyness

The PREFACE.

shynefs *in determining, is to be practifed. The* Underftanding *is to* order *all the inferiour fervices of the lower Faculties; but yet it is to do this only as a* lawful Mafter, *and not as a* Tyrant. *It muft* not incroach *upon their Offices, nor take upon it felf the employments which belong to either of them. It muft* watch *the irregularities of the Senfes, but it muft not go before them, or* prevent *their information. It muft* examine, range, *and* difpofe *of the bank which is laid up in the* Memory : *but it muft be fure to make* diftinction *between the* fober *and* well collected heap, *and the* extravagant Idea's, *and* miftaken Images, *which there it may fometimes light upon. So many are the* links, *upon which the true* Philofophy *depends, of which, if any one be* loofe, *or* weak, *the whole* chain *is in danger of being diffolv'd ; it is to* begin *with the* Hands *and* Eyes, *and to* proceed *on through the* Memory, *to be* continued *by the Reafon ; nor is it to ftop there, but to* come about *to the Hands and Eyes again, and fo, by a* continual paffage round *from one Faculty to another, it is to be maintained in life and ftrength, as much as the body of man is by the* circulation *of the blood through the feveral parts of the body, the* Arms, the Fat, the Lungs, the Heart, and the Head.

If once this method were followed with diligence and attention, there is nothing that lyes within the power of human Wit (*or which is far more effectual*) *of human Induftry, which we might not compafs ; we might not only hope for Inventions to equalize thofe of* Copernicus, Galileo, Gilbert Harvy, *and of others, whofe Names are almoft loft, that were the Inventors of* Gun-powder, *the* Seamans Compafs, Printing, Etching, Graving, Microfcopes, &c. *but multitudes that may far exceed them : for even thofe difcoveries feem to have been the products of fome fuch method, though but imperfect ; What may not be therefore expected from it if thoroughly profecuted?* Talking *and* contention of Arguments *would foon be turn'd into* labours ; *all the* fine dreams *of* Opinions, *and* univerfal metaphyfical natures, *which the luxury of fubtil Brains has devis'd, would quickly vanifh, and give place to* folid Hiftories, Experiments *and* Works. *And as at firft, mankind* fell *by* tafting *of the forbidden Tree of Knowledge, fo we, their Pofterity, may be in part* reftor'd

by

The PREFACE.

by the same way, not only by beholding and contemplating, *but by ta-sting too those fruits of Natural knowledge, that were never yet forbidden.*

From hence the World may be assisted with variety *of Inventions,* new *matter for Sciences may be* collected, *the* old improv'd, *and their* rust *rubb'd away; and as it is by the benefit of Senses that we receive all our Skill in the works of Nature, so they also may be wonderfully benefited by it, and may be guided to an easier and more exact performance of their Offices ; 'tis not unlikely, but that we may find out wherein our Senses are deficient, and as easily find wayes of repairing them.*

The Indeavours of Skilful men have been most conversant about the assistance of the Eye, and many noble Productions have followed upon it ; and from hence we may conclude, that there is a way open'd for advancing the operations, not only of all the other Senses, but even of the Eye it self; that which has been already done ought not to content us, but rather to incourage us to proceed further, and to attempt greater things in the same and different wayes.

'Tis not unlikely, but that there may be yet invented several other helps for the eye, as much exceeding those already found, as those do the bare eye, such as by which we may perhaps be able to discover living Creatures *in the Moon, or other Planets, the figures of the compounding Particles of matter, and the particular* Schematisms *and* Textures *of Bodies.*

And as Glasses *have highly promoted our* seeing, *so 'tis not improbable, but that there may be found many* Mechanical Inventions *to improve our other Senses, of* hearing, smelling, tasting, touching. *'Tis not impossible to hear a* whisper *a* furlongs *distance, it having been already done ; and perhaps the nature of the thing would not make it more impossible, though that furlong should be ten times multiply'd. And though some famous Authors have affirm'd it impossible to hear through the* thinnest *plate of* Muscovy-glass ; *yet I know a way, by which 'tis easie enough to hear one speak through a* wall a yard thick. *It has not been yet thoroughly examin'd, how far* Otocousticons *may be improv'd, nor what other wayes there may be of* quickning *our hearing, or* conveying *sound through other bodies then the* Air: *for that that is not the only* medium, *I can assure the Reader, that I have, by the help of a* distended wire, *propa-gated*

gated the sound to a very considerable distance in an instant, or with as seemingly quick a motion as that of light, at least, incomparably swifter then that, which at the same time was propagated through the Air ; and this not only in a straight line , or direct, but in one bended in many angles.

Nor are the other three so perfect, but that diligence, attention, *and many* mechanical contrivances, *may also highly improve them. For since the sense of* smelling *seems to be made by the* swift passage *of the* Air *(* impregnated *with the steams and* effluvia *of several odorous Bodies) through the grisly* meanders *of the Nose whose surfaces are* cover'd *with a very sensible* nerve , *and* moistned *by a* transuda-*tion from the* processus mamillares *of the Brain , and some ad-joyning* glandules, *and by the moist* steam *of the Lungs, with a Liquor convenient for the reception of those effluvia and by the adhesion and mixing of those steams with that liquor, and thereby affecting the nerve, or perhaps by insinuating themselves into the juices of the brain, after the same manner, as I have in the following Observations intimated, the parts of Salt to pass through the skins of Effs, and Frogs. Since, I say, smelling seems to be made by some such way, 'tis not improbable, but that some con-trivance, for making a great quantity of Air pass quick through the Nose, might as much promote the sense of smelling, as the any wayes hindring that passage does dull and destroy it. Several tryals I have made , both of hindring and promoting this sense, and have succeeded in some according to expectation ; and indeed to me it seems capable of being improv'd , for the judging of the constitutions of many Bodies. Perhaps we may thereby also judge (as other Creatures seem to do) what is wholsome, what poyson ; and in a word, what are the specifick properties of Bodies.*

There may be also some other mechanical wayes found out , of sensibly perceiving the effluvia *of Bodies ; several Instances of which, were it here proper, I could give of Mineral steams and exhalations ; and it seems not impossible, but that by some such wayes improved, may be discovered, what Minerals lye buried under the Earth, without the trouble to dig for them ; some things to confirm this Conjecture may be found in* Agricola, *and other Writers of Minerals, speaking of the Vegetables that are apt to thrive, or pine, in those steams.*

Whether

The PREFACE.

Whether also those steams, which seem to issue out of the Earth, and mix with the Air (and so to precipitate some aqueous *Exhalations, wherewith 'tis impregnated) may not be by some way detected before they produce the effect, seems hard to determine ; yet something of this kind I am able to discover, by an Instrument I contriv'd to shew all the minute variations in the pressure of the Air ; by which I constantly find, that before, and during the time of rainy weather, the pressure of the Air is less, and in* dry *weather , but especially when an* Eastern Wind *(which having past over vast tracts of Land is heavy with Earthy Particles) blows, it is much more, though these changes are varied according to very odd Laws.*

The Instrument is this. I prepare a pretty capaceous Bolt-head A B, with a small stem about two foot and a half long D C ; upon the end of this D I put on a small bended Glass, or brazen *Syphon* D E F (open at D, E and F, but to be closed with cement at F and E, as occasion serves) whose stem F should be about six or eight inches long, but the bore of it not above half an inch diameter, and very even ; these I fix very strongly together by the help of very hard Cement, and then fit the whole Glass A B C D E F into a long Board, or Frame, in such manner, that almost half the head A B may lye buried in a concave Hemisphere cut into the Board R S ; then I place it so on the Board R S, as is exprest in the first Figure of the first Scheme ; and fix it very firm and steady in that posture, so as that the weight of the *Mercury* that is afterwards to be put into it, may not in the least shake or stir it ; then drawing a line X Y on the Frame R T, so that it may divide the ball into two equal parts, or that it may pass, as 'twere, through the center of the ball. I begin from that, and divide all the rest of the Board towards U T into inches, and the inches between the 25 and the end E (which need not be above two or three and thirty inches distant from the line X Y) I subdivide into Decimals ; then stopping the end F with soft Cement, or soft Wax, I invert the Frame, placing the head downwards, and the Orifice E upwards ; and by it, with a small Funnel, I fill the whole Glass with Quicksilver ; then by stopping the small Orifice E with my finger, I oftentimes erect and invert the whole Glass and Frame, and thereby free the Quicksilver and Glass from all the bubbles or parcels of lurking Air ; then inverting it as before, I fill it top full with clear and well strain'd Quicksilver, and having made ready a small ball of pretty hard Cement, by heat made very soft, I press it into the hole E, and thereby stop it very fast ; and to secure this Cement from flying out afterward, I bind over it a piece of Leather, that is spread over in the inside with Cement, and wound about it whilst the Cement is hot: Having thus fastned it, I gently erect again the Glass after this manner : I first let the Frame down edge-wayes, till the edge R V touch the Floor, or ly horizontal ; and then in that edging posture raise the end R S ; this I do , that if there chance to be any Air hidden in the small Pipe E, it may ascend into the Pipe F, and not into the Pipe D C : Having thus erected it, and hung it by the hole Q, or fixt it perpendicularly by any other means, I open the end F,

The PREFACE.

and by a fmall *Syphon* I draw out the *Mercury* fo long, till I find the furface of it A B in the head to touch exactly the line X Y; at which time I immediately take away the *Syphon*, and if by chance it be run fomewhat below the line X Y, by pouring in gently a little *Mercury* at F, I raife it again to its defired height, by this contrivance I make all the fenfible rifing and falling of the *Mercury* to be vifible in the furface of the *Mercury* in the Pipe F, and fcarce any in the head A B. But becaufe there really is fome fmall change of the upper furface alfo, I find by feveral Obfervations how much it rifes in the Ball, and falls in the Pipe F, to make the diftance between the two furfaces an inch greater then it was before; and the meafure that it falls in the Pipe is the length of the inch by which I am to mark the parts of the Tube F, or the Board on which it lyes, into inches and Decimals: Having thus juftned and divided it, I have a large Wheel M N O P, whofe outmoft limb is divided into two hundred equal parts; this by certain fmall Pillars is fixt on the Frame R T, in the manner expreft in the Figure. In the middle of this, on the back fide, in a convenient frame, is placed a fmall Cylinder, whofe circumference is equal to twice the length of one of thofe divifions, which I find anfwer to an inch of afcent, or defcent, of *Mercury*: This Cylinder I, is movable on a very fmall Needle, on the end of which is fixt a very light Index K L, all which are fo pois'd on the Axis, or Needle, that no part is heavier then another: Then about this Cylinder is wound a fmall Clew of Silk, with two fmall fteel Bullets at each end of it G H; one of thefe, which is fomewhat the heavier, ought to be fo big, as freely to move to and fro in the Pipe F; by means of which contrivance, every the leaft variation of the height of the *Mercury* will be made exceeding vifible by the motion to and fro of the fmall Index K L.

But this is but one way of difcovering the effluvia of the Earth mixt with the Air; there may be perhaps many others, witnefs the Hygrofcope, an Inftrument whereby the watery fteams volatile in the Air are difcerned, which the Nofe it felf is not able to find. This I have defcrib'd in the following Tract in the Defcription of the Beard of a wild Oat. Others there are, may be difcovered both by the Nofe, and by other wayes alfo. Thus the fmoak of burning Wood is fmelt, feen, and fufficiently felt by the eyes: The fumes of burning Brimftone are fmelt and difcovered alfo by the deftroying the Colours of Bodies, as by the whitening of a red Rofe: And who knows, but that the Induftry of man, following this method, may find out wayes of improving this fenfe to as great a degree of perfeltion as it is in any Animal, and perhaps yet higher.

'Tis not improbable alfo, but that our tafte may be very much improv'd, either by preparing our taft for the Body, as, after eating bitter things, Wine, or other Vinous liquors, are more fenfibly tafted; or elfe by preparing
paring

paring *Bodies for our taft* ; *as the diffolving of Metals with acid Liquors, make them taftable, which were before altogether infipid* ; *thus* Lead *becomes* fweeter *then Sugar, and* Silver *more* bitter *then Gall,* Copper *and* Iron *of moft* loathfome *tafts. And indeed the bufinefs of this fenfe being to difcover the prefence of diffolved Bodies in Liquors put on the Tongue,or in general to difcover that a fluid body has fome folid body diffolv'd in it, and what they are* ; *whatever contrivance makes this difcovery improves this fenfe. In this kind the mixtures of Chymical Liquors afford many Inftances* ; *as the fweet Vinegar that is impregnated with* Lead *may be difcovered to be fo by the affufion of a little of an* Alcalizate folution : *The bitter liquor of* Aqua fortis *and* Silver *may be difcover'd to be charg'd with that Metal, by laying in it fome plates of Copper :* '*Tis not improbable alfo,but there may be multitudes of other wayes of difcovering the parts diffolv'd, or diffoluble in liquors* ; *and what is this difcovery but a kind of* fecundary tafting.

'*Tis not improbable alfo,but that the fenfe of* feeling *may be highly improv'd, for that being a fenfe that judges of the more* grofs *and* robuft motions *of the* Particles *of* Bodies, *feems capable of being improv'd and affifted very many wayes. Thus for the diftinguifhing of* Heat *and* Cold,*the* Weather-glafs *and* Thermometer, *which I have defcrib'd in this following Treatife, do exceedingly perfeft it* ; *by each of which the leaft variations of heat or cold, which the moft* Acute *fenfe is not able to diftinguifh,are manifefted This is oftentimes further promoted alfo by the help of* Burning-glaffes,*and the like, which colleft and unite the radiating beat. Thus the* roughnefs *and* fmoothnefs *of a Body is made much more fenfible by the help of a* Microfcope, *then by the moft* tender *and* delicate Hand. *Perhaps, a Phyfitian might, by feveral other* tangible *proprieties, difcover the confiitution of a Body as well as by the* Pulfe. *I do but inftance in thefe,to fhew what poffibility there may be of many others, and what probability and hopes there were of finding them,if this method were followed* ; *for the Offices of the five Senfes being to deteft either the* fubtil *and* curious Motions *propagated through all* pellucid *or perfeftly* homogeneous *Bodies* ; *Or the more* grofs *and* vibrative Pulfe *communicated through the* Air *and all other convenient* mediums,*whether fluid or folid : Or the*

effluvia

The PREFACE.

effluvia *of Bodies* diffolv'd *in the* Air ; *Or the* particles *of bodies* diffolv'd *or* diffoluble *in* Liquors, *or the more* quick *and* violent fhaking motion *of* heat *in all or any of thefe: whatfoever does any wayes promote any of thefe kinds of* criteria, *does afford a way of improving fome one fenfe. And what a multitude of thefe would a diligent Man meet with in his inquiries? And this for the helping and promoting the* fenfitive faculty *only.*

Next, as for the Memory, *or* retentive faculty, *we may be fufficiently inftructed from the* written Hiftories *of* civil actions, *what great affiftance may be afforded the Memory, in the committing to writing things obfervable in* natural operations. *If a Phyfitian be therefore accounted the more able in his Faculty, becaufe he has had long experience and practice, the remembrance of which, though perhaps very imperfect, does regulate all his after actions : What ought to be thought of that man, that has not only a perfect* regifter *of his own experience, but is grown* old *with the experience of many hundreds of years, and many thoufands of men.*

And though of late , men, beginning to be fenfible of this convenience, have here and there regiftred and printed fome few Centuries, *yet for the moft part they are fet down very lamely and imperfectly, and, I fear, many times not fo truly, they feeming, feveral of them, to be defign'd more for* Oftentation *then* publique ufe : *For, not to inftance, that they do, for the moft part, omit thofe* Experiences they *have made , wherein their Patients have mifcarried, it is very eafie to be perceiv'd, that they do all along* hyperbolically extol *their own Prefcriptions, and vilifie thofe of others. Notwithftanding all which, thefe kinds of Hiftories are generally efteem'd ufeful, even to the ableft Phyfitian.*

What may not be expected from the rational or deductive Faculty *that is furnifht with fuch* Materials, *and thofe fo readily* adapted, *and rang'd for ufe, that in a moment, as 'twere, thoufands of Inftances, ferving for the* illuftration, determination, *or* invention, *of almoft any inquiry, may be* reprefented *even to the fight ? How neer the nature of* Axioms *muft all thofe* Propofitions *be which are examin'd before fo many* Witneffes ? *And how difficult will it be for any, though never fo fubtil an error in Philofophy, to fcape from being difcover'd, after it has indur'd the* touch, *and fo many other* tryals *?*   d   *What*

The PREFACE.

*What kind of mechanical way, and physical invention also is there re-
quir'd, that might not this way be found out ? The* Invention *of a way to
find the* Longitude *of places is easily perform'd, and that to as great* perfection *as is desir'd, or to as great an* accuratenes *as the* Latitude *of
places can be found at Sea ; and perhaps yet also to a greater certainty
then that has been hitherto found, as I shall very speedily freely manifest to
the world The way of* flying *in the Air seems principally unpracticable,
by reason of the* want *of* strength *in humane mulcles ; if therefore
that could be suppli d, it were, I think, easie to make twenty contrivances to
perform the office of* Wings : *What Attempts also I have made for the
supplying that Defect, and my succeffes therein, which, I think, are wholly
new, and not inconsiderable, I shall in another place relate.*

'Tis not unlikely also, but that Chymifts, *if they followed this method,
might find out their so much fought for* Alkahelt. *What an* univerfal
Menftruum , *which diffolves all forts of* Sulphureous Bodies, *I have
discover d (which has not been before taken notice of as fuch) I have
shewn in the sixteenth Obfervation.*

What a prodigious variety of Inventions *in* Anatomy *has this latter
Age afforded, even in our own Bodies, in the very* Heart, *by which we live,
and the* Brain, *which is the feat of our knowledge of other things ? witnefs
all the excellent* Works *of* Pecquet, Bartholinus, Billius, *and many
others ; and at home, of Doctor* Harvy, *Doctor* Ent, *Doctor* Willis, *Doctor*
Gliffon. *In* Celeftial Obfervations *we have far exceeded all the An-
tients, even the* Chaldeans *and* Egyptians *themfelves, whofe vaft* Plains,
high Towers, *and* clear Air, *did not give them fo great advantages over
us, as we have over them by our* Glaffes. *By the help of which, they have
been very much outdone by the famous* Galileo, Hevelius, Zulichem ;
and our own Countrymen, *Mr.* Rook, *Doctor* Wren, *and the great Ornament of our Church and Nation, the* Lord Bifhop *of* Exeter. *And to fay
no more in* Aerial Difcoveries, *there has been a wonderful progrefs made
by the* Noble Engine *of the moft Illuftrious Mr.* Boyle, *whom it becomes
me to mention with all honour, not only as my particular Patron, but as the
Patron of Philofophy it felf ; which he every day increafes by his Labours, and adorns by his Example.*

The

The PREFACE.

The good fucceſs of all theſe great Men, *and many others, and the now feemingly great* obvioufneſs *of moſt of their and divers other Inventions, which from the beginning of the world have been, as 'twere, trod on, and yet not minded till theſe laſt* inquiſitive *Ages (an Argument that there may be yet behind multitudes of the like) puts me in mind to recommend fuch Studies, and the profecution of them by fuch methods, to the* Gentlemen *of our Nation, whoſe* leiſure *makes them fit to* undertake, *and the plenty of their fortunes to* accompliſh, *extraordinary things in this way. And I do not only propoſe this kind of* Experimental Philoſophy *as a matter of high* rapture *and* delight *of the mind, but even as a* material *and* ſenſi-ble Pleaſure. *So vaſt is the* variety of Objects *which will come under their* Inſpections, *ſo many* different wayes *there are of* handling them, *ſo great is the* ſatisfaction *of* finding *out* new things, *that I dare compare the* contentment *which they will injoy, not only to that of* contemplation, *but even to that which moſt men prefer of the* very Senſes *themſelves.*

And if they will pleaſe to take any incouragement from ſo mean and ſo imperfect endeavours as mine, upon my own experience, *I can aſſure them, without arrogance, That there has not been any inquiry or Pro-blem in* Mechanicks, *that I have hitherto propounded to my ſelf, but by a certain method (which I may on ſome other opportunity explain) I have been able preſently to examine the poſſibility of it ; and if ſo, as eaſily to ex-cogitate divers wayes of performing it : And indeed it is poſſible to do as much by* this method *in* Mechanicks, *as by* Algebra *can be perform'd in* Geometry. *Nor can I at all doubt, but that the ſame method is as ap-plicable to* Phyſical Enquiries , *and as likely to find and reap thence as plentiful a crop of Inventions ; and indeed there ſeems to be no ſubject ſo barren, but may with this good husbandry be highly improv'd.*

Toward the profecution of this method in Phyſical Inquiries, *I have here and there* gleaned *up an* handful *of* Obſervations, *in the collection of moſt of which I made uſe of* Microſcopes, *and ſome other* Glaſſes *and In-ſtruments that improve the ſenſe ; which way I have herein taken , not that there are not multitudes of uſeful and pleaſant Obſervables, yet uncol-lected, obvious enough without the helps of* Art , *but only to promote the uſe of* Mechanical *helps for the Senſes, both in the ſurveying the already viſible*

World,

World, and for the discovery of many others hitherto unknown, and to make us, with the great Conqueror, to be affected that we have not yet overcome one World when there are so many others to be discovered, every considerable improvement of Telescopes *or* Microscopes *producing new Worlds and* Terra-Incognita's *to our view.*

The Glasses I used were of our English make, but though very good of the kind, yet far short of what might be expected, could we once find a way of making Glasses Elliptical, or of some more true shape ; for though both Microscopes, *and* Telescopes, *as they now are, will magnifie an Object about a thousand thousand times bigger then it appears to the naked eye ; yet the Apertures of the Object-glasses are so very small, that very few Rays are admitted, and even of those few there are so many false, that the Object appears* dark *and* indistinct : *And indeed these inconveniences are such, as seem inseparable from Spherical Glasses, even when most exactly made; but the way we have hitherto made use of for that purpose is so imperfect, that there may be perhaps ten wrought before one be made tolerably good, and most of those ten perhaps every one differing in goodneß one from another, which is an Argument, that the way hitherto used is, at least, very uncertain. So that these Glasses have a double defect; the one, that very few of them are exactly true wrought ; the other, that even of those that are best among them, none will admit a sufficient number of Rayes to magnifie the Object beyond a determinate bigneß. Against which Inconveniences the only Remedies I have hitherto met with are these.*

First, for *Microscopes* (where the Object we view is near and within our power) the best way of making it appear bright in the Glaß, is to cast a great quantity of light on it by means of *convex glasses,* for thereby, though the aperture be very small, yet there will throng in through it such multitudes, that an Object will by this means indure to be magnifi'd as much again as it would be without it. The way for doing which is this. I make choice of some Room that has only one window open to the South , and at about three or four foot distance from this Window, on a Table, I place my *Microscope,* and then so place either a round Globe of Water, or a very deep clear *plano convex* Glaß (whose convex side is turn'd towards the Window) that there is a great quantity of Rayes collected and thrown upon the Object : Or if the Sun shine, I place a small piece of oyly Paper very near the Object, between that and the light ; then with a good large Burning-Glaß I so collect and throw the Rayes on the Paper, that there may be a very great quantity of light paß through it to the Object ; yet I so proportion that light, that it

may

The PREFACE.

may not finge or burn the Paper. Inftead of which Paper there may be made ufe of a fmall piece of Looking-glafs plate, one of whofe fides is made rough by being rubb'd on a flat Tool with very fine fand, this will, if the heat be leifurely caft on it, indure a much greater degree of heat, and confequently very much augment a convenient light. By all which means the light of the Sun, or of a Window, may be fo caft on an Object, as to make it twice as light as it would otherwife be without it, and that without any inconvenience of glaring, which the immediate light of the Sun is very apt to create in moft Objects; for by this means the light is fo equally diffufed, that all parts are alike inlightned; but when the immediate light of the Sun falls on it, the reflexions from fome few parts are fo vivid, that they drown the appearance of all the other, and are themfelves alfo, by reafon of the inequality of light, indiftinct, and appear only radiant fpots.

But becaufe the light of the Sun, and alfo that of a Window, is in a continual variation, and fo many Objects cannot be view'd long enough by them to be throughly examin'd; befides that, oftentimes the Weather is fo dark and cloudy, that for many dayes together nothing can be view'd: And becaufe alfo there are many Objects to be met with in the night, which cannot fo conveniently be kept perhaps till the day, therefore to procure and caft a fufficient quantity of light on an Object in the night, I thought of, and often ufed this, Expedient.

I procur'd me a fmall Pedeftal, fuch as is defcrib'd in the fifth Figure of the firft *Scheme* on the fmall Pillar A B, of which were two movable Armes C D, which by means of the Screws E F, I could fix in any part of the Pillar; on the undermoft of thefe I plac'd a pretty large Globe of Glafs G, fill'd with exceeding clear Brine, ftopt, inverted, and fixt in the manner vifible in the Figure; out of the fide of which Arm proceeded another Arm H, with many joynts; to the end of which was faftned a deep plain *Convex glafs* I, which by means of this Arm could be moved to and fro, and fixt in any pofture. On the upper Arm was placed a fmall Lamp K, which could be fo mov'd upon the end of the Arm, as to be fet in a fit pofture to give light through the Ball: By means of this Inftrument duly plac'd, as is expreft in the Figure, with the fmall flame of a Lamp may be caft as great and convenient a light on the Object as it will well indure; and being always conftant, and to be had at any time, I found moft proper for drawing the reprefentations of thofe fmall Objects I had occafion to obferve.

None of all which ways (though much beyond any other hitherto made ufe of by any I know) do afford a fufficient help, but after a certain degree of magnifying, they leave us again in the lurch. Hence it were very defirable, that fome way were thought of for making the Object-glafs of fuch a Figure as would conveniently bear a large Aperture.

As for Telefcopes, *the only improvement they feem capable of, is the increafing of their length; for the Object being remote, there is no thought of giving it a greater light then it has; and therefore to augment the Aperture, the Glafs muft be ground of a very large fphere; for, by that*

means,

means,the longer the Glaß be,the bigger aperture will it bear,if the Glaßes be of an equal goodneß in their kind. Therefore a fix will indure a much larger Aperture then a three foot Glaß; and a fixty foot Glaß will proportionably bear a greater Aperture then a thirty,and will as much excel it alfo as a fix foot does a three foot, as I have experimentally obferv'd in one of that length made by Mr. Richard Reives *here at* London, *which will bear an Aperture above three inches over, and yet make the Object proportionably big and diftinct; whereas there are very few thirty foot Glaßes that will indure an Aperture of more then two inches over. So that for* Telefcopes, *fuppofing we had a very ready way of making their Object Glaßes of exactly fpherical Surfaces, we might, by increafing the length of the Glaß, magnifie the Object to any affignable bigneß. And for performing both thefe, I cannot imagine any way more eafie,and more exact, then by this following Engine,by means of which, any Glaßes,of what length foever,may be fpeedily made.It feems the moft eafie, becaufe with one and the fame Tool may be with care ground an Object Glaß, of any length or breadth requifite, and that with very little or no trouble in fitting the Engine, and without much skill in the Grinder. It feems to be the moft exact, for to the very laft ftroke the Glaß does regulate and rectifie the Tool to its exact Figure; and the longer or more the Tool and Glaß are wrought together, the more exact will both of them be of the defir'd Figure. Further, the motions of the Glaß and Tool do fo croß each other, that there is not one point of eithers Surface,but has thoufands of croß motions thwarting it, fo that there can be no kind of Rings or Gutters made either in the Tool or Glaß.*

The contrivance of the Engine is, only to make the ends of two large *Mandrils* fo to move, that the Centers of them may be at any convenient diftance afunder, and that the *Axis* of the *Mandrils* lying both in the fame plain produc'd, may meet each other in any affignable Angle; both which requifites may be very well perform'd by the Engine defcrib'd in the third Figure of the firft *Scheme*: where A B fignifies the Beam of a Lath fixt perpendicularly or Horizontally, C D the two Poppet heads, fixt at about two foot diftance, E F an Iron *Mandril*,whofe tapering neck F runs in an adapted tapering brafs Collar; the other end E runs on the point of a Screw G; in a convenient place of this is faftned H a pully Wheel, and into the end of it,that comes through the Poppet head C, is fcrewed a Ring of a hollow *Cylinder* K, or fome other conveniently fhap'd Tool, of what wideneß fhall

be

The PREFACE.

be thought moft pr oper for the cize of Glaffes , about which it is to be imploy'd : As, for Object glaffes, between twelve foot and an hundred foot
long , the Ring may be about fix inches over, or indeed fomewhat
more for thofe longer Glaffes. It would be convenient alfo, and not
very chargeable, to have four or five feveral Tools; as one for all Glaffes
between an inch and a foot, one for all Glaffes between a foot and ten foot
long, another for all between ten and an hundred,a fourth for all between a
hundred and a thoufand foot long; and if Curiofity fhall ever proceed fo
far,one for all lengths between a thoufand and ten thoufand foot long ; for
indeed the principle is fuch,that fuppofing the *Mandrils* well made,and of a
good length, and fuppofing great care be ufed in working and polifhing
them,I fee no reafon,but that a Glafs of a thoufand,nay of ten thoufand foot
long, may be as well made as one of ten; for the reafon is the fame,fuppofing
the *Mandrils* and Tools be made fufficiently ftrong, fo that they cannot
bend; and fuppofing the Glafs, out of which they are wrought, be capable
of fo great a regularity in its parts as to refraction : this hollow *Cylinder* K
is to contain the Sand, and by being drove round very quick to and fro by
means of a fmall Wheel,which may be mov'd with ones foot, ferves to grind
the Glafs : The other *Mandril* is fhap'd like this, but it has an even neck inftead of a taper one,and runs in a Collar, that by the help of a Screw, and a
joynt made like M in the Figure, it can be ftill adjuftned to the wearing or
wafting neck : into the end of this *Mandril* is fcrewed a Chock N, on which
with Cement or Glew is faftned the piece of Glafs Q that is to be form'd;
the middle of which Glafs is to be plac'd juft on the edge of the Ring, and
the Lath O P is to be fet and fixt (by means of certain pieces and fcrews,
the manner whereof will be fufficiently evidenc'd by the Figure) in fuch
an Angle as is requifite to the forming of fuch a Sphere as the Glafs is defign'd to be of; the geometrical ground of which being fufficiently plain,
though not heeded before, I fhall, for brevities fake, pafs over. This laft
Mandril is to be made (by means of the former, or fome other Wheel) to
run round very fwift alfo, by which two crofs motions the Glafs cannot
chufe (if care be us'd) but be wrought into a moft exactly fpherical
Surface.

But becaufe we are certain, from the Laws of refraction *(which I
I have experimentally found to be fo,by an Inftrument I fhall prefently defcribe) that the* lines of the angles of Incidence *are proportionate to the* lines of the angles of Refraction, *therefore if Glaffes could
be made of thofe kind of Figures, or fome other, fuch as the moft incomparable* Des Cartes *has invented, and demonftrated in his Philofophical and
Mathematical Works,we might hope for a much greater perfection of Opticks
then can be rationally expected from fpherical ones;for though,cæteris paribus, we find, that the larger the* Telefcope Object Glaffes *are, and the
fhorter thofe of the* Microfcope, *the better they magnifie, yet both of them,*
befide

The PREFACE.

befide fuch determinate dimenfions , are by certain inconveniences rendred unufeful ; for it will be exceeding difficult *to make and* manage *a Tube* above an hundred foot long, *and it will be as difficult to* inlighten *an Objeƈt lefs then an hundred part of an inch diftant from the Objeƈt Glafs.*

I have not as yet made any attempts of that kind, though I know two or three wayes, which, as far as I have yet confidered, feem very probable,and may invite me to make a tryal as foon as I have an opportunity, of which I may hereafter perhaps acquaint the world. In the Interim, I fhall defcribe the Inftrument I even now mention'd, by which the refraƈtion *of all kinds of Liquors may be moft exaƈtly meafur'd, thereby to give the curious an opportunity of making what further tryals of that kind they fhall think requifite to any of their intended tryals ; and to let them fee that the laws of Refraƈtion are not only notional.*

The Inftrument confifted of five Rulers , or long pieces placed together, after the manner expreft in the fecond Figure of the firft *Scheme* , where A B denotes a ftraight piece of wood about fix foot and two inches long, about three inches over , and an inch and half thick , on the back fide of which was hung a fmall plummet by a line ftretcht from top to bottom, by which this piece was fet exaƈtly upright,and fo very firmly fixt ; in the middle of this was made a hole or center, into which one end of a hollow cylindrical brafs Box C C, fafhion'd as I fhall by and by defcribe , was plac'd, and could very eafily and truly be mov'd to and fro ; the other end of this Box being put into, and moving in, a hole made in a fmall arm D D; into this box was faftned the long Ruler E F, about three foot and three or four inches long, and at three foot from the above mention'd Centers P P was a hole E, cut through, and crofs'd with two fmall threads, and at the end of it was fixt a fmall fight G, and on the back fide of it was fixt a fmall Arm H, with a Screw to fix it in any place on the Ruler L M ; this Ruler L M was mov'd on the Center B (which was exaƈtly three foot diftance from the middle Center P) and a line drawn through the middle of it L M, was divided by a Line of cords into fome fixty degrees,and each degree was fubdivided into minutes , fo that putting the crofs of the threads in E upon any part of this divided line, I prefently knew what Angle the two Rules A B and E F made with each other, and by turning the Screw in H, I could fix them in any pofition. The other Ruler alfo R S was made much after the fame manner, only it was not fixt to the hollow cylindrical Box, but,by means of two fmall brafs Armes or Ears, it mov'd on the Centers of it ; this alfo, by means of the crofs threads in the hole S, and by a Screw in K, could be faftned on any divifion of another line of cords of the fame radius drawn on N O. And fo by that means, the Angle made by the two Rulers, A B and R S, was alfo known. The Brafs box C C in the middle was fhap'd very much like the Figure X, that is, it was a cylindrical Box ftopp'd clofe at either end,off of which a part both of the fides and bottomes was cut out, fo
that

The PREFACE.

that the Box, when the Pipe and that was joyne d to it, would contain the Water when fill'd half full, and would likewife, without running over, indure to be inclin'd to an Angle , equal to that of the greateft refraction of Water, and no more,without running over. The Ruler E F was fixt very faft to the Pipe V, fo that the Pipe V directed the length of the RulerE F and the Box and Ruler were mov'd on the Pin T T, fo as to make any defirable Angle with the Ruler A B. The bottom of this Pipe V was ftop'd with a fmall piece of exactly plain Glafs , which was plac'd exactly perpendicular to the Line of direction, or *Axis* of the Ruler E F. The Pins alfo T T were drill'd with fmall holes through the *Axis*,and through thofe holes was ftretcht and faftned a fmall Wire. There was likewife a fmall Pipe of Tin loofly put on upon the end of V, and reaching down to the fight G ; the ufe of which was only to keep any falfe Rayes of light from paffing through the bottom of V, and only admitting fuch to pafs as pierced through the fight G : All things being placed together in the manner defcrib'd in the Figure ; that is, the Ruler A B being fixt perpendicular, I fill'd the Box C C with Water, or any other Liquor, whofe refraction I intended to try , till the Wire paffing through the middle of it were juft covered : then I moved and fixt the Ruler F E at any affignable Angle, and placed the flame of a Candle juft againft the fight G ; and looking through the fight I, I moved the Ruler R S to and fro, till I perceived the light paffing through G to be covered, as 'twere, or divided by the dark Wire paffing through P P: then turning the Screw in K, I fixt it in that pofture : And through the hole S, I obferved what degree and part of it was cut by the crofs threads in S. And this gave me the Angle of Inclination, A P S anfwering to the Angle of Refraction B P E : for the furface of the Liquor in the Box will be alwayes horizontal , and confequently A B will be a perpendicular to it; the Angle therefore A P S will meafure, or be the Angle of Inclination in the Liquor ; next E P B muft be the Angle of Refraction,for the Ray that paffes through the fight G, paffes alfo perpendicularly through the Glafs *Diaphragme* at F, and confequently alfo perpendicularly through the lower furface of the Liquor contiguous to the Glafs, and therefore fuffers no refraction till it meet with the horizontal furface of the Liquor in C C, which is determined by the two Angles.

By means of this *Inftrument* I can with little trouble, and a very fmall quantity of any Liquor, examine, moft accurately, the refraction of it , not only for one inclination, but for all; and thereby am inabled to make very accurate Tables ; feveral of which I have alfo experimentally made,and find, that Oyl of Turpentine has a much greater Refraction then Spirit of Wine , though it be lighter ; and that Spirit of Wine has a greater Refraction then Water, though it be lighter alfo ; but that falt Water alfo has a greater Refraction then frefh, though it be heavier : but Allum water has a lefs refraction then common Water, though heavier alfo So that it feems,as to the refraction made in a Liquor,the fpeci-

f   fick

The PREFACE.

fick gravity is of no efficacy. By this I have also found, that look what pro-portion *the* Sine *of the* Angle *of one* Inclination *has to the* Sine *of the* Angle *of* Refraction, *correspondent to it, the same* proportion *have all the* Sines *of other Inclinations to the* Sines *of their appropriate Refractions.*

My way for measuring how much a Glass magnifies an Object, plac'd at a convenient distance from my eye, is this. Having rectifi'd the *Microscope*, to see the desir'd Object through it very distinctly, at the same time that I look upon the Object through the Glass with one eye, I look upon other Objects at the same distance with my other bare eye; by which means I am able, by the help of a *Ruler* divided into inches and small parts, and laid on the *Pedestal* of the *Microscope*, to cast, as it were, the magnifi'd appearance of the Object upon the Ruler, and thereby exactly to measure the Diameter it appears of through the Glass, which being compar'd with the Diameter it appears of to the naked eye, will easily afford the quantity of its magnify-ing.

The *Microscope*, which for the most part I made use of, was shap'd much like that in the sixth Figure of the first *Scheme*, the Tube being for the most part not above six or seven inches long, though, by reason it had four Draw-ers, it could very much be lengthened, as occasion required; this was con-triv'd with three Glasses; a small Object Glass at A, a thinner Eye Glass about B, and a very deep one about C: this I made use of only when I had oc-casion to see much of an Object at once; the middle Glass conveying a very great company of radiating Pencils, which would go another way, and throwing them upon the deep Eye Glass. But when ever I had occasion to examine the small parts of a Body more accurately, I took out the middle Glass, and only made use of one Eye Glass with the Object Glass, for always the fewer the Refractions are, the more bright and clear the Object appears. And therefore 'tis not to be doubted, but could we make a *Microscope* to have one only refraction, it would, *ceteris paribus*, far excel any other that had a greater number. And hence it is, that if you take a very clear piece of a broken *Venice* Glass, and in a Lamp draw it out into very small hairs or threads, then holding the ends of these threads in the flame, till they melt and run into a small round Globul, or drop, which will hang at the end of the thread; and if further you stick several of these upon the end of a stick with a little sealing Wax, so as that the threads stand upwards, and then on a Whetstone first grind off a good part of them, and afterward on a smooth Metal plate, with a little Tripoly, rub them till they come to be very smooth; if one of these be fixt with a little soft Wax against a small needle hole, prick'd through a thin Plate of Brass, Lead, Pewter, or any other Me-tal, and an Object, plac'd very near, be look'd at through it, it will both magnifie and make some Objects more distinct then any of the great *Micro-scopes*. But because these, though exceeding easily made, are yet very trou-blesome to be us'd, because of their smalness, and the nearness of the Object; therefore to prevent both these, and yet have only two Refractions, I pro-vided me a Tube of Brass, shap'd much like that in the fourth Figure of the first *Scheme*; into the smaller end of this I fixt with Wax a good *plano con-*

vex

The PREFACE.

vex Object Glafs, with the convex fide towards the Object, and into the bigger end I fixt alfo with wax a pretty large plano *Convex* Glafs, with the *convex* fide towards my eye, then by means of the fmall hole by the fide, I fill'd the intermediate fpace between thefe two Glaffes with very clear Water, and with a Screw ftopp'd it in; then putting on a Cell for the Eye, I could perceive an Object more bright then I could when the intermediate fpace was only fill'd with Air, but this, for other inconveniences, I made but little ufe of.

My way for fixing both the Glafs and Object to the Pedeftal moft conveniently was thus : Upon one fide of a round Pedeftal A B, in the fixth Figure of the firft *Scheme*, was fixt a fmall Pillar C C, on this was fitted a fmall Iron Arm D, which could be mov'd up and down, and fixt in any part of the Pillar, by means of a fmall Screw E; on the end of this Arm was a fmall Ball fitted into a kind of focket F, made in the fide of the Brafs Ring G, through which the fmall end of the Tube was fcrew'd; by means of which contrivance I could place and fix the Tube in what pofture I defir'd (which for many Obfervations was exceeding neceffary) and adjuften it moft exactly to any Object.

For placing the Object, I made this contrivance; upon the end of a fmall brafs Link or Staple H H, I fo faftned a round Plate I I, that it might be turn'd round upon its Center K, and going pretty ftiff, would ftand fixt in any pofture it was fet; on the fide of this was fixt a fmall Pillar P, about three quarters of an inch high, and through the top of this was thruft a fmall Iron pin M, whofe top juft ftood over the Center of the Plate; on this top I fixt a fmall Object, and by means of thefe contrivances I was able to turn it into all kind of pofitions, both to my Eye and the Light; for by moving round the fmall Plate on its center, I could move it one way, and by turning the Pin M, I could move it another way, and this without ftirring the Glafs at all, or at leaft but very little : the Plate likewife I could move to and fro to any part of the Pedeftal (which in many cafes was very convenient) and fix it alfo in any Pofition, by means of a Nut N, which was fcrew'd on upon the lower part of the Pillar C C. All the other Contrivances are obvious enough from the draught, and will need no defcription

Now though this were the Inftrument I made moft ufe of, yet I have made feveral other Tryals with other kinds of Microfcopes, *which both for* matter *and* form *were very different from common fpherical Glaffes. I have made a* Microfcope *with one piece of Glaß, both whofe furfaces were* plains. *I have made another only with a* plano concave, *without any kind of refleĉion, divers alfo by means of* refleĉion. *I have made others of* Waters, Gums, Refins, Salts, Arfenick, Oyls, *and with divers other* mixtures of *watery and* oyly Liquors. *And indeed the fubjeĉt is capable of a great variety ; but I find generally none more ufeful then that which is made with* two Glaffes, *fuch as I have already defcrib'd.*

What

The PREFACE.

What the things are I obferv'd, the following defcriptions will manifeft ; in brief, they were either exceeding fmall Bodies, or exceeding fmall Pores, or exceeding fmall Motions, fome of each of which the Reader will find in the following Notes, and fuch, as I prefume, (many of them at leaft) will be new, and perhaps not lefs ftrange : Some fpecimen of each of which Heads the Reader will find in the fubfequent delineations, and indeed of fome more then I was willing there fhould be; which was occafioned by my firft Intentions to print a much greater number then I have fince found time to compleat. Of fuch therefore as I had, I felecti-ed only fome few of every Head, which for fome particulars feem'd moft ob-fervable, rejecting the reft as fuperfluous to the prefent Defign.

What each of the delineated Subjects are, the following defcriptions an-next to each will inform, of which I fhall here, only once for all, add, That in divers of them the Gravers have pretty well follow'd my directions and draughts ; and that in making of them, I indeavoured (as far as I was able) firft to difcover the true appearance, and next to make a plain re-prefentation of it. This I mention the rather, becaufe of thefe kind of Objects there is much more difficulty to difcover the true fhape, then of thofe vifible to the naked eye, the fame Object feeming quite differing, in one pofition to the Light, from what it really is, and may be difcover'd in another. And therefore I never began to make any draught before by many examinations in feveral lights, and in feveral pofitions to thofe lights, I had difcover'd the true form. For it is exceeding difficult in fome Objects, to diftinguifh between a prominency and a depreffion, between a fhadow and a black ftain, or a reflection and a whitenefs in the colour. Befides, the tranfparency of moft Objects renders them yet much more difficult then if they were opacous. The Eyes of a Fly in one kind of light appear almoft like a Lattice, drill'd through with abun-dance of fmall holes ; which probably may be the Reafon, why the Ingeni-ous Dr. Power feems to fuppofe them fuch. In the Sunfhine they look like a Surface cover'd with golden Nails ; in another pofture, like a Sur-face cover'd with Pyramids ; in another with Cones ; and in other po-ftures of quite other fhapes ; but that which exhibits the beft, is the Light collected on the Object, by thofe means I have already defcrib'd.

And

The PREFACE.

And this was undertaken in profecution of the Defign which the ROYAL SOCIETY *has propos'd to it felf. For the Members of the Affembly having before their eys fo many* fatal Inftances *of the errors and falfhoods,in which the greateft part of mankind has fo long wandred, becaufe they rely'd upon the ftrength of humane Reafon alone*, *have begun anew to correct all* Hypothefes *by fenfe, as* Seamen *do their* dead Reckonings *by* Cœleftial Obfervations;*and to this purpofe it has been their principal indeavour to* enlarge *&* ftrengthen *the* Senfes *by* Medicine,*and by fuch* outward Inftruments *as are proper for their particular works. By this means they find fome reafon to fuffect,that thofe effects of Bodies,which have been commonly attributed to* Qualities, *and thofe confefs'd to be* occult, *are perform'd by the fmall* Machines *of* Nature, *which are not to be difcern'd without thefe helps, feeming the meer products of* Motion,Figure,*and* Magnitude; *and that the* Natural Textures, *which fome call the* Plaftick *faculty, may be made in* Looms,*which a greater perfection of* Opticks *may make difcernable by thefe* Glaffes:*fo as now they are no more puzzled about them,then the vulgar are to conceive,how* Tapeftry *or* fiowred Stuffs *are woven. And the ends of all thefe* Inquiries *they intend to be the* Pleafure *of Contemplative minds, but above all,the* eafe *and* difpatch *of the labours of mens hands.They do indeed neglect no opportunity to bring all the* rare things *of* Remote Countries *within the compafs of their knowledge and practice.But they* ftill *acknowledg their* moft ufeful Informations *to arife from* common *things, and from* diverfifying *their moft* ordinary *operations upon them. They do not wholly reject Experiments of meer* light *and* theory ; *but they principally aim at fuch, whofe Applications will* improve *and* facilitate *the prefent way of* Manual Arts. *And though fome men, who are perhaps taken up about lefs honourable Employments, are pleas'd to cenfure their proceedings, yet they can fhew more fruits of their firft three years,wherein they have affembled, then any other* Society *in* Europe *can for a much larger fpace of time. Tis true, fuch undertakings as theirs do commonly meet with fmall incouragement, becaufe men are generally rather taken with the* plaufible *and* difcurfive, *then the* real *and the* folid part *of* Philofophy ; *yet by the good fortune of their inftitution,in an Age of all others the moft* inquifitive,*they have been affifted by the* contribution *and* prefence *of very many of the chief* Nobility *and* Gentry,

g

and

The PREFACE.

and others,who are some of the most confiderable *in their feveral Profeffions.*
But that that yet farther convinces me of the Real efteem *that the more fe-*
rious *part of men have of this* Society,*is, that feveral* Merchants,*men who*
act in earneft(whofe Object is meum *&* tuum,*that great* Rudder *of humane*
affairs)have adventur'd confiderable fums of Money,*to put in practice what*
fome of our Members *have contrived,* and have continued ftedfaft *in their*
good opinions of fuch Indeavours, when not one of a hundred of the vulgar
have believed their undertakings feafable.And it is alfo fit to be added,that
they have one advantage peculiar to themfelves,that very many of their num-
ber are men of Converfe *and* Traffick ; *which is a good Omen, that their*
attempts will bring Philofophy from words *to* action,*feeing the men of Bufi-*
nefs have had fo great a fhare in their firft foundation.

And of this kind I ought not to conceal one particular Generofity,*which more*
nearly concerns my felf.It is the munificence *of* Sir JohnCutler,*in endowing*
a Lecture for the promotion of Mechanick Arts,*to be governed and directed*
*by*This*Society.This*Bounty *I mention for the* Honourablenefs *of the thing it*
felf,and for the expectation which I have of the efficacy *of the* Example ;*for*
it cannot now be objected to them,that their Defigns will be efteemed frivolous
and vain, *when they have fuch a* real Teftimony *of* the Approbation *of*
a Man that *is fuch an* eminent Ornament *of this renowned City, and one,*
who, by the Variety, *and the* happy Succefs, *of his negotiations, has given*
evident proofs, that he is not eafie to be deceiv'd. This Gentleman has well
obferv'd, that the Arts *of life have been too long* imprifon'd *in the dark*
fhops of Mechanicks themfelves,& there hindred from growth,*either by ig-*
norance,or felf-intereft:and he has bravely freed *them from thefe* inconveni-
ences:He hath not only obliged Tradefmen,*but* Trade *it felf:He has done a*
work that is worthy of London, *and has taught the chief City of Commerce*
in the world the right way how Commerce is to be improv'd. We have already
feen many other great figns of Liberality *and a large mind, from the fame*
hand:For by his diligence *about the*Corporation for the Poor;*by his hono-*
rable Subfcriptions *for the rebuilding of* St.Paul's;*by his chearful* Disburf-
ment *for the replanting of* Ireland,*and by many other fuch* publick works,
he has fhewn by what means he indeavours to eftablifh *his Memory ; and*
now by this laft gift he has done that,which became one of the wifeft Citizens

of

The PREFACE.

of our Nation to accomplish, seeing one of the wiseft of our Statefmen, the Lord Verulam, *firſt propounded it.*

But to return to my Subject, from a digreſſion, which, I hope, my Reader will pardon me, seeing the Example is so rare that I can make no more such digreſſions. If theſe my firſt Labours ſhall be any wayes uſeful to inquiring men, I muſt attribute the incouragement and promotion of them to a very Reverend and Learned Perſon, of whom this ought in juſtice to be ſaid, That there is ſcarce any one Invention, which this Nation has produc'd in our Age, but it has ſome way or other been ſet forward by his aſſiſtance. My Reader, I believe, will quickly gheß, that it is Dr. Wilkins that I mean. He is indeed a man born for the good of mankind, and for the honour of his Couutry. In the ſweetneſs of whoſe behaviour, in the calmneſs of his mind, in the unbounded goodneſs of his heart, we have an evident Inſtance, what the true and the primitive unpaſſionate Religion was, before it was ſowred by particular Factions. In a word, his Zeal has been ſo conſtant and effectual in advancing all good and profitable Arts, that as one of the Antient Romans ſaid of Scipio, That he thanked God that he was a Roman; becauſe whereever Scipio had been born, there had been the ſeat of the Empire of the world: So may I thank God, that Dr. Wilkins was an Engliſhman, for whereever he had lived, there had been the chief Seat of generous Knowledge and true Philoſophy. To the truth of this, there are ſo many worthy men living that will ſubſcribe, that I am confident, what I have here ſaid, will not be look'd upon, by any ingenious Reader, as a Panegyrick, but only as a real teſtimony.

By the Advice of this Excellent man I firſt ſet upon this Enterpriſe, yet ſtill came to it with much Reluctancy, becauſe I was to follow the footſteps of ſo eminent a Perſon as Dr. Wren, who was the firſt that attempted any thing of this nature; whoſe original draughts do now make one of the Ornaments of that great Collection of Rarities in the Kings Cloſet. This Honor, which his firſt beginnings of this kind have receiv'd, to be admitted into the moſt famous place of the world, did not ſo much incourage, as the hazard of coming after Dr. Wren did affright me; for of him I muſt affirm, that, ſince the time of Archimedes, there ſcarce ever met in one man, in ſo

great

great a perfection, such a Mechanical Hand, *and so* Philosophical *a* Mind.

But at last, being assured both by Dr. Wilkins, *and* Dr. Wren *himself, that he had given over his intentions of prosecuting it, and not finding that there was any else design'd the pursuing of it, I set upon this undertaking, and was not a little incourag'd to proceed in it, by the Honour the* Royal Society *was pleas'd to favour me with, in approving of those draughts (which from time to time as I had an opportunity of describing) I presented to them. And particularly by the Incitements of divers of those Noble and excellent Persons of it, which were my more especial Friends, who were not less urgent with me for the publishing, then for the prosecution of them.*

After I had almost compleated these Pictures and Observations (having had divers of them ingraven, and was ready to send them to the Press) I was inform'd, that the Ingenious Physitian Dr. Henry Power *had made several* Microscopical *Observations, which had I not afterwards, upon our interchangably viewing each others Papers, found that they were for the most part differing from mine, either in the Subject it self, or in the particulars taken notice of; and that his design was only to print Observations without Pictures, I had even then* suppressed *what I had so far proceeded in. But being further* excited *by several of my Friends, in complyance with their opinions, that it would not be unacceptable to several inquisitive Men, and hoping also, that I should thereby discover something New to the World, I have at length cast in my Mite, into the vast Treasury of* A Philosophical History. *And it is my* hope, *as well as* belief, *that these my* Labours *will be no more comparable to the Productions of many other* Natural Philosophers, *who are now every where busie about greater things; then my little Objects are to be compar'd to the greater and more beautiful* Works of Nature, A Flea, a Mite, a Gnat, *to an* Horse, an Elephant, *or a* Lyon.

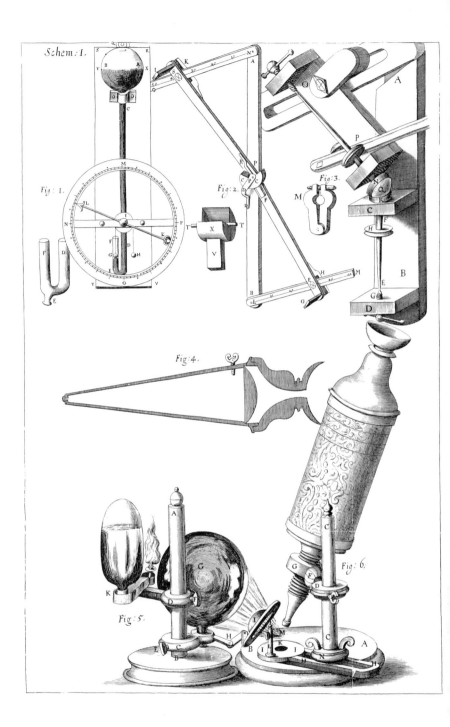

MICROGRAPHIA,

OR SOME

Phyſiological Deſcriptions

OF

MINUTE BODIES,

MADE BY

MAGNIFYING GLASSES;

WITH

OBSERVATIONS and INQUIRIES thereupon.

Obſerv. I. *Of the Point of a ſharp ſmall Needle.*

S in *Geometry*, the moſt natural way of beginning is from a Mathematical *point* ; ſo is the ſame method in Obſervations and *Natural hiſtory* the moſt genuine, ſimple, and inſtructive. We muſt firſt endeavour to make *letters*, and draw *ſingle* ſtrokes true, before we venture to write whole *Sentences*, or to draw large *Pictures*. And in *Phyſical* Enquiries, we muſt endeavour to follow Nature in the more *plain* and *eaſie* ways ſhe treads in the moſt *ſimple* and *uncompounded bodies*, to trace her ſteps, and be acquainted with her manner of walking there, before we venture our ſelves into the multitude of *meanders* ſhe has in *bodies of a more complicated* nature ; leſt, being unable to diſtinguiſh and judge of our way, we quickly loſe both *Nature* our Guide,and *our ſelves* too and are left to wander in the *labyrinth* of groundleſs opinions ; wanting both *judgment*, that *light*, and *experience*, that *clew*, which ſhould direct our proceedings.

We will begin theſe our Inquiries therefore with the Obſervations of Bodies of the moſt *ſimple nature* firſt, and ſo gradually proceed to thoſe of a more *compounded* one. In proſecution of which method, we ſhall begin with a *Phyſical point*; of which kind the *Point of a Needle* is commonly reckon'd for one ; and is indeed, for the moſt part, made ſo ſharp, that the naked eye cannot diſtinguiſh any parts of it : It very eaſily pierces, and makes its way through all kind of bodies ſofter then it ſelf: But if view'd with a very good *Microſcope*, we may find that the *top* of a Needle (though as to the

*Schem.*2, *Fig.*1.

B   ſenſe

fenfe very *fharp*) appears a *broad*,*blunt*, and very *irregular* end; not refembling a Cone, as is imagin'd, but onely a piece of a tapering body, with a great part of the top remov'd, or deficient. The Points of Pins are yet more blunt, and the Points of the moft curious Mathematital Inftruments do very feldome arrive at fo great a fharpnefs; how much therefore can be built upon demonftrations made onely by the productions of the Ruler and Compaffes, he will be better able to confider that fhall but view thofe *points* and *lines* with a *Microfcope*.

Now though this point be commonly accounted the fharpeft (whence when we would exprefs the fharpnefs of a point the moft *fuperlatively*, we fay, As fharp as a Needle) yet the *Microfcope* can afford us hundreds of Inftances of Points many thoufand times fharper: fuch as thofe of the *hairs*, and *briftles*, and *claws* of multitudes of *Infects*; the *thorns*, or *crooks*, or *hairs* of *leaves*, and other fmall vegetables; nay, the ends of the *ftiriæ* or fmall *parallelipipeds* of *Amianthus*, and *alumen plumofum*; of many of which, though the Points are fo fharp as not to be vifible, though view'd with a *Microfcope* (which magnifies the Object, in bulk, above a million of times) yet I doubt not, but were we able *practically* to make *Microfcopes* according to the *theory* of them, we might find hills, and dales, and pores, and a fufficient bredth, or expanfion, to give all thofe parts elbow-room, even in the blunt top of the very Point of any of thefe fo very fharp bodies. For certainly the *quantity* or extenfion of any body may be *Divifible in infinitum*, though perhaps not the *matter*.

But to proceed: The Image we have here exhibited in the firft Figure, was the top of a fmall and very fharp Needle, whofe point *a a* neverthelefs appear'd through the *Microfcope* above a quarter of an inch broad, not round nor flat, but *irregular* and *uneven*; fo that it feem'd to have been big enough to have afforded a hundred armed Mites room enough to be rang'd by each other without endangering the breaking one anothers necks, by being thruft off on either fide. The furface of which, though appearing to the naked eye very fmooth, could not neverthelefs hide a multitude of holes and fcratches and ruggedneffes from being difcover'd by the *Microfcope* to inveft it, feveral of which inequalities (as A,B,C, feem'd *holes* made by fome fmall fpecks of *Ruft*; and D fome *adventitious body*, that ftuck very clofe to it) were *cafual*. All the reft that roughen the furface, were onely fo many marks of the rudenefs and bungling of *Art*. So unaccurate is it, in all its productions, even in thofe which feem moft neat, that if examin'd with an organ more acute then that by which they were made, the more we fee of their *fhape*, the lefs appearance will there be of their *beauty*: whereas in the works of *Nature*, the deepeft Difccveries fhew us the greateft Excellencies. An evident Argument, that he that was the Author of all thefe things, was no other then *Omnipotent*; being able to include as great a variety of parts and contrivances in the yet fmalleft Difcernable Point, as in thofe vafter bodies (which comparatively are called alfo Points) fuch as the *Earth*, *Sun*, or *Planets*. Nor need it feem ftrange that the Earth it felf may be by an *Analogie* call'd a Phyfical Point: For as its body, though now

fo

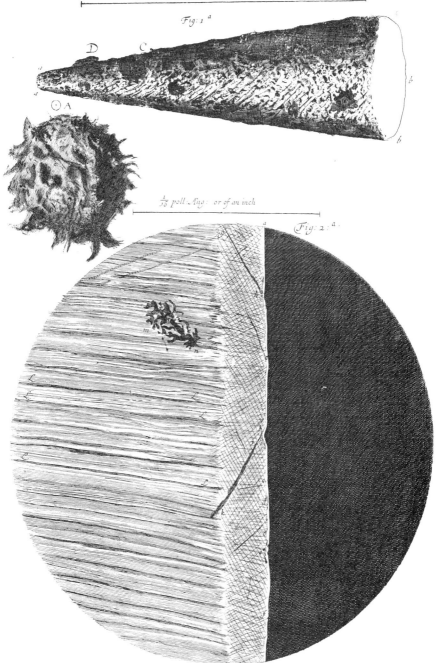

$\frac{1}{10}$ pallicis *Aug:* Or of an inch

Fig: 1.ᵈ

D   C

⊙ A

$\frac{1}{16}$ *poll: Aug:* or of an inch

Fig: 2:ᵈ

fo near us as to fill our eys and fancies with a fenfe of the vaftnefs of it, may by a little Diftance, and fome convenient *Diminifhing* Glaffes, be made vanifh into a fcarce vifible Speck, or Point (as I have often try'd on the *Moon,* and (when not too bright) on the *Sun* it felf.) So, could a Mechanical contrivance fucceffully anfwer our *Theory,* we might fee the leaft fpot as big as the Earth it felf ; and Difcover, as *Des Cartes Diop ch.* alfo conjectures, as great a variety of bodies in the *Moon,* or *Planets,* as in 10. § 9. the *Earth.*

But leaving thefe Difcoveries to future Induftries, we fhall proceed to add one Obfervation more of a *point* commonly fo call'd,that is, the mark of a *full ftop,* or *period.* And for this purpofe I obferved many both *printed* ones and *written* ; and among multitudes I found *few* of them more *round* or *regular* then this which I have delineated in the third figure of the fe-cond Scheme, but *very many* abundantly *more disfigur'd* ; and for the moft part if they feem'd equally round to the eye, I found thofe points that had been made by a *Copper-plate,* and Roll-prefs, to be as misfhapen as thofe which had been made with *Types,* the moft curious and fmothly *engraven ftrokes* and *points,* looking but as fo many *furrows* and *holes,* and their *printed impreffions,* but like *fmutty daubings* on a matt or uneven floor with a blunt extinguifht brand or ftick's end. And as for *points* made with a *pen* they were much *more rugged* and *deformed.* Nay,having view'd certain pieces of exceeding curious writing of the kind (one of which in the bredth of a *two-pence* compris'd *the Lords prayer, the Apoftles Creed, the ten Commandments, and about half a dozen verfes befides of the Bible,* whofe *lines* were fo *fmall* and *near together,* that I was unable to *number* them with my *naked eye,* a very ordinary *Microfcope,* I had then a-bout me, inabled me to fee that what the Writer of it had afferted was *true,* but withall difcover'd of what pitifull *bungling fcribbles* and *fcrawls* it was compos'd, *Arabian* and *China characters* being almoft as well fhap'd ; yet thus much I muft fay for the Man, that it was for the moft part *legible* enough, though in fome places there wanted a good *fantfy* well *prepofeft* to help one through. If this manner of *fmall writing* were made *eafie* and *practicable* (and I think I know fuch a one, but have never yet made tryal of it, whereby one might be inabled to write *a great deale* with *much eafe,* and *accurately* enough in a very *little roome*) it might be of very good ufe to convey *fecret Intelligence* without any danger of *Difcovery* or *miftrufting.* But to come again to the point. The *Irregularities* of it are caufed by three or four *coadjutors,* one of which is, the *uneven furface* of the *paper,* which at beft appears no fmother then a very courfe piece of *fhag'd cloth,* next the *irregularity of the Type* or *Ingraving,* and a third is the *rough Daubing* of the *Printing-Ink* that lies upon the inftrument that makes the impreffion, to all which, add the *variation* made by the Different *lights* and *fhadows,* and you may have fufficient reafon to ghefs that a *point* may appear much more *ugly* then *this,* which I have here prefented, which though it appear'd through the *Microfcope gray,* like a great fplatch of *London* dirt, about three inches over ; yet to the *naked eye* it was *black,* and no bigger then that in the midft of the Circle A. And could I have

found

found Room in this Plate to have inferted an O you fhould have feen that the *letters* were not more diftinct then the *points* of Diftinction, nor a *drawn circle* more exactly *fo*, then we have now fhown a *point* to be a *point*.

Obferv. II. *Of the Edge of a Razor.*

THe fharpeft *Edge* hath the fame kind of affinity to the fharpeft *Point* in Phyficks, as a *line* hath to a *point* in Mathematicks; and therefore the Treaty concerning this, may very properly be annexed to the former. A Razor doth appear to be a Body of a very neat and curious a-fpect, till more clofely viewed by the *Microfcope,* and there we may obferve its very Edge to be of all kind of fhapes, except what it fhould be. For examining that of a very fharp one, I could not find that any part of it had any thing of fharpnefs in it; but it appear'd a rough furface of a very confiderable bredth from fide to fide, the narroweft part not feeming thinner then the back of a pretty thick Knife. Nor is't likely that it fhould appear any otherwife, fince as we juft now fhew'd that a *point* appear'd a *circle,* 'tis rational a *line* fhould be a *parallelogram.*

Now for the drawing this fecond Figure(which reprefents a part of the Edge about half a quarter of an inch long of a Razor well fet) I fo plac'd it between the Object-glafs & the light, that there appear'd a reflection from the very Edge, reprefented by the white line *a b c d e f.* In which you may perceive it to be fomewhat fharper then elfewhere about *d,* to be indented or pitted about *b,* to be broader and thicker about *c,* and unequal and rugged about *e,* and pretty even between *a b* and *e f.* Nor was that part of the Edge *g h i k* fo fmooth as one would imagine fo fmooth bodies as a Hone and Oyl fhould leave it; for befides thofe multitudes of fcratches, which appear to have raz'd the furface *g h i k,* and to crofs each other every way which are not half of them expreft in the Figure, there were feveral great and deep fcratches, or furrows, fuch as *g h* and *i k,* which made the furface yet more rugged, caus'd perhaps by fome fmall Duft cafually falling on the Hone, or fome harder or more flinty part of the Hone it felf. The other part of the Razor *l l,* which is polifh'd on a grinding-ftone, appear'd much rougher then the other, looking almoft like a plow'd field, with many parallels, ridges, and furrows, and a cloddy, as 'twere, or an uneven furface : nor fhall we wonder at the roughneffes of thofe furfaces, fince even in the moft curious wrought Glaffes for *Microfcopes,* and other Optical ufes, I have, when the Sun has fhone well on them, difcover'd their furface to be varioufly raz'd or fcratched, and to confift of an infinite of fmall broken furfaces, which reflect the light of very various and differing colours. And indeed it feems impoffible by Art to cut the furface of any hard and brittle body fmcoth, fince *Putte,* or even the moft curious *Powder* that can be made ufe of, to polifh fuch a body, muft confift of little hard rough particles, and each of them muft cut its way, and confequently leave fome kind of gutter or

furrows

Schem: **III**

Fig: 1

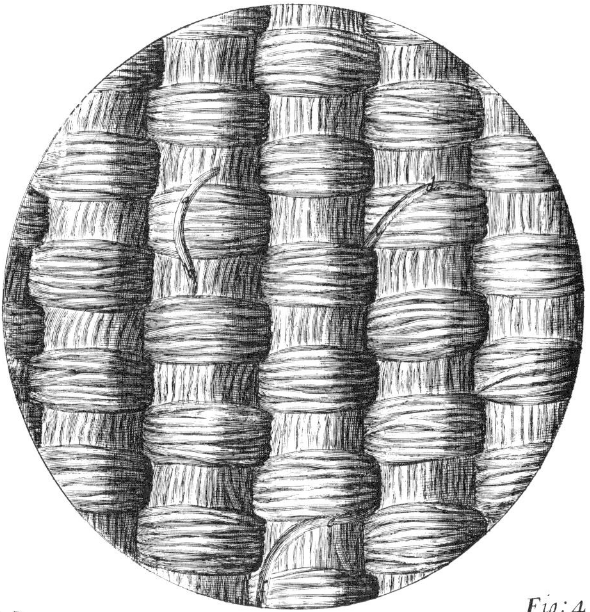

Fig: 3

Fig: 4

A

Fig: 2

C

D

B

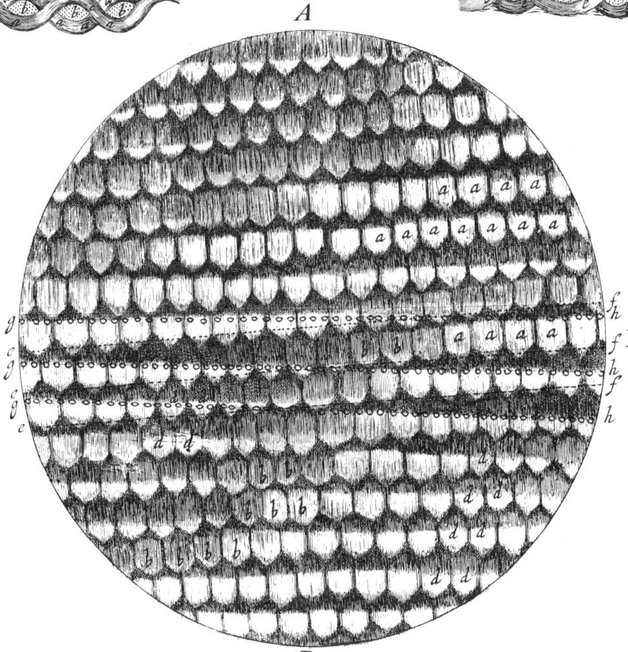

furrow behind it. And though Nature does seem to do it very readily in all kinds of fluid bodies, yet perhaps future obfervators may difcover even thefe alfo rugged; it being very probable, as I elfewhere fhew, that fluid bodies are made up of fmall folid particles varioufly and ftrongly mov'd, and may find reafon to think there is fcarce a furface *in rerum natura* perfectly fmooth. The black fpot *m n*, I ghefs to be fome fmall fpeck of ruft, for that I have oft obferv'd to be the manner of the working of Corrofive Juyces. To conclude, this Edge and piece of a Razor, if it had been really fuch as it appear'd through the *Microfcope*, would fcarcely have ferv'd to cleave wood, much lefs to have cut off the hair of beards, unlefs it were after the manner that *Lucian* merrily relates *Charon* to have made ufe of, when with a Carpenters Axe he chop'd off the beard of a fage Philofopher, whofe gravity he very cautioufly fear'd would indanger the overfetting of his Wherry.

Obferv. III. *Of fine Lawn, or Linnen Cloth.*

Schem. 15.
Fig. 3.

THis is another product of Art, A piece of the fineft Lawn I was able to get, fo curious that the threads were fcarce difcernable by the naked eye, and yet through an ordinary *Microfcope* you may perceive what a goodly piece of *coarfe Matting* it is; what proportionable cords each of its threads are, being not unlike, both in fhape and fize, the bigger and coarfer kind of *fingle Rope-yarn*, wherewith they ufually make *Cables*. That which makes the Lawn fo tranfparent, is by the *Microfcope*, nay by the naked eye, if attentively viewed, plainly enough evidenced to be the multitude of fquare holes which are left between the threads, appearing to have much more hole in refpect of the intercurrent parts then is for the moft part left in a *lattice-window*, which it does a little refemble, onely the croffing parts are round and not flat.

These threads that compofe this fine contexture, though they are as fmall as thofe that conftitute the finer forts of Silks, have notwithftanding nothing of their gloffie, pleafant, and lively reflection. Nay, I have been informed both by the Inventor himfelf, and feveral other eye-witneffes, that though the flax, out of which it is made, has been (by a fingular art, of that excellent Perfon, and Noble Vertuofo, M. *Charls Howard*, brother to the *Duke of Norfolk*) fo curioufly drefs'd and prepar'd, as to appear both to the eye and the touch, full as *fine* and as *gloffie*, and to receive all kinds of colours, as well as Sleave-Silk; yet when this Silken Flax is twifted into threads, it quite lofeth its former lufter, and becomes as plain and bafe a thread to look on, as one of the fame bignefs, made of common Flax.

The reafon of which odd *Phenomenon* feems no other then this; that though the curioufly dreft Flax has its parts fo exceedingly fmall, as to equallize, if not to be much fmaller then the clew of the Silk-worm, efpecially in thinnefs, yet the differences between the figures of the conftituting filaments are fo great, and their fubftances fo various, that whereas

C   thofe

thofe of the *Silk* are *fmall,round, hard, tranfparent*, and to their bignefs
proportionably *ftiff*, fo as each filament preferves its proper *Figure*, and
confequently its vivid *reflection* intire, though twifted into a thread, if
not too hard ; thofe of Flax are *flat, limber, fofter*, and *lefs tranfparent*, and
in twifting into a thread they joyn, and lie fo clofe together, as to lofe their
own, and deftroy each others particular reflections. There feems therefore three Particulars very requifite to make the fo dreft Flax appear Silk
alfo when fpun into threads. Firft, that the fubftance of it fhould be
made more *clear* and *tranfparent*, Flax retaining in it a kind of opacating
brown, or yellow ; and the parts of the whiteft kind I have yet obferv'd
with the *Microfcope* appearing white, like flaw'd Horn or Glafs, rather
then clear, like clear Horn or Glafs. Next that, the filaments fhould each
of them be *rounded*, if that could be done, which yet is not fo very neceffary, if the firft be perform'd, and this third, which is, that each of the
fmall filaments be *ftifned* ; for though they be fquare, or flat, provided
they be *tranfparent* and ftiff, much the fame appearances muft neceffarily
follow. Now, though I have not yet made trial, yet I doubt not, but that
both thefe proprieties may be alfo induc'd upon the Flax, and perhaps too
by one and the fame Expedient, which fome trials may quickly inform any
ingenious attempter of, who from the ufe and profit of fuch an Invention,
may find fufficient argument to be prompted to fuch Inquiries. As for
the *tenacity* of the fubftance of Flax, out of which the thread is made, it
feems much inferiour to that of Silk, the one being a *vegetable*, the
other an *animal* fubftance.   And whether it proceed from the better con-
coction, or the more homogeneous conftitution of *animal* fubftances
above thofe of *vegetables*, I do not here determine ; yet fince I generally find, that *vegetable* fubftances do not equalize the *tenacity* of *animal*, nor thefe the *tenacity* of fome purified *mineral* fubftances ; I am
very apt to think, that the *tenacity* of bodies does not proceed from the
hamous, or *hooked* particles, as the *Epicureans*, and fome modern *Philofophers* have imagin'd ; but from the more exact *congruity* of the conftituent parts, which are contiguous to each other, and fo bulky, as not to
be eafily feparated, or fhatter'd, by any fmall pulls or concuffion of
heat.

Obferv. IV.   *Of fine waled Silk, or Taffety.*

Schem. 3.
Fig. 1.
T His is the appearance of a piece of very fine Taffety-riband in the
bigger magnifying Glafs, which you fee exhibits it like a very convenient fubftance to make Bed-matts, or Door-matts of, or to ferve for Bee-
hives, Corn-fcuttles, Chairs, or Corn-tubs, it being not unlike that kind of
work, wherewith in many parts in *England*, they make fuch Utenfils of
Straw, a little wreathed, and bound together with thongs of Brambles. For
in this Contexture, each little filament, fiber, or clew of the Silk-worm,
feem'd about the bignefs of an ordinary Straw, as appears by the little irregular

regular pieces,*a b,c d*,and *e f*; The *Warp*,or the thread that ran croffing the Riband,appear'd like a fingle Rope of an Inch Diameter; but the *Woof*, or the thread that ran the length of the Riband, appear'd not half fo big. Each Inch of fix-peny-broad Riband appearing no lefs then a piece of Matting Inch and half thick, and twelve foot fquare; a few yards of this, would be enough to floor the long Gallery of the *Loure* at *Paris*. But to return to our piece of Riband : It affords us a not unpleafant ob-ject, appearing like a bundle, or wreath, of very clear and tranfparent *Cylinders*,if the Silk be white, and curioufly ting'd; if it be colour'd,each of thofe fmall horney *Cylinders* affording in fome place or other of them, as vivid a reflection, as if it had been fent from a *Cylinder* of Glafs or Horn. In-fo-much, that the reflections of Red, appear'd as if coming from fo many *Granates*, or *Rubies*. The lovelinefs of the colours of Silks above thofe of hairy Stuffs,or Linnen,confifting as I elfe-where intimate,chiefly in the tranfparency, and vivid reflections from the *Concave*,or inner furface of the *tranfparent Cylinder*, as are alfo the colours of Precious Stones; for moft of the reflections from each of thefe *Cylinders*, come from the *Concave* furface of the air, which is as 'twere the foil that incompaffes the *Cylinder*. The colours with which each of thefe *Cylinders* are ting'd, feem partly to be fuperficial, and fticking to the out-fides of them; and partly, to be imbib'd, or funck into the fubftance of them : for Silk, feeming to be little elfe then a dried thread of Glew, may be fuppos'd to be very eafily relaxt,and foftened, by being fteeped in warm, nay in cold, if pene-trant, juyces or liquors. And thereby thofe tinctures, though they tinge perhaps but a fmall part of the fubftance, yet being fo highly impregnated with the colour, as to be almoft black with it, may leave an impreffion ftrong enough to exhibit the defir'd colour. A pretty kinde of artifi-cial Stuff I have feen, looking almoft like tranfparent Parchment, Horn, or Ifing-glafs, and perhaps fome fuch thing it may be made of, which be-ing tranfparent, and of a glutinous nature, and eafily mollified by keep-ing in water, as I found upon trial, had imbib'd, and did remain ting'd with a great variety of very vivid colours, and to the naked eye, it look'd very like the fubftance of the Silk. And I have often thought, that pro-bably there might be a way found out, to make an artificial glutinous compofition, much refembling, if not full as good, nay better, then that Excrement,or whatever other fubftance it be out of which, the Silk-worm wire-draws his clew. If fuch a compofition were found, it were certain-ly an eafie matter to find very quick ways of drawing it out into fmall wires for ufe. I need not mention the ufe of fuch an Invention,nor the be-nefit that is likely to accrue to the finder,they being fufficiently obvious. This hint therefore,may, I hope, give fome Ingenious inquifitive Perfon an occafion of making fome trials, which if fuccefsfull, I have my aim, and I fuppofe he will have no occafion to be difpleas'd.

Obferv. V.

Obſerv. V. *Of watered Silks, or Stuffs.*

THere are but few *Artificial* things that are worth obſerving with a *Microſcope*; and therefore I ſhall ſpeak but briefly concerning them. For the Productions of art are ſuch rude miſ-ſhapen things, that when view'd with a *Microſcope*,there is little elſe obſervable,but their deformity. The moſt curious Carvings appearing no better then thoſe rude *Ruſſian* Images we find mention'd in *Purchas*, where three notches at the end of a Stick, ſtood for a face. And the moſt ſmooth and burniſh'd ſurfaces appear moſt rough and unpoliſht : So that my firſt Reaſon why I ſhall add but a few obſervations of them, is, their miſ-ſhapen form ; and the next, is their uſeleſsneſs. For why ſhould we trouble our ſelves in the examination of that form or ſhape (which is all we are able to reach with a *Microſcope*) which we know was deſign'd for no higher a uſe, then what we were able to view with our naked eye? Why ſhould we endeavour to diſcover myſteries in that which has no ſuch thing in it? And like *Rabbins* find out *Caballiſms*, and *ænigmâs* in the Figure, and placing of Letters, where no ſuch thing lies hid : whereas in *natural* forms there are ſome ſo ſmall, and ſo curious,and their deſign'd buſineſs ſo far remov'd beyond the reach of our ſight,that the more we magnify the object, the more excellencies and myſteries do appear ; And the more we diſcover the imperfections of our ſenſes, and the Omnipotency and Infinite perfections of the great Creatour. I ſhall therefore onely add one or two Obſervations more of *artificial* things, and then come to the Treaty concerning ſuch matters as are the Productions of a more curious Workman. One of theſe,ſhall be that of a piece of water'd Silk, repreſented in the ſecond Figure of the third *Scheme*,as it appear'd through the leaſt magnifying Glaſs. *A B.* ſignifying the long way of the Stuff,and *C D* the broad way. This Stuff, if the right ſide of it be looked upon, appears to the naked eye, all over ſo waved, undulated, or grain'd, with a curious, though irregular variety of brighter and darker parts, that it adds no ſmall gracefulneſs to the Gloſs of it. It is ſo known a propriety, that it needs but little explication, but it is obſervable, which perhaps every one has not conſidered, that thoſe parts which appear the darker part of the wave, in one poſition to the light, in another appears the lighter,and the contrary;and by this means the undulations become tranſient, and in a continual change,according as the poſition of the parts in reſpect of the incident beams of light is varied. The reaſon of which odd *phænomena*, to one that has but diligently examin'd it even with his naked eye, will be obvious enough. But he that obſerves it with a *Microſcope*, may more eaſily perceive what this *Proteus* is, and how it comes to change its ſhape. He may very eaſily perceive, that it proceeds onely from the variety of the *Reflections* of light, which is caus'd by the various *ſhape of the Particles*, or little protuberant parts of the thread that compoſe the ſurface ; and that thoſe parts of the waves that

appear

appear the brighter, throw towards the eye a multitude of small reflections of light, whereas the darker scarce afford any. The reason of which
reflection, the *Microscope* plainly discovers, as appears by the Figure. In
which you may perceive, that the brighter parts of the surface consist of
an abundance of large and strong reflections, denoted by *a, a, a, a, a,* &c.
for the surfaces of those threads that run the *long way,* are by the Mechanical process of watering, *creas'd* or *angled* in another kind of posture
then they were by the weaving: for by the weaving they are onely *bent
round* the warping threads ; but by the watering, they are *bent with an
angle, or elbow,* that is instead of lying, or being bent *round* the threads,
as in the third Figure, *a, a, a, a, a,* are about *b, b, b* (*b, b, b* representing the
ends, as 'twere, of the cross threads, they are bent about) they are creas'd
on the top of those threads, with an *angle,* as in the fourth Figure, and
that with all imaginable variety ; so that, whereas before they reflected
the light onely from one point of the round surface, as about *c, c, c,* they
now when water'd, reflect the beams from more then half the whole surface, as *d e, d e, d e,* and in other postures they return no reflections at all
from those surfaces. Hence in one posture they compose the brighter
parts of the waves, in another the darker. And these reflections are also
varied, according as the particular parts are variously bent. The reason
of which creasing we shall next examine ; and here we must fetch our information from the Mechanism or manner of proceeding in this operation ;
which, as I have been inform'd, is no other then this.

They double all the Stuff that is to be water'd, that is, they crease it just
through the middle of it, the whole length of the piece, leaving the right
side of the Stuff inward, and placing the two edges, or silvages just upon
one another, and, as near as they can, place the wale so in the doubling of it,
that the wale of the one side may lie very near parallel, or even with the
wale of the other ; for the nearer that posture they lie, the greater will
the watering appear ; and the more obliquely, or across to each other they
lie, the smaller are the waves. Their way for folding it for a great wale
is thus : they take a Pin, and begin at one side of the piece in any wale, and
so moving it towards the other side, thereby direct their hands to the opposite ends of the wale, and then, as near as they can, place the two opposite ends of the same wale together, and so double, or fold the whole
piece, repeating this enquiry with a Pin at every yard or two's distance
through the whole length ; then they sprinkle it with water, and fold it the
longways, placing between every fold a piece of Pastboard, by which
means all the wrong side of the water'd Stuff becomes flat, and with little
wales, and the wales on the other side become the more protuberant ;
whence the creasings or angular bendings of the wales become the more
perspicuous. Having folded it in this manner, they place it with an interjacent Pastboard into an hot Press, where it is kept very violently prest,
till it be dry and stiff ; by which means, the wales of either contiguous
sides leave their own impressions upon each other, as is very manifest by the second Figure, where 'tis obvious enough, that the wale of the
piece *A B C D* runs parallel between the pricked lines *e f, e f, e f,* and as

manifeſt to diſcern the impreſſions upon theſe wales, left by thoſe that were preſt upon them,which lying not exactly parallel with them,but a little athwart them, as is denoted by the lines of,*o o o o,gh. gh,gh,* between which the other wales did lie parallel;they are ſo variouſly,and irregularly creas'd that being put into that ſhape when wet,and kept ſo till they be drie, they ſo ſet each others threads, that the Moldings remain almoſt as long as the Stuff laſts.

Hence it may appear to any one that attentively conſiders the Figure, why the parts of the wale *a, a, a, a, a, a,* ſhould appear bright ; and why the parts *b, b, b, b, b, b,* ſhould appear ſhadowed, or dark; why ſome, as *d,d,d,d,d,d,* ſhould appear partly light,and partly dark : the varieties of which reflections and ſhadows are the only cauſe of the appearance of watering in Silks, or any other kind of Stuffs.

From the variety of reflection, may alſo be deduc'd the cauſe why a ſmall breez or gale of wind ruffling the ſurface of a ſmooth water, makes it appear black ; as alſo,on the other ſide, why the ſmoothing or burniſhing the ſurface of whitened Silver makes it look black ; and multitudes of other phænomena might hereby be ſolv'd, which are too many to be here inſiſted on.

Obſerv. VI. *Of ſmall Glaſs Canes.*

Schem. 4.

THat I might be ſatisfi'd, whether it were not poſſible to make an *Artificial* pore as *ſmall* as any *Natural* I had yet found, I made ſeveral attemps with ſmall *glaſs pipes,* melted in the flame of a Lamp, and then very *ſuddenly* drawn out into a great length. And, by *that means,* without much difficulty, I was able to draw ſome almoſt as ſmall as a *Cobweb,* which yet, with the *Microſcope,* I could plainly perceive to be *perforated,* both by looking on the *ends* of it, and by looking on it *againſt the light* ; which was much the *eaſier way* to determine whether it were ſolid or perforated; for, taking a ſmall pipe of glaſs, and cloſing one end of it, then filling it *half full* of water, and holding it *againſt the light,* I could, by this means, very eaſily find what was the *differing aſpect* of a *ſolid* and a *perforated* piece of glaſs ; and ſo eaſily diſtinguiſh, without ſeeing either end, whether any *Cylinder* of glaſs I look'd on, were a *ſolid ſtick,* or a *hollow cane.* And by this means,I could alſo preſently judge of any ſmall *filament* of glaſs, whether it were *hollow* or *not,* which would have been exceeding tedious to examine by looking on the end. And many ſuch like ways I was fain to make uſe of, in the examining of divers other particulars related in this Book, which would have been no eaſie task to have determined meerly by the more common way of looking on, or viewing the Object. For, if we conſider firſt, the very *faint light* wherewith the object is enlightened, whence many particles appear *opacous,* which when more enlightned, appear very *tranſparent,* ſo that I was fain to *determine* its *tranſparency* by one glaſs, and its *texture* by another Next, the *unmanageableneſs* of moſt *Objects,* by reaſon

of

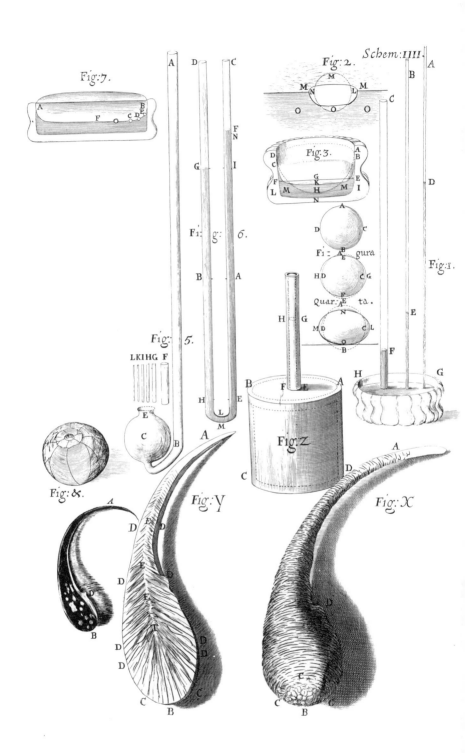

Schem:IIII

Fig:7.

Fig:2.

Fig:3.

F₁ gura

Quar. ta.

Fig:6.

Fig:5.

Fig:Z

Fig:I

Fig:8.

Fig:Y

Fig:X

of their *fmalnefs*, 3. The *difficulty of finding* the defired point, and of *placing* it fo, as to reflect the *light conveniently* for the Inquiry, Laftly, ones being able to view it but with *one eye* at once, they will appear no fmall *obftructions*, nor are they eafily *remov'd* without many *contrivances*. But to proceed, I could not find that water, or fome *deeply ting'd* liquors would in fmall ones rife fo high as one would expect; and the *higheft* I have found it yet rife in any of the pipes I have try'd, was to 21 *inches* above the level of the water in the veffel: for though I found that in the fmall pipes it would *nimbly enter* at firft, and run about 6 or 7 *inches* upwards; yet I found it then to move upwards *fo flow*, that I have not yet had the *patience* to obferve it above that height of 21 *inches* (and that was in a pretty *large Pipe*, in comparifon of thofe I formerly mentioned; for I could obferve the *progrefs* of a *very deep ting'd liquor* in it with my *naked eye*, without much trouble; whereas many of the *other pipes* were fo *very fmall*, that unlefs in a *convenient pofture* to the light, I could not perceive *them* :) But 'tis very probable. that a greater *patience* and *affiduity* may difcover the liquors to *rife*, at leaft to remain *fufpended*, at heights that I fhould be loath now even to *ghefs* at, if at leaft there be any *proportion* kept between the height of the afcending liquor, and the *bignefs of the holes* of the pipes.

An Attempt for the Explication of this Experiment.

My Conjecture, *That the unequal height of the furfaces of the water,* *proceeded from the greater preffure made upon the water by the Air* *without the Pipes* A B C, *then by that within them*; I fhall endeavour to confirm from the truth of the two following *Propofitions*:
Schem. 4.
Fig. 1.
The firft of which is, *That an unequal preffure of the incumbent Air,* *will caufe an unequal height in the water's Surfaces.*
And the fecond is, *That in this experiment there is fuch an unequal* *preffure.*
That the firft is true, the following *Experiment* will evince. For if you take any Veffel fo contrived, as that you can at pleafure either *increafe* or *diminifh* the *preffure* of the Air upon this or that part of the *Superficies* of the *water*, the *equality* of the height of thofe parts will prefently be *loft*; and that part of the *Superficies* that fuftains the *greater preffure*, will be *inferior* to that which undergoes the *lefs*. A fit Veffel for this purpofe, will be an inverted Glafs *Syphon*, fuch an one as is defcribed in the *Sixth Figure*. For if into it you put Water enough to fill it as high as *A B*, and gently blow in at *D*, you fhall *deprefs* the Superficies *B*, and thereby *raife* the oppofite Superficies *A* to a *confiderable height*, and by gently *fucking* you may produce clean *contrary* effects.
Next, That there is fuch an *unequal preffure*, I fhall prove from this, *That there is a much greater incongruity of Air to Glafs, and fome other Bodies,* *then there is of Water to the fame.*

By

By *Congruity*, *I mean a property of a fluid Body, whereby any part of it is readily united with any other part, either of it self, or of any other Similar, fluid, or solid body:* And by *Incongruity a property of a fluid, by which it is hindred from uniting with any dissimilar, fluid, or solid Body.*

This laſt property, any one that hath been obſervingly converſant about fluid Bodies, cannot be ignorant of. For (not now to mention ſeveral *Chymical Spirits* and *Oyls*, which will *very hardly*, if at *all*, be brought to *mix* with one another; infomuch that there may be found ſome 8 or 9, or more, ſeveral diſtinct Liquors, which *ſwimming* one upon another, will not preſently *mix*) we need ſeek no further for Examples of this kind in *fluids*, then to obſerve the *drops of rain* falling through the *air*, and the *bubbles of air* which are by any means conveyed under the ſurface of the *water* ; or a drop of common *Sallet Oyl* ſwimming upon water. In all which, and many more examples of this kind that might be enumerated, the *incongruity* of two *fluids* is eaſily diſcernable. And as for the *Congruity* or *Incongruity* of Liquids, with ſeveral kinds of *firm* Bodies, they have long ſince been taken notice of, and called by the Names of *Drineſs* and *Moiſture* (though theſe two names are not comprehenſive enough, being commonly uſed to ſignifie only the adhering or not adhering of *water* to ſome other *ſolid Bodies*)of this kind we may obſerve that *water* will more readily *wet ſome woods* then *others* ; and that *water*, let fall upon a *Feather*, the whiter ſide of a *Colwort*, and ſome other leaves, or upon almoſt any *duſty*, *unctuous*, or *reſinous* ſuperficies, will not *at all adhere* to them, but eaſily *tumble off* from them, like a ſolid *Bowl* ; whereas, if dropt upon *Linnen*, *Paper*, *Clay*, green *Wood*, &c. it will not be taken off, without leaving ſome part of it behind *adhering* to them. So *Quick-ſilver*, which will very *hardly* be brought to *ſtick* to any *vegetable body*, will *readily adhere* to, and *mingle* with, ſeveral clean *metalline bodies*.

And that we may the better finde what the *cauſe* of *Congruity* and *Incongruity* in bodies is, it will be requiſite to conſider, Firſt, what is the *cauſe* of *fluidneſs* ; And this, *I conceive*, to be nothing elſe but a certain *pulſe* or *ſhake* of *heat* ; for Heat being nothing elſe but a very *briſk* and *vehement agitation* of the parts of a body (as I have elſewhere made *probabable*) the parts of a body are thereby made ſo *looſe* from one another, that they eaſily *move any way*, and become *fluid*. That I may explain this a little by a groſs Similitude, let us ſuppoſe a diſh of ſand ſet upon ſome body that is very much *agitated*, and ſhaken with ſome *quick* and *ſtrong vibrating motion*, as on a *Milſtone* turn'd round upon the under ſtone very violently whilſt it is empty; or on a very ſtiff *Drum*-head, which is vehemently or very nimbly beaten with the Drumſticks. By this means, the ſand in the diſh, which before lay like a *dull* and unactive body, becomes a perfect *fluid* ; and ye can no ſooner make a *hole* in it with your finger, but it is immediately *filled up again*, and the upper ſurface of it *levell'd*. Nor can you *bury* a *light body*, as a piece of Cork under it, but it preſently *emerges* or *ſwims* as 'twere on the top ; nor can you lay a *heavier* on the top of it, as a piece of Lead, but it is immediately *buried*

in

in Sand, and (as 'twere) finks to the bottom. Nor can you make a *hole* in the fide of the Difh, but the fand fhall *run out* of it to a *level*, not an *obvious property* of a fluid body, as fuch, but this dos *imitate* ; and all this meerly caufed by the vehement *agitation* of the conteining veffel ; for by this means, *each* fand becomes to have a *vibrative* or *dancing* motion, fo as no other heavier body can *reft* on it, unlefs *fuftein'd* by fome other on either fide : Nor will it fuffer any Body to be *beneath* it, unlefs it be a *heavier* then it felf. Another Inftance of the ftrange *loofening* nature of a violent jarring Motion, or a ftrong and nimble vibrative one, we may have from a piece of *iron* grated on very ftrongly with a *file* : for if into that a pin be *fcrew'd* fo firm and hard, that though it has a convenient head to it, yet it can by no means be *unfcrew'd* by the fingers ; if, I fay, you attempt to unfcrew this whilft *grated on by the file*, it will be found to undoe and turn very *eafily*. The firft of thefe Examples manifefts, how a body actually *divided* into fmall parts, becomes a *fluid*. And the latter manifefts by what means the agitation of heat fo eafily *loofens* and *unties* the parts of *folid* and *firm* bodies. Nor need we fuppofe heat to be any thing elfe, befides fuch a motion ; for fuppofing we could *Mechanically* produce fuch a one *quick* and *ftrong* enough, we need not fpend *fuel* to *melt* a body. Now, that I do not fpeak this altogether groundlefs, I muft refer the Reader to the Obfervations I have made upon the fhining fparks of Steel, for there he fhall find that the *fame* effects are produced upon fmall chips or parcels of Steel by the *flame*, and by *a quick and violent motion* ; and if the body of *fteel* may be thus melted (as I there fhew it may) I think we have little reafon to doubt that almoft *any other* may not alfo. Every Smith can inform one how quickly both his *File* and the *Iron* grows *hot* with *filing*, and if you *rub* almoft any two *hard* bodies together, they will do the fame : And we know, that a fufficient degree of heat caufes *fluidity*, in fome bodies much fooner, and in others later ; that is, the parts of the body of fome are fo *loofe* from one another, and fo *unapt to cohere*, and fo *minute* and *little*, that a very *fmall* degree of agitation keeps them always in the *ftate of fluidity*. Of this kind, I fuppofe, the *Æther*, that is the *medium* or *fluid* body, in which all other bodies do as it were fwim and move ; and particularly, the *Air*, which feems nothing elfe but a kind of *tincture* or *folution* of terreftrial and aqueous particles *diffolv'd* into it, and agitated by it, juft as the *tincture* of *Cocheneel* is nothing but fome finer *diffoluble* parts of that Concrete lick'd up or *diffolv'd* by the *fluid* water. And from this Notion of it, we may eafily give a more Intelligible reafon how the Air becomes fo capable of *Rarefaction* and *Condenfation*. For, as in *tinctures*, one grain of fome *ftrongly tinging* fubftance may *fenfibly* colour fome hundred thoufand grains of *appropriated* Liquors, fo as every *drop* of it has its proportionate fhare, and be fenfibly ting'd, as I have try'd both with *Logwood* and *Cocheneel* : And as fome few grains of *Salt* is able to infect as great a quantity, as may be found by *precipitations*, though not fo eafily by the *fight* or *afte* ; fo the *Air*, which feems to be but as 'twere a *tincture* or *faline fubftance*, *diffolv'd and agitated by the fluid and agil Æther*, may difperfe perſe and *expand* it ſelf into a *vaſt ſpace,* if it have room enough, and infect,as it were,every part of that ſpace. But,as on the other ſide,if there be but ſome *few grains* of the liquor, it may *extract all* the colour of the tinging ſubſtance, and may *diſſolve* all the Salt, and thereby become *much more impregnated* with thoſe ſubſtances, ſo may *all* the air that ſuf-ficed in a *rarify'd ſtate* to fill ſome *hundred thouſaud* ſpaces of Æther, be compris'd in only *one,*but in a poſition proportionable *denſe.* And though we have not yet found out ſuch *ſtrainers* for Tinctures and Salts as we have for the Air, being yet unable to *ſeparate* them from their diſſolving liquors by any kind of *filtre,* without *præcipitation,* as we are able to *ſe-parate* the Air from the Æther by *Glaſs,* and ſeveral other bodies. And though we are yet unable and ignorant of the ways of *præcipitating* Air out of the Æther as we can Tinctures, and Salts out of ſeveral *diſſolvents;* yet neither of theſe ſeeming *impoſſible* from the nature of the things, nor ſo *improbable* but that ſome happy future induſtry may find out ways to effect them ; nay, further, ſince we find that Nature *does really perform* (though by what means we are not certain) both theſe actions, namely, by *præcipitating* the Air in Rain and Dews, and by ſupplying the Streams and Rivers of the World with freſh water, *ſtrain'd* through ſecret ſub-terraneous Caverns: And ſince, that in very many other *proprieties* they do ſo exactly *ſeem* of the *ſame nature ;* till further obſervations or tryals do inform us of the *contrary,* we may *ſafely enough conclude* them of the *ſame kind.* For it ſeldom happens that any two natures have ſo ma-ny properties *coincident* or the *ſame,* as I have obſerv'd Solutions and Air to have, and to be *different* in the reſt. And therefore I think it nei-ther *impoſſible, irrational,* nay nor *difficult* to be able to *predict* what is *likely* to happen in other particulars alſo, beſides thoſe which *Obſervation* or *Experiment* have declared thus or thus; eſpecially, if the *circum-ſtances* that do often very much conduce to the variation of the effects be duly *weigh'd* and *conſider'd.* And indeed, were there not a *probability* of this, our *inquiries* would be *endleſs,* our *tryals vain,* and our greateſt *in-ventions* would be nothing but the meer *products* of *chance,* and not of *Reaſon ;* and, like *Mariners* in an Ocean, deſtitute both of a *Compaſs* and the ſight of the *Celeſtial guids,* we might indeed, *by chance,* Steer *directly* towards our deſired Port, but 'tis *a thouſand to one* but we *miſs* our aim. But to proceed, we may hence alſo give a plain reaſon, how the Air comes to be *darkned* by *clouds,* &c. which are nothing but a kind of *precipitati-on,* and how thoſe *precipitations* fall down in *Showrs.* Hence alſo could I very eaſily, and I think truly, deduce the cauſe of the curious *ſixangu-lar figures* of Snow, and the appearances of *Haloes, &c.* and the ſudden *thickning* of the Sky with Clouds, and the *vaniſhing* and *diſappearing* of thoſe Clouds again; for all theſe things may be very eaſily *imitated* in a *glaſs of liquor,*with ſome ſlight *Chymical preparations* as I have often try'd, and may ſomewhere elſe more largely relate, but have not now time to ſet them down. But to proceed, there are other bodies that conſiſt of particles more *Groſs,* and of a more *apt* figure for *coheſion,* and this re-quires a *ſomewhat greater* agitation ; ſuch, I ſuppoſe ☿. *fermented vinous*

Spirits

Spirits, feveral *Chymical Oils*, which are much of kin to thofe Spirits, &c. Others yet require a *greater*, as *water*, and fo others *much greater*, for almoft infinite degrees: For, I fuppofe there are very *few* bodies in the world that may not be made *aliquatenus* fluid, by *fome* or *other* degree of agitation or heat.

Having therefore in fhort fet down my Notion of a Fluid body, I come in the next place to confider what *Congruity* is; and this, as I faid before, being a *Relative property* of a fluid, whereby it may be faid to be *like* or *unlike* to this or that other body, whereby it *does* or *does not mix* with this or that body. We will again have recourfe to our former Experiment, though but a rude one; and here if we mix in the difh *feveral kinds* of fands, fome of *bigger*, others of *lefs* and finer bulks, we fhall find that by the agitation the *fine fand* will *eject* and *throw out* of it felf all thofe *bigger* bulks of fmall *ftones* and the like, and thofe will be *gathered* together all into *one* place; and if there be *other* bodies in it of other natures, thofe alfo will be *feparated* into a place by themfelves, and *united* or *tumbled* up together. And though this do not come up to the *higheft property* of *Congruity*, which is a *Cohæfion* of the parts of the fluid together, or a kind of *attraction* and *tenacity*, yet this does as 'twere *fhadow* it out, and fomewhat refemble it; for juft after the fame manner, I fuppofe the *pulfe* of heat to *agitate* the fmall parcels of matter, and thofe that are of a *like bignefs*, and *figure*, and *matter*, will *hold*, or *dance* together, and thofe which are of a *differing* kind will be *thruft* or *fhov'd* out from between them; for particles that are all *fimilar*, will, like fo many *equal mufical ftrings* equally *ftretcht*, vibrate together in a kind of *Harmony* or *unifon*; whereas others that are *diffimilar*, upon what account foever, unlefs the difproportion be otherwife counter-ballanc'd, will, like fo many *ftrings out of tune* to thofe unifons, though they have the fame agitating *pulfe*, yet make quite *differing* kinds of *vibrations* and *repercuffions*, fo that though they may be both mov'd, yet are their *vibrations* fo *different*, and fo *untun'd*, as 'twere to each other, that they *crofs* and *jar* againft each other, and confequently, *cannot agree* together, but *fly back* from each other to their fimilar particles. Now, to give you an inftance how the *difproportion* of fome bodies in one refpect, may be *counter-ballanc'd* by a *contrary difproportion* of the fame body in another refpect, whence we find that the fubtil *vinous fpirit* is *congruous*, or does readily *mix* with *water*, which in many properties is of a very *differing nature*, we may confider that a *unifon* may be made either by two *ftrings* of the fame *bignefs*, *length*, and *tenfion*, or by two ftrings of the fame *bignefs*, but of *differing length*, and a *contrary differing tenfion*; or 3*ly.* by two ftrings of *unequal length* and *bignefs*, and of a *differing tenfion*, or of *equal length*, and *differing bignefs* and *tenfion*, and feveral other fuch varieties. To which *three* properties in *ftrings*, will correfpond *three* proprieties alfo in *fand*, or the particles of bodies, their *Matter* or *Subftance*, their *Figure* or *Shape*, and their *Body* or *Bulk*. And from the *varieties* of thefe *three*, may arife *infinite varieties* in fluid bodies, though all agitated by the *fame pulfe* or *vibrative* motion. And there may be as many ways of making Harmonies and

and Difcords with thefe, as there may be with *mufical ftrings*. Having therefore feen what is the caufe of Congruity or Incongruity, thofe relative properties of fluids, we may, from what has been faid, very eafily colleét, what is the *reafon* of thofe Relative proprieties alſo between *fluid bodies* and *folid*; for fince all bodies confift of *particles* of fuch a *Subftance*, *Figure*, and *Bulk*; but in fome they are *united* together more *firmly* then to be *loofened* from each other by every *vibrative* motion (though I imagine that there is no body in the world, but that fome degree of agitation may, as I hinted before, agitate and loofen the particles fo as to make them fluid) thofe *cohering* particles may *vibrate* in the fame manner almoft as thofe that are *loofe* and become *unifons* or *difcords*, as I may fo fpeak, to them. Now that the *parts* of all *bodies*, though never fo *folid*, do yet *vibrate*, I think we need go no further for proof, then that *all* bodies have fome *degrees* of *heat* in them, and that there has not been yet found any thing *perfeétly cold*: Nor can I believe indeed that there is any fuch thing in Nature, as a body whofe particles are at *reft*, or *lazy* and *unaétive* in the great *Theatre* of the *World*, it being quite *contrary* to the grand *Oeconomy* of the Univerfe. We fee therefore what is the reafon of the *fympathy* or uniting of fome bodies together, and of the *antipathy* or flight of others from each other : For *Congruity* feems nothing elfe but a *Sympathy*, and *Incongruity* an *Antipathy* of bodies; hence *fimilar* bodies once *united* will not *eafily part*, and *diffimilar* bodies once *difjoyn'd* will not *eafily unite* again; from hence may be very eafily deduc'd the reafon of the *fufpenfion* of *water* and *Quick-filver* above their ufual *ftation*, as I fhall more at large anon fhew.

Thefe properties therefore (alwayes the concomitants of fluid bodies) produce thefe following vifible *Effeéts :*

Firft, They *unite* the parts of a fluid to its *fimilar* Solid, or keep them *feparate* from its *diffimilar*. Hence *Quick-filver* will (as we noted before) *ftick* to *Gold*, *Silver*, *Tin*, *Lead*, &c. and *unite* with them : but *roul* off from *Wood*, *Stone*, *Glafs*, &c. if never fo little fcituated out of its *horizontal level*; and *water* that will *wet falt* and *diffolve* it, will *flip* off from *Tallow*, or the like, without at all *adhering*; as it may likewife be obferved to do upon a *dufty* fuperficies. And next they caufe the parts of *homogeneal fluid* bodies readily to *adhere* together and *mix*, and of *heterogeneal*, to be exceeding *averfe* thereunto. Hence we find, that *two* fmall *drops* of *water*, on any fuperficies they can roul on, will, if they chance to touch each other, *readily unite* and *mix* into one 3^d *drop* : The like may be obferved with two fmall *Bowls* of *Quick-filver* upon a Table or Glafs, provided their furfaces be not *dufty*; and with two drops of *Oyl* upon fair water, *&c*. And further, *water* put unto *wine*, *falt water*, *vinegar*, *fpirit* of *wine*, or the like, does immediately (efpecially if they be fhaken together) *difperfe* it felf all over them. Hence, on the contrary, we alfo find, that *Oyl of Tartar* poured upon *Quick-filver*, and *Spirit of Wine* on that *Oyl*, and *Oyl of Turpentine* on that *Spirit*, and *Air* upon that *Oyl*, though they be ftopt clofely up into a Bottle, and *fhaken* never fo much, they will by no means long fuffer any of their bigger parts to be *united* or included

cluded within any of the other Liquors(by which recited Liquors,may be plainly enough reprefented the four *Peripatetical Elements*, and the more fubtil *Æther* above all.) From this property'tis, that a drop of *water* does not mingle with, or vanifh into *Air*, but is *driven* (by that Fluid equally protruding it on every fide) and forc't into as little a fpace as it can poffibly be contained in, namely, into a *Round Globule*. So likewife a little *Air* blown under the *water*, is *united* or thruft into a *Bubble* by the ambient water. And a parcel of *Quick-filver* enclofed with *Air*, *Water*, or almoft any other *Liquor*, is *formed* into a *round Ball*.

Now the caufe why all thefe included Fluids, newly mentioned, or as many others as are wholly included within a heterogeneous fluid, are not *exactly* of a *Spherical Figure* (feeing that if caufed by thefe Principles only, it could be of no other) muft proceed from fome other kind of *preffure* againft the two oppofite flatted fides. This *adventitious* or *accidental preffure* may proceed from *divers caufes*, and accordingly muft *diverfifie* the Figure of the included heterogeneous fluid : For feeing that a body may be included either with a fluid only, or only with a folid, or partly with a fluid, and partly with a folid, or partly with one fluid, and partly with another ; there will be found a very great variety of the terminating *furfaces*, much differing from a *Spherical*, according to the various refiftance or preffure that belongs to each of thefe encompaffing bodies.

Which Properties may in general be deduced from two heads, *viz. Motion*, and *Reft*. For, either this Globular Figure is altered by a *natural Motion*, fuch as is *Gravity* ; or a *violent*, fuch as is any *accidental motion* of the fluids, as we fee in the *wind* ruffling up the water,and the *purlings* of *Streams*, and *foaming* of *Catarracts*, and the like. Or thirdly, By the *Reft*, *Firmnefs* and *Stability* of the ambient *Solid*. For if the including *Solid* be of an *angular* or any other *irregular* Form, the included *fluid* will be near of the *like*,as a Pint-*Pot* full of *water*,or a *Bladder* full of *Air*. And next, if the including or included fluid have a greater *gravity* one than another,then will the *globular* Form be depreft into an *Elliptico-fpherical* : As if, for example, we fuppofe the Circle *A B C D*, in the *fourth Figure*, to reprefent a *drop of water*, *Quick-filver*, or the like, included with the *Air* or the like , which fuppofing there were no *gravity* at all in either of the *fluids*, or that the *contained* and *containing* were of the *fame weight*, would be *equally compreft* into an exactly *fpherical* body (the ambient fluid *forcing equally* againft every fide of it.) But fuppofing either a greater *gravity* in the included , by reafon whereof the parts of it being *preft* from *A* towards *B*, and thereby the whole put into *motion* , and that *motion* being *hindred* by the *refiftance* of the *fubjacent* parts of the ambient, the *globular* Figure *A D B C* will be *depreft* into the *Elliptico-fpherical*, *E G F H*. For the fide *A* is *detruded* to *E* by the *Gravity*, and *B* to *F* by the *refiftance* of the fubjacent medium : and therefore *C* muft neceffarily be thruft to *G*; and *D* to *H*. Or elfe, fuppofing a greater *gravity* in the *ambient*, by whofe more then ordinary *preffure* againft the under fide of the included globule ; *B* will be forced to *F*, and by its *refiftance* of

E   the

the motion *upwards*, the fide *A* will be *deprefs* to *E*, and therefore *C* being thruft to *G* and *D* to *H*; the *globular* Figure by this means alfo will be made an *Elliptico-fpherical.* Next if a fluid be included *partly* with one, and *partly* with another fluid, it will be found to be fhaped *diverfly*, according to the proportion of the *gravity* and *incongruity* of the 3 *fluids* one to another : As in the *fecond Figure*, let the upper *M M M* be *Air*,the middle *L M N O* be common *Oyl*, the lower *O O O* be *Water*, the *Oyl* will be form'd, not into a *fpherical* Figure , fuch as is reprefented by the *pricked Line*, but into fuch a Figure as *L M N O*, whofe fide *L M N* will be of a flatter *Elliptical* Figure, by reafon of the great difproportion between the *Gravity* of *Oyl* and *Air*, and the fide *L O M* of a rounder, becaufe of the fmaller difference between the weight of *Oyl* and *Water.* Laftly,The *globular* Figure will be changed,if the *ambient* be partly *fluid* and partly *folid.* And here the termination of the incompafled *fluid* towards the incompaffing is fhap'd according to the proportion of the congruity or incongruity of the *fluids* to the *folids* , and of the gravity and incongruity of the *fluids* one to another. As fuppofe the fubjacent *medium* that hinders an included fluids defcent,be a *folid* , as let K I, in the *fourth Figure*, reprefent the fmooth fuperficies of a *Table* ; E G F H, a parcel of *running Mercury*; the fide G F H will be more flatted , according to the proportion of the incongruity of the *Mercury* and *Air* to the *Wood*,and of the *gravity* of *Mercury* and *Air* one to another ; The fide G E H will likewife be a little more depreft by reafon the fubjacent parts are now at reft,which were before in motion.

　　Or further in the *third Figure*, let A I L D reprefent an including *folid* medium of a cylindrical fhape (as fuppofe a fmall *Glafs Jar*) Let F G E M M reprefent a contain'd *fluid*, as water; this towards the bottom and fides, is figured according to the concavity of the *Glafs :* But its upper *Surface*, (which by reafon of its gravity, (not confidering at all the Air above it, and fo neither the congruity or incongruity of either of them to the Glafs) fhould be terminated by part of a *Sphere* whofe diameter fhould be the fame with that of the earth, which to our fenfe would appear a ftraight *Line*, as F G E. Or which by reafon of its having a greater congruity to Glafs than Air has, (not confidering its Gravity) would be thruft into a *concave Sphere*, as C H B, whofe diameter would be the fame with that of the concavity of the Veffel :) Its upper Surface, I fay, by reafon of its having a greater gravity then the Air, and having likewife a greater congruity to Glafs then the Air has, is terminated, by a *concave Elliptico-fpherical* Figure, as C K B. For by its congruity it eafily conforms it felf, and adheres to the Glafs, and conftitutes as it were one containing body with it, and therefore fhould thruft the contained Air on that fide it touches it,into a *fpherical* Figure, as B H C, but the motion of Gravity depreffing a little the Corners B and C, reduces it into the aforefaid Figure C K B. Now that it is the greater congruity of one of the two *contiguous fluids*,then of the other,to the containing *folid*,that caufes the feparating furfaces to be thus or thus figured : And that it is not becaufe this or that figured furface is more proper, natural, or peculiar to one of thefe fluid bodies, then to the other, will appear from this; that the fame *fluids* will by being put into differing *folids*, change their *furfaces*. For the fame water, which in a Glafs or wooden Veffel will have a concave furface upwards, and will rife higher in a fmaller then a greater Pipe, the fame water, I fay, in the fame Pipes greafed over or oyled, will produce quite contrary effects; for it will have a *protuberant* and *convex* furface upwards, and will not rife fo high in fmall, as in bigger Pipes : Nay, in the very fame folid Veffel, you may make the very fame two contiguous *Liquids* to alter their Surfaces; for taking a fmall Wine-glafs, or fuch like Veffel, and pouring water gently into it, you fhall perceive the *furface* of the water all the way *concave*, till it rife even with the top, when you fhall find it (if you gently and carefully pour in more) to grow very *protuberant* and *convex*; the reafon of which is plain, for that the *folid* fides of the containing body are no longer extended, to which the water does more readily adhere then the air; but it is henceforth to be included with air, which would reduce it into a *hemifphere*, but by reafon of its *gravity*, it is flatted into an *Oval*. *Quickfilver* alfo which to *Glafs* is more incongruous then *Air* (and thereby being put into a *Glafs-pipe*, will not adhere to it, but by the more *congruous air* will be forced to have a very *protuberant* furface, and to rife higher in a greater then a leffer Pipe) this *Quickfilver* to clean *Metal*, efpecially to *Gold, Silver, Tin, Lead*, &c. *Iron* excepted, is more *congruous* then *Air*, and will not only ftick to it, but have a *concave* Surface like *water*, and rife higher in a lefs, then in a greater Pipe.

In all thefe Examples it is evident, that there is an *extraordinary* and *adventitious force*, by which the *globular* Figure of the contained *heterogeneous* fluid is altered; neither can it be imagined, how it fhould otherwife be of any other Figure then *Globular* : For being by the *heterogeneous* fluid equally *protruded* every way, whatfoever part is *protuberant*, will be thereby *depreft*. From this caufe it is, that in its effects it does very much refemble a *round Spring* (fuch as a *Hoop*.) For as in a *round Spring* there is required an additional *preffure* againft two oppofite fides, to reduce it into an *Oval* Form, or to force it in between the fides of a *Hole*, whofe *Diameter* is lefs then that of the *Spring*, there muft be a confiderable force or *protrufion* againft the *concave* or inner fide of the *Spring*; So to alter this *fpherical* conftitution of an included fluid body, there is required more preffure againft oppofite fides to reduce it into an *Oval*; and, to prefs it into an *Hole* lefs in *Diameter* then it felf, it requires a greater *protrufion* againft all the other fides. What degrees of force are requifite to reduce them into longer and longer *Ovals*, or to prefs them into lefs and lefs *holes*, I have not yet experimentally calculated; but thus much by experiment I find in general, that there is always required a greater preffure to clofe them into longer *Ovals*, or protude them into fmaller *holes*. The neceffity and reafon of this, were it requifite, I could eafily explain : but being not fo neceffary, and requiring more room and time then I have for it at prefent, I fhall here omit it; and proceed to fhew, that this may be prefently found true, if Experiment be made with a

round

round Spring (the way of making which trials is *obvious* enough.) And with the fluid bodies of *Mercury, Air, &c,* the way of trying which, will be fomewhat more difficult ; and therefore I fhall in brief defcribe it. He therefore that would try with *Air,* muft firft be provided of a *Glafs-pipe,* made of the fhape of that in the *fifth Figure,* whereof the fide A B, reprefents a ftraight *Tube* of about three foot long, C, reprefents another part of it, which confifts of a *round Bubble*; fo ordered, that there is left a *paffage* or *hole* at the top, into which may be faftened with *cement* feveral *fmall Pipes* of determinate *cylindrical* cavities : as let the *hollow* of

F.		$\frac{1}{4}$	
G.		$\frac{1}{6}$	
H.		$\frac{1}{8}$	
I.	be	$\frac{1}{12}$	of an inch.
K.		$\frac{1}{16}$	
L.		$\frac{1}{24}$	
M.		$\frac{1}{32}$	
&c			

There may be added as many more, as the Experimenter fhall think fit, with holes continually decreafing by known quantities, fo far as his fenfes are able to help him ; I fay, fo far, becaufe there may be made *Pipes* fo fmall that it will be impoffible to perceive the *perforation* with ones naked eye, though by the help of a *Microfcope,* it may eafily enough be perceived : Nay, I have made a *Pipe* perforated from end to end, fo fmall, that with my naked eye I could very hardly fee the body of it, infomuch that I have been able to knit it up into a knot without breaking : And more accurately examining one with my *Microfcope,* I found it not fo big as a fixteenth part of one of the fmaller hairs of my head which was of the fmaller and finer fort of hair, fo that fixteen of thefe *Pipes* bound faggot-wife together, would but have equalized one fingle hair ; how fmall therefore muft its *perforation* be ? It appearing to me through the *Microfcope* to be a proportionably *thick-fided Pipe.*

To proceed then, for the trial of the Experiment, the Experimenter muft place the *Tube* A B, perpendicular, and fill the *Pipe* F (cemented into the hole E) with water, but leave the *bubble* C full of *Air,* and then gently pouring in water into the Pipe A B, he muft obferve diligently how high the water will rife in it before it protrude the *bubble* of Air C, through the narrow paffage of F, and denote exactly the height of the *Cylinder* of water, then cementing in a fecond Pipe as G, and filling it with water ; he may proceed as with the former, denoting likewife the height of the *Cylinder* of water, able to protrude the *bubble* C through the paffage of G, the like may he do with the next *Pipe,* and the next, &c. as far as he is able : then comparing the feveral heights of the *Cylinders,* with the feveral *holes* through which each *Cylinder* did force the air (having due regard to the *Cylinders* of water in the fmall *Tubes*) it will be very eafie to determine, what force is requifite to prefs the *Air* into fuch and fuch *a hole,* or (to apply it to our prefent experiment) how

how much of the preſſure of the *Air* is taken off by its ingreſs into ſmaller and ſmaller *holes.* From the application of which to the entring of the *Air* into the bigger *hole* of the *Veſſel*, and into the ſmaller *hole* of the *Pipe*, we ſhall clearly find, that there is a greater preſſure of the air upon the water in the *Veſſel* or greater *pipe*, then there is upon that in the leſſer *pipe*: For ſince the preſſure of the *air* every way is found to be equal, that is, as much as is able to preſs up and ſuſtain a *Cylinder* of *Quickſilver* of two foot and a half high, or thereabouts; And ſince of this preſſure ſo many more degrees are required to force the *Air* into a ſmaller then into a greater *hole* that is full of a more congruous fluid. And laſtly, ſince thoſe degrees that are requiſite to preſs it in, are thereby taken off from the *Air* within, and the *Air* within left with ſo many degrees of preſſure leſs then the *Air* without ; it will follow, that the *Air* in the leſs *Tube* or *pipe*, will have leſs preſſure againſt the ſuperficies of the *water* therein, then the *Air* in the bigger: which was the minor Propoſition to be proved.

The Concluſion therefore will neceſſarily follow, *viz.* That *this unequal preſſure of the Air cauſed by its ingreſs into unequal holes, is a cauſe ſufficient to produce this effect*, *without the help of any other concurrent* ; and therefore is probably the principal (if not the only) cauſe of theſe *Phænomena.*

This therefore being thus explained, there will be divers *Phænomena* explicable thereby, as, the riſing of *Liquors* in a *Filtre*, the riſing of *Spirit of Wine, Oyl, melted Tallow, &c.* in the *Week* of a *Lamp*, (though made of ſmall *Wire, Threeds* of *Asbeſtus, Strings* of *Glaſs*, or the like) the riſing of *Liquors* in a *Spunge*, piece of *Bread, Sand, &c.* perhaps alſo the aſcending of the *Sap* in *Trees* and *Plants*, through their ſmall, and ſome of them *imperceptible pores*, (of which I have ſaid more, on another occaſion) at leaſt the paſſing of it out of the earth into their roots. And indeed upon the conſideration of this Principle, multitudes of other uſes of it occurr'd to me, which I have not yet ſo well examined and digeſted as to propound for *Axioms*, but only as *Queries* and *Conjectures* which may ſerve as *hints* toward ſome further *diſcoveries.*

As firſt, Upon the conſideration of the *congruity* and *incongruity* of Bodies, as to *touch*, I found alſo the like *congruity* and *incongruity* (if I may ſo ſpeak) as to the *Tranſmitting* of the *Raies* of Light: For as in this regard, *water* (not now to mention other Liquors) ſeems nearer of affinity to *Glaſs* then *Air*, and *Air* then *Quickſilver* : whence an *oblique Ray* out of *Glaſs*, will paſs into *water* with very little *refraction* from the *perpendicular*, but none out of *Glaſs* into *Air*, excepting a *direct*, will paſs without a very great refraction from the perpendicular, nay any oblique Ray under thirty degrees, will not be admitted into the Air at all. And *Quickſilver* will neither admit oblique or direct, but reflects all ; ſeeming, as to the tranſmitting of the Raies of Light, to be of a quite differing conſtitution, from that of *Air, Water, Glaſs, &c.* and to reſemble moſt thoſe opacous and ſtrong reflecting bodies of Metals: So alſo as to the property of coheſion or congruity, Water ſeems to keep the ſame order, being more congruous to Glaſs then Air, and Air then Quickſilver.

A Second thing (which was hinted to me, by the conſideration of the included fluids globular form, cauſed by the protruſion of the ambient heterogeneous fluid) was, whether the *Phænomena* of gravity might not by this means be explained, by ſuppoſing the *Globe* of Earth, Water, and Air to be included with a *fluid*, heterogeneous to all and each of them, ſo ſubtil, as not only to be every where *interſperſed* through the *Air*, (or rather the *air* through it) but to *pervade* the bodies of *Glaſs*, and even the *cloſeſt Metals*, by which means it may endeavour to *detrude* all earthly bodies as far from it as it can ; and partly thereby, and partly by other of its properties may move them towards the Center of the Earth. Now that there is ſome ſuch fluid, I could produce many Experiments and Reaſons, that do ſeem to prove it : But becauſe it would ask ſome time and room to ſet them down and explain them, and to conſider and anſwer all the Objections (many whereof I foreſee) that may be alledged againſt it ; I ſhall at preſent proceed to other *Queries*, contenting my ſelf to have here only given a hint of what I may ſay more elſwhere.

A Third *Query* then was, Whether the *heterogeneity* of the *ambient fluid* may not be accounted a *ſecondary cauſe* of the *roundneſs* or *globular form* of the *greater bodies* of the world, ſuch as are thoſe of the *Sun, Stars,* and *Planets*, the *ſubſtance* of each of which ſeems altogether *heterogeneous* to the *circum-ambient fluid æther*? And of this I ſhall ſay more in the Obſervation of the Moon.

A Fourth was, Whether the *globular form* of the *ſmaller parcels* of matter here upon the *Earth*, as that of *Fruits, Pebbles,* or *Flints* , &c. (which ſeem to have been a *Liquor* at firſt) may not be cauſed by the *heterogeneous ambient fluid*. For thus we ſee that melted *Glaſs* will be naturally formed into a *round Figure*; ſo likewiſe any ſmall Parcel of any *fuſible body*, if it be perfectly encloſed by the *Air*, will be driven into a *globular* Form ; and, when cold, will be found a *ſolid Ball*. This is plainly enough manifeſted to us by their way of making *ſhot* with the *drops of Lead*; which being a very pretty curioſity, and known but to a very few, and having the liberty of publiſhing it granted me, by that *Eminent Virtuoſo* Sir *Robert Moray*, who brought in this Account of it to the *Royal Society*, I have here tranſcribed and inſerted.

To make ſmall ſhot of different ſizes ; Communicated by his Highneſs *P. R.*

*T*Ake *Lead out of the Pig what quantity you pleaſe, melt it down, ſtir and clear it with an iron Ladle , gathering together the blackiſh parts that ſwim at top like ſcum, and when you ſee the colour of the clear Lead to be greeniſh, but no ſooner, ſtrew upon it* Auripigmentum

pigmentum *powdered according to the quantity of Lead, about as much as will lye upon a half Crown piece will ferve for eighteen or twenty pound weight of fome forts of Lead* ; *others will require more, or lefs.* After the Auripigmentum *is put in, ftir the Lead well, and the* Auripigmentum *will flame: when the flame is over, take out fome of the Lead in a Ladle having a lip or notch in the brim for convenient pouring out of the Lead, and being well warmed amongft the melted Lead, and with a ftick make fome fingle drops of Lead trickle out of the Ladle into water in a Glafs, which if they fall to be round and without tails, there is* Auripigmentum *enough put in, and the temper of the heat is right, otherwife put in more.* Then lay two bars of Iron (*or fome more proper Iron-tool made on purpofe*) *upon a Pail of water, and place upon them a round Plate of Copper, of the fize and figure of an ordinary large Pewter or Silver Trencher, the hollow whereof is to be about three inches over, the bottom lower then the brims about half an inch, pierced with thirty, forty, or more fmall holes; the fmaller the holes are, the fmaller the fhot will be; and the brim is to be thicker then the bottom, to conferve the heat the better.*

The bottom of the Trencher being fome four inches diftant frum the water in the Pail, lay upon it fome burning Coles, to keep the Lead melted upon it. Then with the hot Ladle take Lead off the Pot where it *ftands melted, and pour it foftly upon the burning Coles over the bottom of the Trencher, and it will immediately run through the holes into the water in fmall round drops.* Thus pour on new Lead ftill as faft as *it runs through the Trencher till all be done* ; *blowing now and then the Coles with hand-Bellows, when the Lead in the Trencher cools fo as to ftop from running.*

Whilft one pours on the Lead, another muft, with another Ladle, thrufted four or five inches under water in the Pail, catch from time to time fome of the fhot, as it drops down, to fee the fize of it, and whether there be any faults in it. The greateft care is to keep the Lead upon *the Trencher in the right degree of heat* ; *if it be too cool, it will not run through the Trencher, though it ftand melted upon it* ; *and this is to*

be

be helped by blowing the Coals a little , or pouring on new Lead that is hotter : but the cooler the Lead, the larger the Shot; and the hotter, the smaller ; when it is too hot, the drops will crack and fly ; then you must stop pouring on new Lead, and let it cool ; and so long as you observe the right temper of the heat, the Lead will constantly drop into very round Shot, without so much as one with a tail in many pounds.

When all is done, take your Shot out of the Pail of water, and put it in a Frying-pan over the fire to dry them , which must be done warily, still shaking them that they melt not ; and when they are dry you may separate the small from the great , in Pearl Sives made of Copper or Lattin let into one another, into as many sizes as you please. But if you would have your Shot larger then the Trencher makes them , you may do it with a Stick , making them trickle out of the Ladle, as hath been said.

If the Trencher be but toucht a very little when the Lead stops from going through it, and be not too cool, it will drop again , but it is better not to touch it at all. At the melting of the Lead take care that there be no kind of Oyl, Grease, or the like, upon the Pots, or Ladles, or Trencher.

The Chief cause of this Globular Figure of the Shot, seems to be the Auripigmentum ; for, as soon as it is put in among the melted Lead, it loses its shining brightness, contracting instantly a grayish film or skin upon it, when you scum it to make it clean with the Ladle. So that when the Air comes at the falling drop of the melted Lead, that skin constricts them every where equally : but upon what account, and whether this be the true cause, is left to further disquisition,

Much after this same manner, when the Air is exceeding cold through which it passes, do we find the drops of Rain, falling from the Clouds, congealed into round Hail-stones by the freezing Ambient.

To which may be added this other known Experiment, That if you gently let fall a drop of *water* upon small *sand* or *dust*, you shall find, as it were, an artificial *round stone* quickly generated. I cannot upon this occasion omit the mentioning of the strange kind of *Grain*, which I have observed in a *stone* brought from *Kettering* in *Northamptonshire*, and therefore called by Masons *Kettering-stone*, of which see the Description.

Which

Which brings into my mind what I long since obferved in the fiery Sparks that are ftruck out of a Steel. For having a great defire to fee what was left behind, after the Spark was gone out, I purpofely ftruck fire over a very white piece of Paper, and obferving diligently where fome confpicuous fparks went out, I found a very little black fpot no bigger then the point of a Pin, which through a *Microfcope* appeared to be a perfectly round Ball, looking much like a polifht ball of Steel, infomuch that I was able to fee the Image of the window reflected from it. I cannot here ftay (having done it more fully in another place) to examine the particular Reafons of it, but fhall only hint, that I imagine it to be fome fmall parcel of the Steel, which by the violence of the motion of the ftroke (moft of which feems to be impreft upon thofe fmall parcels) is made fo glowing hot, that it is melted into a *Vitrum*, which by the ambient Air is thruft into the form of a Ball.

A Fifth thing which I thought worth Examination was, Whether the motion of all kind of Springs, might not be reduced to the Principle whereby the included *heterogeneous fluid* feems to be moved ; or to that whereby two Solids, as Marbles, or the like, are thruft and kept together by the *ambient fluid*.

A Sixth thing was, Whether the Rifing and Ebullition of the Water out of Springs and Fountains (which lie much higher from the Center of the Earth then the Superficies of the Sea, from whence it feems to be derived) may not be explicated by the rifing of Water in a fmaller Pipe: For the Sea-water being ftrained through the Pores or Crannies of the Earth, is, as it were, included in little Pipes, where the preffure of the Air has not fo great a power to refift its rifing: But examining this way, and finding in it feveral difficulties almoft irremovable, I thought upon a way that would much more naturally and conceivably explain it, which was by this following Experiment: I took a Glafs-Tube, of the form of that defcribed in the fixth Figure, and chufing two *heterogeneous fluids*, fuch as Water and Oyl, I poured in as much Water as filled up the Pipes as high as A B, then putting in fome Oyl into the Tube A C, I depreft the fuperficies A of the Water to E, and B I raifed to G, which was not fo high perpendicularly as the fuperficies of the Oyl F, by the fpace F I, wherefore the proportion of the gravity of thefe two Liquors was as G H to F E.

This Experiment I tried with feveral other Liquors, and particularly with frefh Water and Salt (which I made by diffolving Salt in warm Water) which two though they are nothing heterogeneous, yet before they would perfectly mix one with another, I made trial of the Experiment: Nay, letting the Tube wherein I tried the Experiment remain for many dayes, I obferved them not to mix ; but the fuperficies of the frefh was rather more then lefs elevated above that of the Salt. Now the proportion of the gravity of Sea-water, to that of River-water, according to *Stevinus* and *Varenius*, and as I have fince found pretty true by making trial my felf, is as 46. to 45. that is, 46. Ounces of the falt Water will take up no more room then 45. of the fresh. Or reciprocally 45 pints of salt-water weigh as much as 46 of fresh.

But I found the proportion of Brine to fresh Water to be near 13 to 12: Suppofing therefore G H M to reprefent the Sea, and F I the height of the Mountain above the Superficies of the Sea, F M a Cavern in the Earth, beginning at the bottom of the Sea, and terminated at the top of the Mountain, L M the Sand at the bottom, through which the Water is as it were ftrained, fo as that the frefher parts are only permitted to tranfude, and the faline kept back ; if therefore the proportion of G M to F M be as 45 to 46, then may the Cylinder of Salt-water G M make the Cylinder of Frefh-water to rife as high as E, and to run over at N. I cannot here ftand to examine or confute their Opinion, who make the depth of the Sea, below its Superficies, to be no more perpendicularly meafured then the height of the Mountains above it: 'Tis enough for me to fay, there is no one of thofe that have afferted it, have experimentally known the perpendicular of either; nor fhall I here determine, whether there may not be many other caufes of the feparation of the frefh water from the falt, as perhaps fome parts of the Earth through which it is to pafs, may contain a Salt, that mixing and uniting with the Sea-falt, may precipitate it ; much after the fame manner as the *Alkalizate* and *Acid Salts* mix and precipitate each other in the preparation of *Tartarum Vitriolatum.* I know not alfo whether the exceeding cold (that muft neceffarily be) at the bottom of the Water, may not help towards this feparation, for we find, that warm Water is able to diffolve and contain more Salt, then the fame cold; infomuch that Brines ftrongly impregnated by heat, if let cool, do fuffer much of their Salt to fubfide and cryftallize about the bottom and fides. I know not alfo whether the exceeding preffure of the parts of the Water one againft another, may not keep the Salt from defcending to the very bottom, as finding little or no room to infert it felf between thofe parts, protruded fo violently together, or elfe fqueeze it upwads into the fuperiour parts of the Sea, where it may more eafily obtain room for it felf, amongft the parts of the Water, by reafon that there is more heat and lefs preffure. To this Opinion I was fomewhat the more induced by the relations I have met with in *Geographical Writers*, of drawing frefh Water from the bottom of the Sea, which is falt above. I cannot now ftand to examine, whether this natural perpetual motion may not artificially be imitated : Nor can I ftand to anfwer the Objections which may be made againft this my Suppofition: As, Firft, How it comes to pafs, that there are fometimes falt Springs much higher then the Superficies of the Water? And, Secondly, Why Springs do not run fafter and flower, according to the varying height made of the Cylinder of Sea-water, by the ebbing and flowing of the Sea ?

As to the Firft, In fhort, I fay, the frefh Water may receive again a faline Tincture near the Superficies of the Earth, by paffing through fome falt *Mines*, or elfe many of the faline parts of the Sea may be kept back, though not all.

And

And as to the Second, The fame *Spring* may be fed and fupplyed by divers *Caverns*, coming from very far diftant parts of the *Sea*, fo as that it may in one place be *high*, in another *low water*; and fo by that means the *Spring* may be equally fupply'd at all times. Or elfe the *Cavern* may be fo ftraight and narrow, that the water not having fo ready and free paffage through it, cannot upon fo fhort and quick mutations of preffure, be able to produce any fenfible effect at fuch a diftance. Befides that, to confirm this *hypothefis*, there are many *Examples* found in *Natural Hiftorians*, of *Springs* that do ebb and flow like the Sea : As particularly, thofe recorded by the Learned *Camden*, and after him by *Speed*, to be found in this *Ifland* : One of which, they relate to be on the Top of a Mountain, by the fmall Village *Kilken* in *Flintfhire*, *Maris æmulus qui ftatis temporibus fuas evomit & reforbet Aquas* ; Which at certain times rifeth and falleth after the manner of the Sea. A Second in *Caermardenfhire*, near *Caermarden*, at a place called *Cantred Bichan* ; *Qui (ut fcribit Giraldus) naturali die bis undis deficiens, & toties exuberans, marinas imitatur inftabilitates* ; That twice in four and twenty hours ebbing and flowing, refembleth the unftable motions of the Sea. The *Phænomena* of which two may be eafily made out, by fuppofing the *Cavern*, by which they are fed, to arife from the bottom of the next Sea. A Third, is a Well upon the River *Ogmore* in *Glamorganfhire*, and near unto *Newton*, of which *Camden* relates himfelf to be certified, by a Letter from a Learned Frier d of his that obferved it, *Fons abeft hinc, &c.* The Letter is a little too long to be inferted, but the fubftance is this ; That this Well ebbs and flows quite contrary to the flowing and ebbing of the Sea in thofe parts : for 'tis almoft empty at Full Sea, but full at Low water. This may happen from the Channel by which it is fupplied, which may come from the bottom of a Sea very remote from thofe parts, and where the Tides are much differing from thofe of the approximate fhores. A Fourth, lies in *Weftmorland*, near the River *Loder* ; *Qui inftar Euripi fæpius in die reciprocantibus undis fluit & refluit*, which ebbs and flows many times a day. This may proceed from its being fupplyed from many Channels, coming from feveral parts of the Sea, lying fufficiently diftant afunder to have the times of High-water differing enough one from the other ; fo as that whenfoever it fhall be High water over any of thofe places, where thefe Channels begin, it fhall likewife be fo in the Well ; but this is but a fuppofition.

A Seventh *Query* was, Whether the *diffolution* or mixing of feveral bodies, whether fluid or folid, with faline or other Liquors, might not partly be attributed to this Principle of the congruity of thofe bodies and their diffolvents ? As of Salt in Water, Metals in feveral *Menftruums*, Unctuous Gums in Oyls, the mixing of Wine and Water, *&c.* And whether *precipitation* be not partly made from the fame Principle of Incongruity ? I fay *partly*, becaufe there are in fome Diffolutions, fome other Caufes concurrent.

I fhall laftly make a much more feemingly ftrange and unlikely *Query* ; and that is, Whether this Principle, well examined and explained, may not be found a *co-efficient* in the moſt conſiderable Operations of Nature? As in thoſe of *Heat*, and *Light*, and conſequently, of *Rarefaction* and *Condenſation*, *Hardneſs*, and *Fluidneſs*, *Perſpicuity* and *Opacouſneſs*, *Refractions* and *Colours*, *&c.* Nay, I know not whether there may be many things done in Nature, in which this may not (be ſaid to) have a Finger? This I have in ſome other paſſages of this Treatiſe further enquired into and ſhewn, that as well *Light* as *Heat* may be cauſed by *corroſion*, which is applicable to *congruity*, and conſequently all the reſt will be but *ſubſequents*: In the mean time I would not willingly be guilty of that *Error*, which the thrice Noble and Learned *Verulam* juſtly takes notice of, as ſuch, and calls *Philoſophiæ Genus Empiricum*, *quod in paucorum Experimentorum Anguſtiis & Obſcuritate fundatum eſt*. For I neither conclude from one ſingle Experiment, nor are the Experiments I make uſe of all made upon one Subject : Nor wreſt I any Experiment to make it *quadrare* with any preconceiv'd Notion. But on the contrary, I endeavour to be converſant in divers kinds of Experiments, and all and every one of thoſe Trials, I make the Standards or Touchſtones, by which I try all my former Notions, whether they hold out in weight, and meaſure, and touch, *&c.* For as that Body is no other then a Counterfeit Gold, which wants any one of the Proprieties of Gold, (ſuch as are the Malleableneſs, Weight, Colour, Fixtneſs in the Fire, Indiſſolubleneſs in *Aqua fortis*, and the like) though it has all the other ; ſo will all thoſe Notions be found to be falſe and deceitful, that will not undergo all the Trials and Teſts made of them by Experiments. And therefore ſuch as will not come up to the deſired *Apex* of Perfection, I rather wholly reject and take new, then by piecing and patching, endeavour to retain the old, as knowing ſuch things at beſt to be but lame and imperfect. And this courſe I learned from Nature ; whom we find neglectful of the old Body, and ſuffering its Decaies and Infirmities to remain without repair, and altogether ſollicitous and careful of perpetuating the *Species* by new *Individuals*. And it is certainly the moſt likely way to erect a glorious Structure and Temple to *Nature*, ſuch as ſhe will be found (by any *zealous Votary*) to reſide in ; to begin to build a new upon a ſure Foundation of Experiments.

But to digreſs no further from the conſideration of the *Phænomena*, more immediately explicable by this Experiment, we ſhall proceed to ſhew, That, as to the riſing of Water in a *Filtre*, the reaſon of it will be manifeſt to him, that does take notice, that a *Filtre* is conſtituted of a great number of ſmall long ſolid bodies, which lie ſo cloſe together, that the Air in its getting in between them, doth loſe of its preſſure that it has againſt the *Fluid* without them, by which means the Water or Liquor not finding a reſiſtance between them as is able to counter-ballance the preſſure on its ſuperficies without, is raiſed upward, till it meet with a preſſure of the Air which is able to hinder it. And as to the Riſing of Oyl, melted Tallow, Spirit of Wine, *&c.* in the Week of a Candle or Lamp, it is evident, that it differs in nothing from the former, ſave only in this, that in a *Filtre* the Liquor deſcends and runs away by another part ; and in the Week the Liquor is diſperſed and carried away by the

Flame ;

Flame; fomething there is afcribable to the Heat, for that it may rarifie the more volatil and fpirituous parts of thofe combuftible Liquors, and fo being made lighter then the Air, it may be protruded upwards by that more ponderous fluid body in the Form of Vapours; but this can be afcribed to the afcenfion of but a very little, and moft likely of that only which afcends without the Week. As for the Rifing of it in a Spunge, Bread, Cotton. &c. above the fuperficies of the fubjacent Liquor; what has been faid about the *Filtre* (if confidered) will eafily fuggeft a reafon, confidering that all thefe bodies abound with fmall holes or pores.

From this fame Principle alfo (*viz. the unequal preffure of the Air against the unequal fuperficies of the water*) proceeds the caufe of the acceffion or incurfion of any floating body againft the fides of the containing Veffel, or the *appropinquation* of two floating bodies, as *Bubbles, Corks, Sticks, Straws, &c.* one towards another. As for inftance, Take a Glafs-jar, fuch as A B in the feventh *Figure*, and filling it pretty near the top with water, throw into it a fmall round piece of Cork, as C, and plunge it all over in water, that it be wet, fo as that the water may rife up by the fides of it, then placing it any where upon the fuperficies, about an inch, or one inch and a quarter from any fide, and you fhall perceive it by degrees to make *perpendicularly* toward the neareft part of the fide, and the nearer it approaches, the fafter to be moved; the reafon of which *Phænomenon* will be found no other then this, that the Air has a greater preffure againft the middle of the *fuperficies*, then it has againft thofe parts that approach nearer, and are *contiguous* to the fides. Now that the preffure is greater, may (as I fhewed before in the explication of the third *Figure*) be evinced from the flatting of the water in the middle, which arifes from the gravity of the under *fluid*: for fince, as I fhewed before, if there were no gravity in the under *fluid*, or that it were equal to that of the upper, the terminating Surface would be *fpherical*, and fince it is the additional preffure of the gravity of water that makes it fo flat, it follows, that the preffure upon the middle muft be greater then towards the fides. Hence the Ball having a ftronger preffure againft that fide of it which refpects the middle of the *fuperficies*, then againft that which refpects the *approximate* fide, muft neceffarily move towards that part, from whence it finds leaft refiftance, and fo be *accelerated*, as the refiftance decreafes. Hence the more the water is raifed under that part of its way it is paffing above the middle, the fafter it is moved : And therefore you will find it to move fafter in E then in D, and in D then in C. Neither could I find the floating fubftance to be moved at all, until it were placed upon fome part of the *fuperficies* that was fenfibly elevated above the height of the middle part. Now that this may be the true caufe, you may try with a blown Bladder, and an exactly round Ball upon a very fmooth fide of fome pliable body, as *Horn* or *Quickfilver*. For if the Ball be placed under a part of the Bladder which is upon one fide of the middle of its preffure, and you prefs ftrongly againft the Bladder, you fhall find the Ball moved from the middle towards the fides.

Having

Having therefore fhewn the reafon of the motion of any float towards the fides, the reafon of the incurfion of any two floating bodies will eafily appear : For the rifing of the water againft the fides of either of them, is an Argument fufficient, to fhew the preffure of the Air to be there lefs, then it is further from it, where it is not fo much elevated ; and therefore the reafon of the motion of the other toward it, will be the fame as towards the fide of the Glafs; only here from the fame reafon, they are mutually moved toward each other, whereas the fide of the Glafs in the former remains fixt. If alfo you gently fill the Jar fo full with water, that the water is *protuberant* above the fides, the fame piece of Cork that before did haften towards the fides, does now fly from it as faft towards the middle of the Superficies; the reafon of which will be found no other then this, that the preffure of the Air is ftronger againft the fides of the Superficies G and H, then againft the middle I ; for fince, as I fhewed before, the Principle of congruity would make the terminating Surface Spherical, and that the flatting of the Surface in the middle is from the abatement of the waters preffure outwards, by the contrary indeavour of its gravity ; it follows that the preffure in the middle muft be lefs then on the fides; and therefore the confecution will be the fame as in the former. It is very odd to one that confiders not the reafon of it, to fee two floating bodies of wood to approach each other, as though they were indued with fome magnetical vigour ; which brings into my mind what I formerly tried with a piece of Cork or fuch like body, which I fo ordered, that by putting a little ftick into the fame water, one part of the faid Cork would approach and make toward the ftick, whereas another would difcede and fly away, nay it would have a kind of verticity, fo as that if the *Æquator* (as I may fo fpeak) of the Cork were placed towards the ftick, if let alone, it would inftantly turn its appropriate Pole toward it, and then run a-tilt at it: and this was done only by taking a dry Cork, and wetting one fide of it with one fmall ftroak ; for by this means gently putting it upon the water, it would deprefs the fuperficies on every fide of it that was dry, and therefore the greateft preffure of the Air, being near thofe fides caufed it either to chafe away, or elfe to fly off from any other floating body, whereas that fide only, againft which the water afcended, was thereby able to attract.

It remains only, that I fhould determine how high the Water or other Liquor may by this means be raifed in a fmaller Pipe above the Superficies of that without it, and at what height it may be fuftained : But to determine this, will be exceeding difficult, unlefs I could certainly know how much of the Airs preffure is taken off by the fmalnefs of fuch and fuch a Pipe, and whether it may be wholly taken off, that is, whether there can be a hole or pore fo fmall, into which Air could not at all enter, though water might with its whole force ; for were there fuch, 'tis manifeft, that the water might rife in it to fome five or fix and thirty Englifh Foot high. I know not whether the capillary Pipes in the bodies of fmall Trees, which we call their *Microfcopical pores*, may not be fuch ; and whether the congruity of the fides of the Pore may not yet draw the juyce
even

even higher then the Air was able by its bare preſſure to raiſe it : For, Congruity is a principle that not only unites and holds a body joyned to it, but, which is more, attracts and draws a body that is very near it, and holds it above its uſual height.

And this is obvious even in a drop of water ſuſpended under any Simílar or Congruous body : For, beſides the ambient preſſure that helps to keep it ſuſtein'd, there is the Congruity of the bodies that are contiguous. This is yet more evident in Tenacious and Glutinous bodies ; ſuch as Gummous Liquors, Syrups, Pitch, and Roſin melted, &c. Tar, Turpentine, Balſom, Bird-lime, &c. for there it is evident, that the Parts of the tenacious body, as I may ſo call it, do ſtick and adhere ſo cloſely together, that though drawn out into long and very ſlender Cylinders, yet they will not eaſily relinquiſh one another ; and this, though the bodies be aliquatenus fluid, and in motion by one another ; which, to ſuch as conſider a fluid body only as its parts are in a confuſed irregular motion, without taking in alſo the congruity of the parts one among another, and incongruity to ſome other bodies, does appear not alittle ſtrange. So that beſides the incongruity of the ambient fluid to it, we are to conſider alſo the congruity of the parts of the contein'd fluid one with another.

And this Congruity (that I may here a little further explain it) is both a Tenaceous and an Attractive power ; for the Congruity, in the Vibrative motions, may be the cauſe of all kind of attraction, not only Electrical, but Magnetical alſo, and therefore it may be alſo of Tenacity and Glutinouſneſs. For, from a perfect congruity of the motions of two diſtant bodies, the intermediate fluid particles are ſeparated and droven away from between them, and thereby thoſe congruous bodies are, by the incompaſſing mediums, compell'd and forced neerer together ; wherefore that attractiveneſs muſt needs be ſtronger, when, by an immediate contact, they are forc'd to be exactly the ſame : As I ſhew more at large in my Theory of the Magnet. And this hints to me the reaſon of the ſuſpenſion of the Mercury many inches, nay many feet, above the uſual ſtation of 30 inches. For the parts of Quick-ſilver, being ſo very ſimilar and congruous to each other, if once united, will not eaſily ſuffer a divulſion : And the parts of water, that were any wayes heterogeneous, being by exantlation or rarefaction exhauſted, the remaining parts being alſo very ſimilar, will not eaſily part neither. And the parts of the Glaſs being ſolid, are more difficultly disjoyn'd ; and the water, being ſomewhat ſimilar to both, is, as it were, a medium to unite both the Glaſs and the Mercury together. So that all three being united, and not very diſſimilar, by means of this contact, if care be taken that the Tube in erecting be not ſhogged, the Quickſilver will remain ſuſpended, notwithſtanding its contrary indeavour of Gravity, a great height above its ordinary Station ; but if this immediate Contact be removed, either by a meer ſeparation of them one from another by the force of a ſhog, whereby the other becomes imbodied between them, and licks up from the ſurface ſome agil parts, and ſo hurling them makes them air ; or elſe

by

by fome fmall heterogeneous agil part of the Water, or Air, or Quick-filver, which appears like a bubble, and by its jumbling to and fro there is made way for the *heterogeneous Æther* to obtrude it felf between the Glafs and either of the other Fluids, the Gravity of *Mercury precipitates* it downward with very great violence ; and if the Veffel that holds the reftagnating *Mercury* be convenient, the *Mercury* will for a time *vibrate* to and fro with very large *reciprocations*, and at laft will remain kept up by the preffure of the external Air at the height of neer thirty inches. And whereas it may be objected, that it cannot be, that the meer imbodying of the *Æther* between thefe bodies can be the caufe,fince the *Æther* ha-ving a free paffage alwayes, both through the Pores of the Glafs, and through thofe of the Fluids, there is no reafon why it fhould not make a feparation at all times whilft it remains fufpended, as when it is violently dif-joyned by a fhog. To this I anfwer, That though the *Æther* paffes between the Particles, that is, through the Pores of bodies, fo as that any chafme or fepa tion being made, it has infinite paffages to admit its en-try into it, yet fuch is the tenacity or attractive virtue of Congruity, that till it be overcome by the meer ftrength of Gravity, or by a fhog affifting that Conatus of Gravity, or by an agil Particle, that is like a leaver agi-tated by the *Æther* ; and thereby the parts of the congruous fubftances are feparated fo far afunder, that the ftrength of congruity is fo far wea-kened,as not to be able to reunite them, the parts to be taken hold ofbe-ing removed out of the attractive Sphere, as I may fo fpeak, of the con-gruity ; fuch, I fay, is the tenacity of congruity, that it retains and holds the almoft contiguous Particles of the Fluid, and fuffers them not to be feparated, till by meer force that attractive or retentive faculty be over-come: But the feparation being once made beyond the Sphere of the attractive activity of congruity, that virtue becomes of no effect at all, but the *Mercury* freely falls downwards till it meet with a refiftance from the preffure of the *ambient* Air, able to refift its gravity, and keep it for-ced up in the Pipe to the height of about thirty inches.

Thus have I gently raifed a Steel *pendulum* by a Loadftone to a great Angle,till by the fhaking of my hand I have chanced to make a fepara-tion between them, which is no fooner made, but as if the Loadftone had retained no attractive virtue, the *Pendulum* moves freely from it towards the other fide. So vafta difference is there between the attractive vir-tue of the *Magnet* when it acts upon a contiguous and upon a disjoyned body: and much more muft there be between the attractive virtues of congruity upon a contiguous and disjoyned body ; and in truth the attra-ctive virtue is fo little upon a body disjoyned. that though I have with a *Microfcope* obferved very diligently, whether there were any extraordi-nary *protuberance* on the fide of a drop of water that was exceeding neer to the end of a green ftick,but did not touch it, I could not perceive the leaft; though I found, that as foon as ever it toucht it the whole drop would prefently unite it felf with it ; fo that it feems an abfolute con-tact is requifite to the exercifing of the tenacious faculty of congruity.

Obſerv. VII. *Of ſome* Phænomena *of Glaſs drops.*

THeſe *Glaſs Drops* are ſmall ¦parcels of coarſe green Glaſs taken out of the Pots that contain the *Metal* (as they call it) in fuſion, upon the end of an Iron Pipe ; and being exceeding hot, and thereby of a kind of ſluggiſh fluid Conſiſtence, are ſuffered to drop from thence into a Bucket of cold Water, and in it to lye till they be grown ſenſibly cold.

Some of theſe I broke in the open air, by ſnapping off a little of the ſmall ſtem with my fingers, others by cruſhing it with a ſmall pair of Ply-ers ; which I had no ſooner done, then the whole bulk of the drop flew violently, with a very briſk noiſe, into multitudes of ſmall pieces, ſome of which were as ſmall as duſt, though in ſome there were remaining pieces pretty large, without any flaw at all, and others very much flaw'd, which by rubbing between ones fingers was eaſily reduced to duſt ; theſe di-ſperſed every way ſo violently, that ſome of them pierced my skin. I could not find, either with my naked Eye, or a *Microſcope*, that any of the broken pieces were of a regular figure, nor any one like another, but for the moſt part thoſe that flaw'd off in large pieces were prettily bran-ched.

The ends of others of theſe drops I nipt off whilſt all the bodies and ends of them lay buried under the water, which, like the former, flew all to pieces with as briſk a noiſe, and as ſtrong a motion.

Others of theſe I tried to break, by grinding away the blunt end, and though I took a ſeemingly good one, and had ground away neer two thirds of the Ball, yet would it not fly to pieces, but now and then ſome ſmall rings of it would ſnap and fly off, not without a briſk noiſe and quick motion, leaving the Surface of the drop whence it flew very pretti-ly branched or creaſed, which was eaſily diſcoverable by the *Microſcope*. This drop, after I had thus ground it, without at all impairing the remnant that was not ground away, I cauſed to fly immediately all into ſand upon the nipping off the very tip of its ſlender end.

Another of theſe drops I began to grind away at the ſmaller end, but had not worn away on the ſtone above a quarter of an inch before the whole drop flew with a briſk crack into ſand or ſmall duſt ; nor would it have held ſo long, had there not been a little flaw in the piece that I ground away, as I afterwards found.

Several others of theſe drops I covered over with a thin but very tuff skin of *Icthyocolla*, which being very tough and very tranſparent, was the moſt convenient ſubſtance for theſe tryals that I could imagine, having dipt, I ſay, ſeveral of theſe drops in this tranſparent Glue whilſt hot, and ſuffering them to hang by a ſtring tied about the end of them till they were cold, and the skin pretty tough ; then wrapping all the body of the

drop

drop (leaving out only the very tip) in fine fupple Kids-leather very clofely,I nipped off the fmall top, and found, as I expected, that notwith-ftanding this skin of Glue, and the clofe wrapping up in Leather, upon the breaking of the top, the drop gave a crack like the reft, and gave my hand a pretty brisk impulfe: but yet the skin and leather was fo ftrong as to keep the parts from flying out of their former pofture; and, the skin being tranfparent, I found that the drop retained exactly its former fi-gure and polifh, but was grown perfectly opacous and all over flaw'd, all thofe flaws lying in the manner of rings, from the bottom or blunt end, to the very top or fmall point. And by feveral examinations with a *Micro-fcope*, of feveral thus broken, I found the flaws, both within the body of the drop, and on the outward furface, to lye much in this order.

Let A B in the Figure X of the fourth Scheme reprefent the drop cafed over with *Icthyocolla* or *Ifinglafs*,and (by being ordered as is before pre-fcribed) crazed or flawed into pieces, but by the skin or cafe kept in its former figure, and each of its flawed parts preferved exactly in its due pofture; the outward appearance of it fomewhat plainly to the naked eye, but much more confpicuous if viewed with a fmall fenfs appeared much after this fhape. That is, the blunt end B for a pretty breadth, namely, as far as the Ring C C C feemed irregularly flawed with divers clefts, which all feemed to tend towards the Center of it, being, as I af-terwards found, and fhall anon fhew in the defcription of the figure Y, the Bafis, as it were, of a Cone, which was terminated a little above the middle of the drop, all the reft of the Surface from C C C to A was flawed with an infinite number of fmall and parallel Rings, which as they were for the moft part very round, fo were they very thick and clofe together, but were not fo exactly flaw'd as to make a perfect Ring, but each circular part was by irregular cracks flawed likewife into multitudes of irregular flakes or tiles; and this order was obferved likewife the whole length of the neck,

Now though I could not fo exactly cut this *conical Body* through the *Axis*, as is reprefented by the figure Y; yet by *anatomizing*, as it were, of feveral, and taking notice of divers particular circumftances, I was in-formed, that could I have artificially divided a flaw'd drop through the *Axis* or *Center*, I fhould with a *Microfcope* have found it to appear much of this form, where A fignifies the *Apex*, and B the blunt end, C C the Cone of the Bafis, which is terminated at T the top or end of it, which feems to be the very middle of the blunt end, in which, not only the co-nical body of the Bafis C C is terminated, but as many of the parts of the drop as reach as high as D D.

And it feemed to be the head or beginning of a Pith, as it were, or a a part of the body which feemed more fpungy then the reft, and much more irregularly flawed, which from T afcended by E E, though lefs vi-fible, into the fmall neck towards A. The Grain, as it were, of all the flaws, that from all the outward Surface A D C C D A, was much the fame,as is reprefented by the black ftrokes that meet in the middle D T, D T, D E, D E, &c.

Nor is this kind of Grain, as I may call it, peculiar to Glafs drops thus quenched; for (not to mention *Coperas-ftones*, and divers other *Marchafites* and *Minerals*, which I have often taken notice of to be in the very fame manner flaked or grained, with a kind of Pith in the middle) I have obferved the fame in all manner of caft Iron, efpecially the coarfer fort, fuch as Stoves, and Furnaces, and Backs, and Pots are made of : For upon the breaking of any of thofe Subftances it is obvious to obferve, how from the out-fides towards the middle, there is a kind of Radiation or Grain much refembling this of the Glafs-drop; but this Grain is moft confpicuous in Iron-bullets, if they be broken: the fame *Phænomena* may be produced by cafting *regulus* of *Antimony* into a Bullet-mold, as alfo with *Glafs of Antimony*, or with almoft any fuch kind of *Vitrified fubftance*, either caft into a cold Mold or poured into Water.

Others of thefe Drops I heat red hot in the fire, and then fuffered them to cool by degrees. And thefe I found to have quite loft all their *fulminating* or flying quality, as alfo their hard, brittle and fpringy texture ; and to emerge of a much fofter temper, and much eafier to be broken or fnapt with ones finger; but its ftrong and brittle quality was quite deftroyed, and it feemed much of the fame confiftence with other green Glafs well nealed in the Oven.

The Figure and bignefs of thefe for the moft part was the fame with that of the Figure Z; that is, all the furface of them was very fmooth and polifht, and for the moft part round, but very rugged or knobbed about D, and all the length of the ftem was here and there pitted or flatted. About D, which is at the upper part of the drop under that fide of the ftem which is concave, there ufually was made fome one or more little Hillocks or Prominences. The drop it felf, before it be broken, appears very tranfparent, and towards the middle of it, to be very full of fmall Bubbles, of fome kind of aerial fubftance, which by the refraction of the outward furface appear much bigger then really they are ; and this may be in good part removed, by putting the drop under the furface of clear Water, for by that means moft part of the refraction of the convex Surface of the drop is deftroyed, and the bubbles will appear much fmaller. And this, by the by, minds me of the appearing magnitude of the *aperture* of the *iris*, or *pupil* of the eye, which though it appear, and be therefore judged very large, is yet not above a quarter of the bignefs it appears of, by the *lenticular* refraction of the *Cornea*.

The caufe of all which *Phænomena* I imagine to be no other then this, That the Parts of the Glafs being by the exceffive heat of the fire kept off and feparated one from another, and thereby put into a kind of fluggifh fluid confiftence, are fuffered to drop off with that heat or agitation remaining in them, into cold Water; by which means the outfides of the drop are prefently cool'd and *crufted*, and are thereby made of a loofe texture, becaufe the parts of it have not time to fettle themfelves leifurely together, and fo to lie very clofe together : And the innermoft parts of the drop, retaining ftill much of their former heat and agitations, remain

of

of a loofe texture alfo, and, according as the cold ftrikes inwards from the bottom and fides, are quenched, as it were, and made rigid in that very pofture wherein the cold finds them. For the parts of the *cruft* being already hardened, will not fuffer the parts to fhrink any more from the outward Surface inward ; and though it fhrink a little by reafon of the fmall parcels of fome Aerial fubftances difperfed through the matter of the Glafs, yet that is not neer fo much as it appears (as I juft now hinted ;) nor if it were, would it be fufficient for to confolidate and condenfe the body of Glafs into a *tuff* and clofe *texture*, after it had been fo exceffively rarified by the heat of the glafs-Furnace.

But that there may be fuch an expanfion of the aerial fubftance contained in thofe little *blebbs* or bubbles in the body of the drop, this following Experiment will make more evident.

Take a fmall Glafs-Cane about a foot long, feal up one end of it *hermetically*, then put in a very fmall bubble of Glafs, almoft of the fhape of an Effence-viol with the open mouth towards the fealed end, then draw out the other end of the Pipe very fmall, and fill the whole Cylinder with water, then fet this Tube by the Fire till the Water begin to boyl, and the Air in the bubble be in good part rarified and driven out, then by fucking at the fmalling Pipe, more of the Air or vapours in the bubble may be fuck'd out, fo that it may fink to the bottom ; when it is funk to the bottom, in the flame of a Candle, or Lamp, nip up the flender Pipe and let it cool : whereupon it is obvious to obferve, firft, that the Water by degrees will fubfide and fhrink into much lefs room : Next, that the Air or vapours in the Glafs will expand themfelves fo, as to buoy up the little Glafs : Thirdly, that all about the infide of the Glafs-pipe there will appear an infinite number of fmall bubbles, which as the Water grows colder and colder will fwell bigger and bigger, and many of them buoy themfelves up and break at the top.

From this *Difceding* of the heat in Glafs drops, that is, by the quenching or cooling Irradiations propagated from the Surface upwards and inwards, by the lines C T, C T, D T, D E, &c. the bubbles in the drop have room to expand themfelves a little, and the parts of the Glafs contract themfelves ; but this operation being too quick for the fluggifh parts of the Glafs, the contraction is performed very unequally and irregularly, and thereby the Particles of the Glafs are bent, fome one way, and fome another, yet fo as that moft of them draw towards the Pith or middle T E E E, or rather from that outward : fo that they cannot *extricate* or unbend themfelves, till fome part of T E E E be broken and loofened, for all the parts about that are placed in the manner of an Arch, and fo till their hold at T E E E be loofened they cannot fly afunder, but uphold, and fhelter, and fix each other much like the ftones in a Vault, where each ftone does concurre to the ftability of the whole Fabrick, and no one ftone can be taken away but the whole Arch falls. And wherefoever any of thofe radiating wedges D T D, &c. are removed, which are the component parts of this Arch, the whole Fabrick prefently falls to

pieces ;

pieces; for all the Springs of the feveral parts are fet at liberty, which immediately extricate themfelves and fly afunder every way; each part by its fpring contributing to the darting of it felf and fome other contiguous part. But if this drop be heat fo hot as that the parts by degrees can unbend themfelves, and be fettled and annealed in that pofture, and be then fuffered gently to fubfide and cool; The parts by this nealing lofing their fpringinefs, conftitute a drop of a more foft but lefs brittle texture, and the parts being not at all under a flexure, though any part of the middle or Pith T E E E be broken, yet will not the drop at all fly to pieces as before.

This Conjecture of mine I fhall indeavour to make out by explaining each particular Affertion with *analogous* Experiments: The Affertions are thefe.

Firft, That the parts of the Glafs, whilft in a fluid Confiftence and hot, are more rarified, or take up more room, then when hard and cold.

Secondly, That the parts of the drop do fuffer a twofold contraction.

Thirdly, That the dropping or quenching the glowing metal in the Water makes it of a hard, fpringing, and rarified texture.

Fourthly, That there is a flexion or force remaining upon the parts of the Glafs thus quenched, from which they indeavour to extricate themfelves.

Fifthly, That the Fabrick of the drop, that is able to hinder the parts from extricating themfelves, is *analogus* to that of an Arch.

Sixthly, That the fudden flying afunder of the parts proceeds from their fpringinefs.

Seventhly, That a gradual heating and cooling does anneal or reduce the parts of Glafs to a texture that is more loofe, and eafilier to be broken, but not fo brittle.

That the firft of thefe is true may be gathered from this, That *Heat is a property of a body arifing from the motion or agitation of its parts*; and therefore whatever body is thereby toucht muft neceffarily receive fome part of that motion, whereby its parts will be fhaken and agitated, and fo by degrees free and extricate themfelves from one another, and each part fo moved does by that motion *exert a conatus* of *protruding* and difplacing all the adjacent Particles. Thus Air included in a veffel, by being heated will burft it to pieces. Thus have I broke a Bladder held over the fire in my hand, with fuch a violence and noife, that it almoft made me deaf for the prefent, and much furpaffed the noife of a Musket: The like have I done by throwing into the fire fmall glafs Bubbles hermetically fealed, with a little drop of Water included in them. Thus Water alfo, or any other Liquor, included in a convenient veffel, by being warmed, manifeftly expands it felf with a very great violence, fo as to break the ftrongeft veffel, if when heated it be narrowly imprifoned in it.

This

This is very manifeſt by the *ſealed Thermometers*, which I have, by ſeveral tryals, at laſt brought to a great certainty and tendernełs : for I have
made ſome with ſtems above four foot long, in which the expanding Liquor would ſo far vary, as to be very neer the very top in the heat of Summer, and prety neer the bottom at the coldeſt time of the Winter. The
Stems I uſe for them are very thick, ſtraight, and even Pipes of Glaſs, with
a very ſmall *perforation*, and both the head and body I have made on
purpoſe at the Glaſs-houſe, of the ſame metal whereof the Pipes are
drawn : theſe I can eaſily in the flame of a Lamp, urged with the blaſt of
a pair of Bellows, ſeal and cloſe together, ſo as to remain very firm, cloſe
and even ; by this means I joyn on the body firſt, and then fill both it and
a part of the ſtem, proportionate to the length of the ſtem and the
warmth of the ſeaſon I fill it in, with the beſt rectified *ſpirit of Wine* highly *ting'd* with the lovely colour of *Cocheneel,* which I deepen the more
by pouring ſome drops of common *ſpirit of Urine,* which muſt not be
too well rectified, becauſe it will be apt to make the Liquor to curdle
and ſtick in the ſmall perforation of the ſtem. This Liquor I have upon
tryal found the moſt tender of any ſpirituous Liquor, and thoſe are much
more ſenſibly affected with the variations of heat and cold then other more
flegmatick and ponderous Liquors, and as capable of receiving a deep
tincture, and keeping it, as any Liquor whatſoever ; and (which makes
it yet more acceptable) is not ſubject to be frozen by any cold yet
known. When I have thus filled it, I can very eaſily in the forementioned flame of a Lamp ſeal and joyn on the head of it.

Then, for graduating the ſtem, I fix that for the beginning of my diviſion where the ſurface of the liquor in the ſtem remains when the
ball is placed in common diſtilled water, that is ſo cold that it juſt begins
to freeze and ſhoot into flakes ; and that mark I fix at a convenient place
of the ſtem, to make it capable of exhibiting very many degrees of cold,
below that which is requiſite to freeze water : the reſt of my diviſions,
both above and below this (which I mark with a [o] or nought) I place
according to the Degrees of *Expanſion,* or *Contraction* of the Liquor in
proportion to the bulk it had when it indur'd the newly mention'd freezing cold. And this may be very eaſily and accurately enough done by
this following way ; Prepare a Cylindrical veſſel of very thin plate Braſs
or Silver, A B C D of the figure Z ; the Diameter A B of whoſe cavity
let be about two inches, and the depth B C the ſame ; let each end be
cover'd with a flat and ſmooth plate of the ſame ſubſtance, cloſely ſoder'd
on, and in the midſt of the upper cover make a pretty large hole E F,
about the bigneſs of a fifth part of the Diameter of the other ; into this
faſten very well with cement a ſtraight and even Cylindrical pipe of Glaſs,
E F G H, the Diameter of whoſe cavity let be exactly one tenth of the
Diameter of the greater Cylinder. Let this pipe be mark'd at G H with
a Diamant, ſo that G from E may be diſtant juſt two inches, or the ſame
height with that of the cavity of the greater Cylinder, then divide the
length E G exactly into 10 parts, ſo the capacity of the hollow of each
of theſe diviſions will be $\frac{1}{1000}$ part of the capacity of the greater Cylinder.

der. This veſſel being thus prepared, the way of marking and graduating the *Thermometers* may be very eaſily thus performed :
Fill this Cylindrical veſſel with the ſame liquor wherewith the *Thermometers* are fill'd, then place both it and the *Thermometer* you are to
graduate, in water that is ready to be frozen, and bring the ſurface of the
liquor in the *Thermometer* to the firſt marke or [o]; then ſo proportion
the liquor in the Cylindrical veſſel, that the ſurface of it may juſt be at
the lower end of the ſmall glaſs-Cylinder ; then very gently and gradually warm the water in which both the *Thermometer* and this Cylindrical
veſſel ſtand, and as you perceive the ting'd liquor to riſe in both ſtems,
with the point of a Diamond give ſeveral marks on the ſtem of the *Thermometer* at thoſe places, which by comparing the expanſion in both
Stems, are found to correſpond to the diviſions of the cylindrical veſſel,
and having by this means marked ſome few of theſe diviſions on the
Stem, it will be very eaſie by theſe to mark all the reſt of the Stem,
and accordingly to aſſign to every diviſion a proper character.
A *Thermometer*, thus marked and prepared, will be the fitteſt Inſtrument to make a Standard of heat and cold that can be imagined. For
being ſealed up, it is not at all ſubject to variation or waſting, nor is it liable to be changed by the varying preſſure of the Air, which all other
kind of *Thermometers* that are open to the Air are liable to. But to proceed.
This property of Expanſion with Heat, and Contraction with Cold, is
not peculiar to Liquors only, but to all kind of ſolid Bodies alſo, eſpecially Metals, which will more manifeſtly appear by this Experiment.
Take the Barrel of a Stopcock of Braſs, and let the Key, which is well
fitted to it, be riveted into it, ſo that it may ſlip, and be eaſily turned round,
then heat this Cock in the fire, and you will find the Key ſo ſwollen, that
you will not be able to turn it round in the Barrel ; but if it be ſuffered
to cool again, as ſoon as it is cold it will be as movable, and as eaſie to be
turned as before.
This Quality is alſo very obſervable in *Lead, Tin, Silver, Antimony,
Pitch, Roſin, Bees-wax, Butter*, and the like; all which, if after they be melted
you ſuffer gently to cool, you ſhall find the parts of the upper Surface
to ſubſide and fall inwards, loſing that plumpneſs and ſmoothneſs it had
whilſt in fuſion. The like I have alſo obſerved in the cooling of *Glaſs
of Antimony*. which does very neer approach the nature of Glaſs,
But becauſe theſe are all Examples taken from other materials then
Glaſs, and argue only, that poſſibly there may be the like property alſo in
Glaſs, not that really there is ; we ſhall by three or four Experiments indeavour to manifeſt that alſo.
And the Firſt is an Obſervation that is very obvious even in theſe very
drops, to wit, that they are all of them terminated with an unequal or irregular Surface, eſpecially about the ſmaller part of the drop, and the
whole length of the ſtem ; as about D, and from thence to A, the whole
Surface, which would have been round if the drop had cool'd leiſurely,
is, by being quenched haſtily, very irregularly flatted and pitted ; which
I

I suppose proceeds partly from the Waters unequally cooling and pressing the parts of the drop, and partly from the self-contracting or subsiding quality of the substance of the Glass: For the vehemency of the heat of the drop causes such sudden motions and bubbles in the cold Water, that some parts of the Water bear more forcibly against one part then against another, and consequently do more suddenly cool those parts to which they are contiguous.

A Second Argument may be drawn from the Experiment of cutting Glasses with a hot Iron. For in that Experiment the top of the Iron heats, and thereby rarifies the parts of the Glass that lie just before the crack, whence each of those agitated parts indeavouring to expand its self and get elbow-room, thrusts off all the rest of the contiguous parts, and consequently promotes the crack that was before begun.

A Third Argument may be drawn from the way of producing a crack in a sound piece or plate of Glass, which is done two wayes, either First, by suddenly heating a piece of Glass in one place more then in another. And by this means *Chymists* usually cut off the necks of Glass-bodies, by two kinds of Instruments, either by a glowing hot round Iron-Ring, which just incompasses the place that is to be cut, or else by a *Sulphur'd* Threed, which is often wound about the place where the separation is to be made, and then fired. Or Secondly, A Glass may be cracked by cooling it suddenly in any place with Water, or the like, after it has been all leisurely and gradually heated very hot. Both which *Phænomena* seem manifestly to proceed from the *expansion* and contraction of the parts of the Glass, which is also made more probable by this circumstance which I have observed, that a piece of common window-glass being heated in the middle very suddenly with a live Coal or hot Iron, does usually at the first crack fall into pieces, whereas if the Plate has been gradually heated very hot, and a drop of cold Water and the like be put on the middle of it, it only flaws it, but does not break it asunder immediately.

A Fourth Argument may be drawn from this Experiment; Take a Glass-pipe, and fit into it a solid stick of Glass, so as it will but just be moved in it. Then by degrees heat them whilst they are one within another, and they will grow stiffer, but when they are again cold, they will be as easie to be turned as before. This Expansion of Glass is more manifest in this Experiment.

Take a stick of Glass of a considerable length, and fit it so between the two ends or screws of a Lath, that it may but just easily turn, and that the very ends of it may be just toucht and susteined thereby; then applying the flame of the Candle to the middle of it, and heating it hot, you will presently find the Glass to stick very fast on those points, and not without much difficulty to be convertible on them, before that by removing the flame for a while from it, it be suffered to cool, and then you will find it as easie to be turned round as at the first.

From all which Experiments it is very evident, that all those Bodies, and particularly Glass, suffers an Expansion by Heat, and that a very considerable fiderable one, whilft they are in a ftate of Fufion. For *Fluidity*, as I elfewhere mention, *being nothing but an effect of a very ftrong and quick fhaking motion, whereby the parts are, as it were, loofened from each other, and confequently leave an interjacent fpace or vacuity*; it follows, that all thofe fhaken Particles muft neceffarily take up much more room then when they were at reft, and lay quietly upon each other. And this is further confirmed by a Pot of *boyling Alabafter*, which will manifeftly rife a fixth or eighth part higher in the Pot, whilft it is boyling, then it will remain at, both before and after it be boyled. The reafon of which odd *Phænomenon* (to hint it here only by the way) is this, that there is in the curious powder of Alabafter, and other calcining Stones, a certain watery fubftance, which is fo fixt and included with the folid Particles, that till the heat be very confiderable they will not fly away; but after the heat is increafed to fuch a degree, they break out every way in vapours, and thereby fo fhake and loofen the fmall corpufles of the Powder from each other, that they become perfectly of the nature of a fluid body, and one may move a ftick to and fro through it, and ftir it as eafily as water, and the vapours burft and break out in bubbles juft as in boyling water, and the like; whereas, both before thofe watery parts are flying away, and after they are quite gone; that is, before and after it have done boyling, all thofe effects ceafe, and a ftick is as difficultly moved to and fro in it as in fand, or the like. Which Explication I could eafily prove, had I time; but this is not a fit place for it.

To proceed therefore, I fay, that the dropping of this expanded Body into cold Water, does make the parts of the Glafs fuffer a double contraction : The firft is, of thofe parts which are neer the Surface of the Drop. For Cold, as I faid before, contracting Bodies, that is, *by the abatement of the agitating faculty the parts falling neerer together*; the parts next adjoying to the Water muft needs lofe much of their motion, and impart it to the Ambient-water (which the Ebullition and commotion of it manifefts) and thereby become a folid and hard cruft, whilft the innermoft parts remain yet fluid and expanded; whence, as they grow cold alfo by degrees, their parts muft neceffarily be left at liberty to be condenfed, but becaufe of the hardnefs of the outward cruft, the contraction cannot be admitted that way; but there being many very fmall, and before inconfpicuous, bubbles in the fubftance of the Glafs, upon the fubfiding of the parts of the Glafs, the agil fubftance contained in them has liberty of expanding it felf a little, and thereby thofe bubbles grow much bigger, which is the fecond Contraction. And both thefe are confirmed from the appearance of the Drop it felf: for as for the outward parts, we fee, firft, that it is irregular and fhrunk, as it were, which is caufed by the yielding a little of the hardened Skin to a Contraction, after the very outmoft Surface is fettled; and as for the internal parts, one may with ones naked Eye perceive abundance of very confpicuous bubbles, and with the *Microfcope* many more.

The Confideration of which Particulars will eafily make the Third Pofition probable, that is, that the parts of the drop will be of a very hard, though of a rarified Texture; for if the outward parts of the Drop, by reafon of its hard cruft, will indure very little Contraction, and the agil Particles, included

ded in thofe bubbles, by the lofing of their agitation, by thedecreafe of the Heat,lofe alfo moft part of their Spring and Expanfive power;it follows (the withdrawing of the heat being very fudden) that the parts muft be left in a very loofe Texture, and by reafon of the implication of the parts one about another,which from their fluggifhnes and glutinoufnefs I fuppofe to be much after the manner of the fticks in a Thorn-bufh,or a Lock of Wool;It will follow, I fay, that the parts will hold each other very ftrongly together,and indeavour to draw each other neerer together, and confequently their Texture muft be very hard and ftiff, but very much rarified.

And this will make probable my next Pofition, That *the parts of the Glafs are under a kind of tenfion or flexure,out of which they indeavour to extricate and free themfelves*,and thereby all the parts draw towards the Center or middle, and would, if the outward parts would give way, as they do when the outward parts cool leifurely (as in baking of Glaffes) contract the bulk of the drop into a much lefs compafs. For fince.as I proved before,the Internal parts of the drop, when fluid,were of a very rarified Texture and,as it were,tos'd open like a Lock of Wool,and if they were fuffered leifurely to cool, would be again preft, as it were,clofe together: And fince that the heat,which kept them bended and open, is removed , and yet the parts not fuffered to get as neer together as they naturally would ; It follows,that the Particles remain under a kind of *tenfion* and *flexure* , and confequently have an indeavour to free themfelves from that *bending* and *diftenfion*, which they do, as foon as either the tip be broken, or as foon as by a leifurely heating and cooling, the parts are nealed into another pofture.

And this will make my next Pofition probable,that *the parts of the Glafs drops are contignated together in the form of an Arch*,and cannot any where yield or be drawn inwards,till by the removing of fome one part of it(as it happens in the removing one of the ftones of an Arch)the whole Fabrick is fhatter'd,and falls to pieces,and each of the Springs is left at liberty,fuddenly to extricate it felf: for fince I have made it probable,that the internal parts of the Glafs have a contractive power inwards, and the external parts are incapable of fuch a Contraction,and the figure of it being fpherical;it follows,that the fuperficial parts muft bear againft each other , and keep one another from being condens'd into a lefs room, in the fame manner as the ftones of an Arch conduce to the upholding each other in that Figure.And this is made more probable by another Experiment which was communicated to me by an excellent Perfon,whofe extraordinary Abilities in all kind of Knowledg, efpecially in that of Natural things,and his generous Difpofition in communicating,incouraged me to have recourfe to him on many occafions. The Experiment was this : Small Glafs-balls (about the bignefs of that reprefented in the *Figure &.*) would,upon rubbing or fcratching the inward Surface, fly all infunder, with a pretty brisk noife ; whereas neither before nor after the inner Surface had been thus fcratcht, did there appear any flaw or crack. And putting the pieces of one of thofe broken ones together again , the flaws appeared much after the manner of the black lines on the Figure, &. Thefe Balls were fmall, but exceeding thick bubbles of Glafs , which being crack'd off from the *Puntilion* whilft very hot , and fo fuffered to cool without nealing them in

the

the Oven over the Furnace, do thereby (being made of white Glaſs, which cools much quicker then green Glaſs, and is thereby made much brittler) acquire a very *porous* and very brittle *texture*: ſo that if with the point of a Needle or Bodkin, the inſide of any of them be rubbed prety hard, and then laid on a Table, it will, within a very little while, break into many pieces with a brisk noiſe, and throw the parts above a ſpan aſunder on the Table: Now though the pieces are not ſo ſmall as thoſe of a *fulminating* drop, yet they as plainly ſhew, that the outward parts of the Glaſs have a great *Conatus* to fly aſunder, were they not held together by the *tenacity* of the parts of the inward Surface: for we ſee as ſoon as thoſe parts are crazed by hard rubbing, and thereby their tenacity ſpoiled, the ſpringineſs of the more outward parts quickly makes a divulſion, and the broken pieces will, if the concave Surface of them be further ſcratcht with a Diamond, fly again into ſmaller pieces.

From which preceding conſiderations it will follow Sixthly, That the ſudden flying aſunder of the parts as ſoon as this Arch is any where diſordered or broken, proceeds from the ſpringing of the parts; which, indeavouring to *extricate* themſelves as ſoon as they get the liberty, they perform it with ſuch a quickneſs, that they throw one another away with very great violence: for the Particles that compoſe the Cruſt have a *Conatus* to lye further from one another, and therefore as ſoon as the external parts are looſened they dart themſelves outward with great violence, juſt as ſo many Springs would do, if they were detained and faſtened to the body, as ſoon as they ſhould be ſuddenly looſened; and the internal parts drawing inward, they contract ſo violently, that they rebound back again and fly into multitude of ſmall ſhivers or ſands. Now though they appear not, either to the naked Eye, or the *Microſcope*, yet I am very apt to think there may be abundance of ſmall flaws or cracks, which, by reaſon the ſtrong reflecting Air is not got between the *contiguous* parts, appear not. And that this may be ſo, I argue from this, that I have very often been able to make a crack or flaw, in ſome convenient pieces of Glaſs, to appear and diſappear at pleaſure, according as by preſſing together, or pulling aſunder the contiguous parts, I excluded or admitted the ſtrong reflecting Air between the parts: And it is very probable, that there may be ſome Body, that is either very rarified Air, or ſomething *analogous* to it, which fills the bubbles of theſe drops; which I argue, firſt, from the roundneſs of them, and next, from the vivid reflection of Light which they exhibite: Now though I doubt not, but that the Air in them is very much rarified, yet that there is ſome in them, to ſuch as well conſider this Experiment of the diſappearing of a crack upon the *extruding* of the Air, I ſuppoſe it will ſeem more then probable.

The Seventh and laſt therefore that I ſhall prove, is, *That the gradual heating and cooling of theſe ſo extended bodies does reduce the parts of the Glaſs to a looſer and ſofter temper.* And this I found by heating them, and keeping them for a prety while very red hot in a fire; for thereby I found them to grow a little lighter, and the ſmall Stems to be very eaſily broken and ſnapt any where, without at all making the drop fly; whereas before they were fo exceeding hard, that they could not be broken without much difficulty; and upon their breaking the whole drop would fly in pieces with very great violence. The Reafon of which laft feems to be, that the leifurely heating and cooling of the parts does not only waft fome part of the Glafs it felf, but ranges all the parts into a better order, and gives each Particle an opportunity of *relaxing* its felf, and confequently neither will the parts hold fo ftrongly together as before, nor be fo difficult to be broken : The parts now more eafily yielding, nor will the other parts fly in pieces, becaufe the parts have no bended Springs. The *relaxation* alfo in the temper of hardned Steel, and hammered Metals, by nealing them in the fire, feems to proceed from much the fame caufe. For both by quenching fuddenly fuch Metals as have *vitrified* parts interfpers'd, as Steel has, and by hammering of other kinds that do not fo much abound with them, as Silver, Brafs, &c. the parts are put into and detained in a bended pofture, which by the agitation of Heat are fhaken, and loofened, and fuffered to unbend themfelves.

Obferv. VIII. *Of the fiery Sparks ftruck from a Flint or Steel.*

Schem. 5.

IT is a very common Experiment, by ftriking with a Flint againft a Steel, to make certain fiery and fhining Sparks to fly out from between thofe two compreffing Bodies. About eight years fince, upon cafually reading the Explication of this odd *Phænomenon*, by the moft Ingenious *Des Cartes*, I had a great defire to be fatisfied, what that Subftance was that gave fuch a fhining and bright Light : And to that end I fpread a fheet of white Paper, and on it, obferving the place where feveral of thefe Sparks feemed to vanifh, I found certain very fmall, black, but gliftering Spots of a movable Subftance, each of which examining with my *Mifcrocope*, I found to be a fmall round *Globule*; fome of which, as they looked pretty fmall, fo did they from their Surface yield a very bright and ftrong reflection on that fide which was next the Light; and each look'd almoft like a pretty bright Iron-Ball, whofe Surface was pretty regular, fuch as is reprefented by the Figure A. In this I could perceive the Image of the Window pretty well, or of a Stick, which I moved up and down between the Light and it. Others I found, which were, as to the bulk of the Ball, pretty regularly round, but the Surface of them, as it was not very fmooth, but rough, and more irregular, fo was the reflection from it more faint and confufed. Such were the Surfaces of B. C. D. and E. Some of thefe I found cleft or cracked, as C, others quite broken in two and hollow, as D. which feemed to be half the hollow fhell of a Granado, broken irregularly in pieces. Several others I found of other fhapes; but that which is reprefented by E, I obferved to be a very big Spark of Fire, which went out upon one fide of the Flint that I ftruck fire withall, to

which

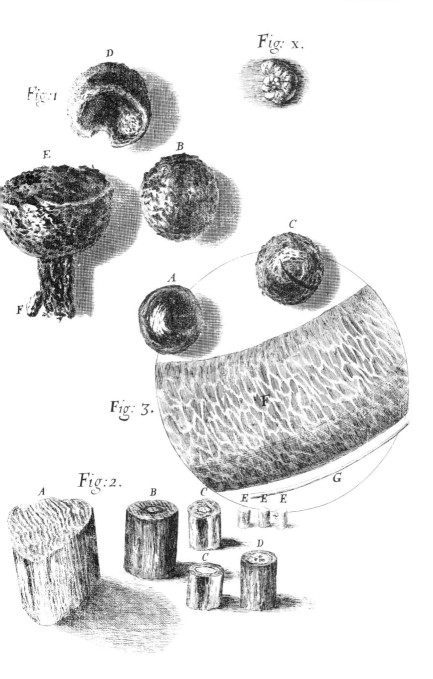

Fig: X.

Fig: 1

D

E.

B

F

A

C

Fig: 3.

F

G

E E E

Fig: 2.

A

B

C

C

D

which it ftuck by the root F, at the end of which fmall Stem was faften-
ed-on a *Hemifphere*, or half a hollow Ball, with the mouth of it open from
the ftemwards, fo that it looked much like a Funnel, or an old fafhioned
Bowl without a foot. This night, making many tryals and obfervations
of this Experiment, I met, among a multitude of the Globular ones which
I had obferved, a couple of Inftances, which are very remarkable to the
confirmation of my *Hypothefis*.

And the Firft was of a pretty big Ball faftened on to the end of a fmall
fliver of Iron, which *Compofitum* feemed to be nothing elfe but a long thin
chip of Iron, one of whofe ends was melted into a fmall round Globul; the
other end remaining unmelted and irregular, and perfectly Iron.

The Second Inftance was not lefs remarkable then the Firft; for I
found, when a Spark went out, nothing but a very fmall thin long fliver
of Iron or Steel, unmelted at either end. So that it feems, that fome of
thefe Sparks are the flivers or chips of the Iron *vitrified*, Others are only the flivers melted into Balls without vitrification, And the third kind
are only fmall flivers of the Iron, made red-hot with the violence of the
ftroke given on the Steel by the Flint.

He that fhall diligently examine the *Phænomena* of this Experiment,
will, I doubt not, find caufe to believe, that the reafon I have heretofore
given of it, is the true and genuine caufe of it, namely, That *the fpark
appearing fo bright in the falling, is nothing elfe but a fmall piece of the Steel
or Flint, but moft commonly of the Steel, which by the violence of the ftroke
is at the fame time fever'd and heatt red-hot, and that fometimes to fuch a
degree, as to make it melt together into a fmall Globule of Steel; and fometimes alfo is that heat fo very intenfe, as further to melt it and vitrifie it; but
many times the heat is fo gentle, as to be able to make the fliver only red hot,
which notwithftanding falling upon the tinder* (that is only a very curious
fmall Coal made of the fmall threads of Linnen burnt to coals and
char'd) *it eafily fets it on fire*. Nor will any part of this *Hypothefis* feem
ftrange to him that confiders, Firft, that either hammering, or filing, or
otherwife violently rubbing of Steel, will prefently make it fo hot as to
be able to burn ones fingers. Next, that the whole force of the ftroke
is exerted upon that fmall part where the Flint and Steel firft touch: For
the Bodies being each of them fo very hard, the puls cannot be far communicated, that is, the parts of each can yield but very little, and therefore the violence of the concuffion will be *exerted* on that piece of Steel
which is cut off by the Flint. Thirdly, that the filings or fmall parts of
Steel are very apt, as it were, to take fire, and are prefently red hot, that
is, there feems to be a very *combuftible fulphureous* Body in Iron or Steel,
which the Air very readily preys upon, as foon as the body is a little violently heated.

And this is obvious in the filings of Steel or Iron caft through the flame
of a Candle; for even by that fudden *tranfitus* of the fmall chips of Iron,
they are heat red hot, and that *combuftible fulphureous* Body is prefently prey'd upon and devoured by the *aereal* incompaffing *Menftruum*,
whofe office in this Particular I have fhewn in the Explication of Char-
cole.   And

And in profecution of this Experiment, having taken the filings of Iron and Steel, and with the point of a Knife caft them through the flame of a Candle, I obferved where fome confpicuous fhining Particles fell, and looking on them with my *Microfcope*, I found them to be nothing elfe but fuch round Globules, as I formerly found the Sparks ftruck from the Steel by a ftroke to be, only a little bigger ; and fhaking together all the filings that had fallen upon the fheet of Paper underneath, and obferving them with the *Microfcope*. I found a great number of fmall Globules, fuch as the former, though there were alfo many of the parts that had remained untoucht, and rough filings or chips of Iron. So that, it feems, Iron does contain a very *combuftible fulphureous* Body, which is, in all likelihood, one of the caufes of this *Phænomenon*, and which may be perhaps very much concerned in the bufinefs of its hardening and tempering : of which fomewhat is faid in the Defcription of *Mufcovy-glafs*.

So that, thefe things confidered, we need not trouble our felves to find out what kind of Pores they are, both in the Flint and Steel, that contain the *Atoms of fire*, nor how thofe *Atoms* come to be hindred from running all out, when a dore or paffage in their Pores is made by the concuffion : nor need we trouble our felves to examine by what *Prometheus* the Element of Fire comes to be fetcht down from above the Regions of the Air, in what Cells or Boxes it is kept, and what *Epimetheus* lets it go : Nor to confider what it is that caufes fo great a conflux of the atomical Particles of Fire, which are faid to fly to a flaming Body, like Vultures or Eagles to a putrifying Carcafs, and there to make a very great pudder. Since we have nothing more difficult in this *Hypothefis* to conceive, firft, as to the kindling of Tinder, then how a large Iron-bullet, let fall red or glowing hot upon a heap of Small-coal, fhould fet fire to thofe that are next to it firft : Nor fecondly, is this laft more difficult to be explicated, then that a Body, as Silver for Inftance, put into a weak *Menftruum*, as unrectified *Aqua fortis* fhould, when it is put in a great heat, be there diffolved by it, and not before ; which *Hypothefis* is more largely explicated in the Defcription of Charcoal. To conclude, we fee by this Inftance, how much Experiments may conduce to the regulating of *Philofophical notions*. For if the moft Acute *Des Cartes* had applied himfelf experimentally to have examined what fubftance it was that caufed that fhining of the falling Sparks ftruck from a Flint and a Steel, he would certainly have a little altered his *Hypothefis*, and we fhould have found, that his Ingenious Principles would have admitted a very plaufible Explication of this *Phænomenon* ; whereas by not examining fo far as he might, he has fet down an Explication which Experiment do's contradict.

But before I leave this Defcription, I muft not forget to take notice of the Globular form into which each of thefe is moft curioufly formed. And this *Phænomenon*, as I have elfewhere more largely fhewn, proceeds from a propriety which belongs to all kinds of fluid Bodies more or lefs, and is caufed by the Incongruity of the Ambient and included Fluid, which fo acts and modulates each other, that they acquire, as neer as is poffible, possible, a *sperical* or *globular* form, which propriety and several of the *Phænomena* that proceed from it, I have more fully explicated in the sixth Observation.

One Experiment, which does very much illustrate my present Explication, and is in it self exceeding pretty, I must not pass by : And that is a way of making small *Globules* or *Balls* of Lead, or Tin, as small almost as these of Iron or Steel, and that exceeding easily and quickly, by turning the filings or chips of those Metals also into perfectly round *Globules*. The way, in short, as I received it from the *Learned Physitian Doctor* I. G. is this;

Reduce the Metal you would thus shape, into exceeding fine filings, the finer the filings are, the finer will the Balls be : *Stratifie* these filings with the fine and well dryed powder of quick Lime in a *Crucible* proportioned to the quantity you intend to make : When you have thus filled your *Crucible*, by continual *stratifications* of the filings and powder, so that, as neer as may be, no one of the filings may touch another, place the *Crucible* in a *gradual fire*, and by degrees let it be brought to a heat big enough to make all the filings, that are mixt with the quick Lime, to melt, and no more ; for if the fire be too hot, many of these filings will joyn and run together ; whereas if the heat be proportioned , upon washing the Lime-dust in fair Water , all those small filings of the Metal will subside to the bottom in a most curious powder , consisting all of exactly round *Globules*, which, if it be very fine, is very excellent to make Hourglasses of.

Now though quick Lime be the powder that this direction makes choice of, yet I doubt not, but that there may be much more convenient ones found out, one of which I have made tryal of, and found very effectual ; and were it not for discovering, by the mentioning of it, another Secret , which I am not free to impart, I should have here inserted it.

Observ. IX. *Of the Colours observable in Muscovy Glass, and other thin Bodies.*

MOscovy-glass, or *Lapis specularis*, is a Body that seems to have as many Curiosities in its Fabrick as any common Mineral I have met with : for first , It is transparent to a great thickness : Next, it is compounded of an infinite number of thin flakes joyned or generated one upon another so close & smooth, as with many hundreds of them to make one smooth and thin Plate of a transparent flexible substance, which with care and diligence may be slit into pieces so exceedingly thin as to be hardly perceivable by the eye, and yet even those, which I have thought the thinnest, I have with a good *Microscope* found to be made up of many other Plates, yet thinner ; and it is probable, that, were our *Microscopes*
much

much better, we might much further difcover its divifibility. Nor are thefe flakes only regular as to the fmoothnefs of their Surfaces; but thirdly, In many Plates they may be perceived to be terminated naturally with edges of the figure of a *Rhomboeid*. This Figure is much more confpicuous in our Englifh talk, much whereof is found in the Lead Mines, and is commonly called *Spar*, and *Kauck*, which is of the fame kind of fubftance with the *Selenitis*, but is feldom found in fo large flakes as that is, nor is it altogether fo tuff, but is much more clear and tranfparent, and much more curioufly fhaped, and yet may be cleft and flak'd like the other *Selenitis*. But fourthly, this ftone has a property, which in refpect of the *Microfcope*, is more notable, and that is, that it exhibits feveral appearances of Colours, both to the naked Eye, but much more confpicuoufly to the *Microfcope*; for the exhibiting of which, I took a piece of *Mufcovy-glafs*, and fplitting or cleaving it into thin Plates, I found that up and down in feveral parts of them I could plainly perceive feveral white fpecks or flaws, and others diverfly coloured with all the Colours of the *Rainbow*; and with the *Microfcope* I could perceive, that thefe Colours were ranged in rings that incompaffed the white fpeck or flaw, and were round or irregular, according to the fhape of the fpot which they terminated; and the pofition of Colours, in refpect of one another, was the very fame as in the *Rainbow*. The confecution of thofe Colours from the middle of the fpot outward being Blew, Purple, Scarlet, Yellow, Green; Blew, Purple, Scarlet, and fo onwards, fometimes half a fcore times repeated, that is, there appeared fix, feven, eight, nine or ten feveral coloured rings or lines, each incircling the other, in the fame manner as I have often feen a very *vivid Rainbow* to have four or five feveral Rings of Colours, that is, accounting all the Gradations between Red and Blew for one: But the order of the Colours in thefe Rings was quite contrary to the primary or innermoft *Rainbow*, and the fame with thofe of the fecondary or outermoft Rainbow; thefe coloured Lines or *Irifes*, as I may fo call them, were fome of them much brighter then others, and fome of them alfo very much broader, they being fome of them ten, twenty, nay, I believe, neer a hundred times broader then others; and thofe ufually were broadifh which were neereft the center or middle of the flaw. And oftentimes I found, that thefe Colours reacht to the very middle of the flaw, and then there appeared in the middle a very large fpot, for the moft part, all of one colour, which was very vivid, and all the other Colours incompaffing it, gradually afcending, and growing narrower towards the edges, keeping the fame order, as in the *fecundary Rainbow*, that is, if the middle were Blew, the next incompaffing it would be a Purple, the third a Red, the fourth a Yellow, *&c.* as above; if the middle were a Red, the next without it would be a Yellow, the third a Green, the fourth a Blew, and fo onward,. And this order it alwayes kept whatfoever were the middle Colour.

There was further obfervable in feveral other parts of this Body, many Lines or Threads, each of them of fome one peculiar Colour, and thofe fo exceedingly bright and vivid, that it afforded a very pleafant object

through

through the *Microscope*. Some of these *threads* I have observed also to be pieced or made up of several short lengths of differently coloured *ends* (as I may so call them) as a line appearing about two inches long through the *Microscope*, has been compounded of about half an inch of a Peach colour, ⅛ of a lovely Grass-green, ¼ of an inch more of a bright Scarlet,and the rest of the line of a Watchet blew. Others of them were much otherwise coloured; the variety being almost infinite. Another thing which is very observable, is, that if you find any place where the colours are very broad and conspicuous to the naked eye, you may, by pressing that place with your finger, make the colours change places,and go from one part to another.

There is one *Phænomenon* more, which may, if care be used, exhibit to the beholder, as it has divers times to me, an exceeding pleasant, and not less instructive Spectacle; And that is, if curiosity and diligence be used, you may so split this admirable Substance, that you may have pretty large Plates (in comparison of those smaller ones which you may observe in the Rings) that are perhaps an ⅛ or a ⅙ part of an inch over, each of them appearing through the *Microscope* most curiously, intirely, and uniformly adorned with some one vivid colour : this, if examined with the *Microscope*, may be plainly perceived to be in all parts of it equally thick. Two, three, or more of these lying one upon another, exhibit oftentimes curious compounded colours, which produce such a *Compositum*, as one would scarce imagine should be the result of such *ingredients*: As perhaps a *faint yellow* and a *blew* may produce a very *deep purple*. But when anon we come to the more strict examination of these *Phænomena*, and to inquire into the causes and reasons of these productions,we shall,I hope, make it more conceivable how they are produced, and shew them to be no other then the natural and necessary effects arising from the peculiar union of concurrent causes.

These *Phænomena* being so various, and so truly admirable, it will certainly be very well worth our inquiry, to examine the causes and reasons of them,and to consider, whether from these causes demonstratively evidenced, may not be deduced the true causes of the production of all kind of Colours. And I the rather now do it, instead of an Appendix or Digression to this History, then upon the occasion of examining the Colours in Peacocks, or other Feathers, because this Subject, as it does afford more variety of particular Colours, so does it afford much better wayes of examining each circumstance. And this will be made manifest to him that considers, first, that this laminated body is more simple and regular then the parts of Peacocks feathers, this consisting only of an indefinite number of plain and smooth Plates, heaped up, or *incumbent* on each other. Next, that the parts of this body are much more manageable, to be divided or joyned, then the parts of a Peacocks feather,or any other substance that I know. And thirdly, because that in this we are able from a colourless body to produce several coloured bodies, affording all the variety of Colours imaginable : And several others, which the subsequent Inquiry will make manifest.

To begin therefore, it is manifeſt from ſeveral circumſtances, that the material cauſe of the *apparition* of theſe ſeveral Colours, is ſome *Lamina* or Plate of a tranſparent or pellucid body of a thickneſs very determinate and proportioned according to the greater or leſs refractive power of the *pellucid* body. And that this is ſo, abundance of Inſtances and particular Circumſtances will make manifeſt.

As *firſt*, if you take any ſmall piece of the *Muſcovy-glaſs*, and with a Needle, or ſome other convenient Inſtrument, cleave it oftentimes into thinner and thinner *Laminæ*, you ſhall find, that till you come to a determinate thinneſs of them, they ſhall all appear tranſparent and colourleſs, but if you continue to ſplit and divide them further, you ſhall find at laſt, that each Plate, after it comes to ſuch a determinate thickneſs, ſhall appear moſt lovely ting'd or imbued with a determinate colour. If *further*, by any means you ſo flaw a pretty thick piece, that one part does begin to cleave a little from the other, and between thoſe two there be by any means gotten ſome pellucid *medium*, thoſe *laminated* pellucid bodies that fill that ſpace, ſhall exhibit ſeveral Rainbows or coloured Lines, the colours of which will be diſpoſed and ranged according to the various thickneſſes of the ſeveral parts of that Plate. That this is ſo, is yet *further* confirmed by this Experiment.

Take two ſmall pieces of ground and poliſht Looking-glaſs-plate, each about the bigneſs of a ſhilling, take theſe two dry, and with your fore-fingers and thumbs preſs them very hard and cloſe together, and you ſhall find, that when they approach each other very near, there will appear ſeveral *Iriſes* or coloured Lines, in the ſame manner almoſt as in the *Muſcovy-glaſs*; and you may very eaſily change any of the Colours of any part of the interpoſed body, by preſſing the Plates cloſer and harder together, or leaving them more lax; that is, a part which appeared coloured with a red, may be preſently ting'd with a yellow, blew, green, purple, or the like, by altering the appropinquation of the terminating Plates. Now that air is not neceſſary to be the interpoſed body, but that any other tranſparent fluid will do much the ſame, may be tryed by wetting thoſe approximated Surfaces with Water, or any other tranſparent Liquor, and proceeding with it in the ſame manner as you did with the Air; and you will find much the like effect, only with this difference, that thoſe compreſt bodies, which differ moſt, in their refractive quality, from the compreſſing bodies, exhibit the moſt ſtrong and vivid tinctures. Nor is it neceſſary, that this *laminated* and *ting'd* body ſhould be of a fluid ſubſtance, any other ſubſtance, provided it be thin enough and tranſparent, doing the ſame thing : this the *Laminæ* of our *Muſcovy-glaſs* hint; but it may be confirm'd by multitudes of other Inſtances.

And firſt, we ſhall find, that even Glaſs it ſelf may, by the help of a Lamp, be blown thin enough to produce theſe *Phænomena* of Colours: which *Phænomena* accidentally happening, as I have been attempting to frame ſmall Glaſſes with a Lamp, did not a little ſurprize me at firſt, having never heard or ſeen any thing of it before; though afterwards comparing it with the *Phænomena*, I had often

obſerved

obferved in thofe Bubbles which Children ufe to make with Soap-water, I did the lefs wonder ; efpecially when upon Experiment I found, I was able to produce the fame *Phænomena* in thin Bubbles made with any other tranfparent Subftance. Thus have I produced them with Bubbles of *Pitch, Rofin, Colophony,Turpentine, Solutions* of feveral *Gums,* as *Gum-Arabick* in water ; any *glutinous* Liquor,as *Wort, Wine,Spirit of Wine, Oyl of Turpentine, Glare of Snails,* &c.

It would be needlefs to enumerate the feveral Inftances , thefe being enough to fhew the generality or univerfality of this propriety. Only I muft not omit, that we have inftances alfo of this kind even in metalline Bodies and animal ; for thofe feveral Colours which are obferved to follow each other upon the polifht furface of hardned Steel, when it is by a fufficient degree of heat gradually tempered or foftened , are produced from nothing elfe but a certain thin *Lamina* of a *vitrum* or *vitrified* part of the Metal,which by that degree of heat, and the concurring action of the ambient Air,is driven out and fixed on the furface of the Steel.

And this hints to me a very probable (at leaft, if not the true) caufe of the hardning and tempering of Steel, which has not, I think, been yet given,nor, that I know of been fo much as thought of by any. And that is this,that the hardnefs of it arifes from a greater proportion of a vitrified Subftance interfperfed through the pores of the Steel. And that the tempering or foftning of it arifes from the proportionate or fmaller parcels of it left within thofe pores. This will feem the more probable , if we confider thefe Particulars.

Firft, That the pure parts of Metals are of themfelves very *flexible* and *tuff* ; that is, will indure bending and hammering,and yet retain their continuity.

Next, That the Parts of all vitrified Subftances, as all kinds of Glafs, the *Scoria* of Metals, *&c.* are very hard, and alfo very brittle, being neither *flexible* nor *malleable,* but may by hammering or beating be broken into fmall parts or powders.

Thirdly,That all Metals (excepting Gold and Silver , which do not fo much with the bare fire, unlefs affifted by other faline Bodies) do more or lefs *vitrifie* by the ftrength of fire, that is, are corroded by a faline Subftance, which I elfewhere fhew to be the true caufe of fire ; and are thereby, as by feveral other *Menftruums,*converted into *Scoria* ; And this is called, *calcining* of them, by Chimifts. Thus Iron and Copper by heating and quenching do turn all of them by degrees into *Scoria,* which are evidently *vitrified* Subftances , and unite with Glafs , and are eafily *fufible* ; and when cold, very hard, and very brittle.

Fourthly, That moft kind of *Vitrifications* or *Calcinations* are made by Salts, uniting and incorporating with the metalline Particles. Nor do I know any one *calcination* wherein a *Saline* body may not, with very great probability, be faid to be an agent or coadjutor.

Fifthly, That Iron is converted into Steel by means of the incorporation of certain falts, with which it is kept a certain time in the fire.

Sixthly, That any Iron may, in a very little time, be *cafe hardned*, as the Trades-men call it, by cafing the iron to be hardned with clay, and putting between the clay and iron a good quantity of a mixture of *Urine*, *Soot*, *Sea-falt*, and *Horfes hoofs* (all which contein great quantities of Saline bodies) and then putting the cafe into a good ftrong fire, and keeping it in a confiderable degree of heat for a good while, and afterwards heating, and quenching or cooling it fuddenly in cold water.

Seventhly, That all kind of vitrify'd fubftances, by being fuddenly cool'd, become very hard and brittle. And thence arifes the pretty *Phænomena* of the Glafs Drops, which I have already further explained in its own place.

Eighthly, That thofe metals which are not fo apt to vitrifie, do not acquire any hardnefs by quenching in water, as Silver, Gold, &c.

Thefe confiderations premis'd, will, I fuppofe, make way for the more eafie reception of this following Explication of the *Phænomena* of hardned and temper'd Steel. That Steel is a fubftance made out of Iron, by means of a certain proportionate *Vitrification* of feveral parts, which are fo curioufly and proportionately mixt with the more tough and unalter'd parts of the Iron, that when by the great heat of the fire this vitrify'd fubftance is melted, and confequently rarify'd, and thereby the pores of the Iron are more open, if then by means of dipping it in cold water it be fuddenly cold, and the parts hardned, that is, ftay'd in that fame degree of *Expanfion* they were in when hot, the parts become very hard and brittle, and that upon the fame account almoft as fmall parcels of glafs quenched in water grow brittle, which we have already explicated. If after this the piece of Steel be held in fome convenient heat, till by degrees certain colours appear upon the furface of the brightned metal, the very hard and brittle tone of the metal, by degrees relaxes and becomes much more tough and foft; namely, the action of the heat does by degrees loofen the parts of the Steel that were before ftreached or fet *atilt* as it were, and ftayed open by each other, whereby they become relaxed and fet at liberty, whence fome of the more brittle interjacent parts are thruft out and melted into a thin skin on the furface of the Steel, which from no colour increafes to a deep Purple, and fo onward by thefe *gradations* or confecutions, *White, Yellow, Orange, Minium, Scarlet, Purple, Blew, Watchet*, &c. and the parts within are more conveniently, and proportionately mixt; and fo they gradually fubfide into a texture which is much better proportion'd and clofer joyn'd, whence that rigidneffe of parts ceafes, and the parts begin to acquire their former *du£til-nefs*.

Now, that 'tis nothing but the vitrify'd metal that fticks upon the furface of the colour'd body, is evident from this, that if by any means it be fcraped and rubb'd off, the metal underneath it is white and clear; and if it be kept longer in the fire, fo as to increafe to a confiderable thicknefs, it may, by blows, be beaten off in flakes. This is further confirm'd by this obfervable, that that Iron or Steel will keep longer from rufting which is covered with this vitrify'd cafe : Thus alfo Lead will, by degrees, be

all

all turn'd into a litharge; for that colour which covers the top being fcum'd or fhov'd afide, appears to be nothing elfe but a litharge or vitrify'd Lead.

This is obfervable alfo in fome fort, on Brafs, Copper, Silver, Gold, Tin, but is moft confpicuous in Lead: all thofe Colours that cover the furface of the Metal being nothing elfe, but a very thin vitrifi'd part of the heated Metal.

The other Inftance we have, is in Animal bodies, as in Pearls, Mother of Pearl-fhels, Oyfter-fhels, and almoft all other kinds of ftony fhels whatfoever. This have I alfo fometimes with pleafure obferv'd even in Mufcles and Tendons. Further, if you take any glutinous fubftance and run it exceedingly thin upon the furface of a fmooth glafs or a polifht metaline body, you fhall find the like effects produced: and in general, wherefoever you meet with a tranfparent body thin enough, that is terminated by reflecting bodies of differing refractions from it, there will be a production of thefe pleafing and lovely colours.

Nor is it neceffary, that the two *terminating* Bodies fhould be both of the fame kind, as may appear by the *vitrified Laminæ* on *Steel, Lead,* and other Metals, one furface of which *Laminæ* is contiguous to the furface of the Metal, the other to that of the Air.

Nor is it neceffary, that thefe colour'd *Laminæ* fhould be of an even thicknefs, that is, fhould have their edges and middles of equal thicknefs, as in a Looking-glafs-plate, which circumftance is only requifite to make the Plate appear all of the fame colour; but they may refemble a *Lens,* that is, have their middles thicker then their edges; or elfe a *double concave,* that is, be thinner in the middle then at the edges; in both which cafes there will be various coloured rings or lines, with differing confecutions or orders of Colours; the order of the firft from the middle outwards being Red, Yellow, Green, Blew, &c. And the latter quite contrary.

But further, it is altogether neceffary, that the Plate, in the places where the Colours appear, fhould be of a determinate thicknefs: Firft, It muft not be more then fuch a thicknefs, for when the Plate is increafed to fuch a thicknefs, the Colours ceafe; and befides, I have feen in a thin piece of *Mufcovy-glafs,* where the two ends of two Plates, which appearing both fingle, exhibited two diftinct and differing Colours; but in that place where they were united, and conftituted one double Plate (as I may call it) they appeared tranfparent and colourlefs. Nor, Secondly, may the Plates be *thinner* then fuch a determinate *cize;* for we alwayes find, that the very outmoft Rim of thefe flaws is terminated in a white and colourlefs Ring.

Further, in this Production of Colours there is no need of a determinate Light of fuch a bignefs and no more, nor of a determinate pofition of that Light, that it fhould be on this fide, and not on that fide; nor of a terminating fhadow, as in the Prifme, and Rainbow, or Water-ball: for we find, that the Light in the open Air, either in or out of the Sun-beams, and within a Room, either from one or many Windows, produces much

the

the fame effect : only where the Light is brighteft, there the Colours are moft *vivid*. So does the light of a Candle, collected by a Glafs-ball. And further, it is all one whatever fide of the coloured Rings be towards the light ; for the whole Ring keeps its proper Colours from the middle outwards in the fame order as I before related, without varying at all, upon changing the pofition of the light.

But above all it is moft obfervable, that here are all kind of Colours generated in a *pellucid* body, where there is properly no fuch refraction as *Des Cartes* fuppofes his *Globules* to acquire a *verticity* by : For in the plain and even Plates it is manifeft, that the fecond refraction (according to *Des Cartes* his Principles in the *fifth Section of the eighth Chapter of his Meteors*) does regulate and reftore the fuppofed *turbinated Globules* unto their former uniform motion. This Experiment therefore will prove fuch a one as our *thrice excellent Verulam* calls *Experimentum Crucis*, ferving as a Guide or Land-mark, by which to direct our courfe in the fearch after the true caufe of Colours. Affording us this particular negative Information, that for the production of Colours there is not neceffary either a great refraction, as in the Prifme ; nor Secondly, a determination of Light and fhadow, fuch as is both in the Prifme and Glafs-ball. Now that we may fee likewife what affirmative and pofitive Inftruction it yields, it will be neceffary, to examine it a little more particularly and ftrictly ; which that we may the better do, it will be requifite to premife fomewhat in general concerning the nature of Light and Refraction.

And firft for Light, it feems very manifeft, that there is no luminous Body but has the parts of it in motion more or lefs.

Firft, That all kind of *fiery burning Bodies* have their parts in motion, I think, will be very eafily granted me. That the *fpark* ftruck from a Flint and Steel is in a rapid agitation, I have elfewhere made probable. And that the Parts of *rotten Wood, rotten Fifh*, and the like, are alfo in motion, I think, will as eafily be conceded by thofe, who confider, that thofe parts never begin to fhine till the Bodies be in a ftate of putrefaction ; and that is now generally granted by all, to be caufed by the motion of the parts of putrifying bodies. That the *Bononian ftone* fhines no longer then it is either warmed by the Sun-beams, or by the flame of a Fire or of a Candle, is the general report of thofe that write of it, and of others that have feen it. And that heat argues a motion of the internal parts, is (as I faid before) generally granted.

But there is one Inftance more, which was firft fhewn to the *Royal Society* by Mr. *Clayton* a worthy Member thereof, which does make this Affertion more evident then all the reft : And that is, That a *Diamond* being *rub'd, ftruck*, or *heated* in the dark, fhines for a pretty while after, fo long as that motion, which is imparted by any of thofe Agents, remains (in the fame manner as a Glafs, rubb'd, ftruck, or (by a means which I fhall elfewhere mention) heated, yields a found which lafts as long as the *vibrating* motion of that *fonorous* body) feveral Experiments made on which Stone, are fince publifhed in a Difcourfe of Colours, by the truly
honou-

MICROGRAPHIA.

honourable Mr. *Boyle.* What may be faid of thofe *Ignes fatui* that appear in the night, I cannot fo well affirm, having never had the opportunity to examine them my felf, nor to be inform'd by any others that had obferv'd them: And the relations of them in Authors are fo imperfect, that nothing can be built on them. But I hope I fhall be able in another place to make it at leaft very probable, that there is even in thofe alfo a Motion which caufes this effect. That the fhining of *Sea-water* proceeds from the fame caufe, may be argued from this, That it fhines not till either it be beaten againft a Rock, or be fome other wayes broken or agitated by Storms, or Oars, or other *percuffing* bodies. And that the Animal *Energyes* or Spirituous *agil* parts are very active in *Cats eyes* when they fhine, feems evident enough, becaufe their eyes never fhine but when they look very intenfly either to find their prey, or being hunted in a dark room, when they feek after their adverfary, or to find a way to efcape. And the like may be faid of the fhining *Bellies of Gloworms,* fince 'tis evident they can at pleafure either increafe or extinguifh that Radiation.

It would be fomewhat too long a work for this place *Zetetically* to examine, and pofitively to prove, what particular kind of motion it is that muft be the efficient of Light; for though it be a motion, yet 'tis not every motion that produces it, fince we find there are many bodies very violently mov'd, which yet afford not fuch an effect; and there are other bodies, which to our other fenfes, feem not mov'd fo much, which yet fhine. Thus Water and quick-filver, and moft other liquors heated, fhine not; and feveral hard bodies, as Iron, Silver, Brafs. Copper, Wood, &c. though very often ftruck with a hammer, fhine not prefently, though they will all of them grow exceeding hot; whereas rotten Wood, rotten Fifh, Sea water, Gloworms, &c. have nothing of tangible heat in them, and yet (where there is no ftronger light to affect the Senfory) they fhine fome of them fo Vividly, that one may make a fhift to read by them.

It would be too long, I fay, here to infert the difcurfive progrefs by which I inquir'd after the proprieties of the motion of Light, and therefore I fhall only add the refult.

And, Firft, I found it ought to be exceeding *quick,* fuch as thofe motions of *fermentation* and *putrefaction,* whereby, certainly, the parts are exceeding nimbly and violently mov'd; and that, becaufe we find thofe motions are able more minutely to fhatter and divide the body, then the moft violent heats or *menftruums* we yet know. And that fire is nothing elfe but fuch a *diffolution* of the Burning body, made by the moft *univerfal menftruum* of all *fulphureous bodies,* namely, the Air, we fhall in an other place of this Tractate endeavour to make probable. And that, in all extreamly hot fhining bodies, there is a very quick motion that caufes Light, as well as a more robuft that caufes Heat, may be argued from the celerity wherewith the bodyes are diffolv'd.

Next, it muft be a *Vibrative motion.* And for this the newly mention'd *Diamond* affords us a good argument; fince if the motion of the parts did
not

not return the Diamond muſt after many rubbings decay and be waſted: but we have no reaſon to ſuſpect the latter, eſpecially if we conſider the exceeding difficulty that is found in cutting or wearing away a Diamond. And a Circular motion of the parts is much more improbable, ſince, if that were granted, and they be ſuppos'd irregular and Angular parts, I ſee not how the parts of the Diamond ſhould hold ſo firmly together, or remain in the ſame ſenſible dimenſions, which yet they do. Next, if they be *Globular*, and mov'd only with a *turbinated* motion, I know not any cauſe that can impreſs that motion upon the *pellucid medium*, which yet is done. Thirdly, any other *irregular* motion of the parts one amongſt another, muſt neceſſarily make the body of a fluid conſiſtence, from which it is far enough. It muſt therefore be a *Vibrating* motion.

And Thirdly, That it is a very *ſhort vibrating motion*, I think the inſtances drawn from the ſhining of Diamonds will alſo make probable. For a Diamond being the hardeſt body we yet know in the World, and conſequently the leaſt apt to yield or bend, muſt conſequently alſo have its *vibrations* exceeding ſhort.

And theſe, I think, are the three principal proprieties of a motion, requiſite to produce the effect call'd Light in the Object.

The next thing we are to conſider, is the way or manner of the *trajection* of this motion through the interpos'd pellucid body to the eye: And here it will be eaſily granted,

Firſt, That it muſt be a body *ſuſceptible* and *impartible* of this motion that will deſerve the name of a Tranſparent. And next, that the parts of ſuch a body muſt be *Homogeneous*, or of the ſame kind. Thirdly, that the conſtitution and motion of the parts muſt be ſuch, that the appulſe of the luminous body may be communicated or propagated through it to the greateſt imaginable diſtance in the leaſt imaginable time; though I ſee no reaſon to affirm, that it muſt be in an inſtant : For I know not any one Experiment or obſervation that does prove it. And, whereas it may be objected, That we ſee the Sun riſen at the very inſtant when it is above the ſenſible Horizon, and that we ſee a Star hidden by the body of the Moon at the ſame inſtant, when the Star, the Moon, and our Eye are all in the ſame line ; and the like Obſervations, or rather ſuppoſitions, may be urg'd. I have this to anſwer, That I can as eaſily deny as they affirm; for I would fain know by what means any one can be aſſured any more of the Affirmative, then I of the Negative. If indeed the propagation were very ſlow, 'tis poſſible ſomething might be diſcovered by Eclypſes of the Moon ; but though we ſhould grant the progreſs of the light from the Earth to the Moon, and from the Moon back to the Earth again to be full two Minutes in performing, I know not any poſſible means to diſcover it ; nay, there may be ſome inſtances perhaps of Horizontal Eclypſes that may ſeem very much to favour this ſuppoſition of the ſlower progreſſion of Light then moſt imagine. And the like may be ſaid of the Eclypſes of the Sun, &c. But of this only by the by. Fourthly, That the motion is propagated every way through an *Homogeneous*

geneous

geneous medium by *direct* or *straight* lines extended every way like Rays from the center of a Sphere. Fifthly, in an *Homogeneous medium* this motion is propagated every way with *equal velocity*, whence neceffarily every *pulfe* or *vitration* of the luminous body will generate a Sphere, which will continually increafe, and grow bigger, juft after the fame manner (though indefinitely fwifter) as the waves or rings on the furface of the water do fwell into bigger and bigger circles about a point of it, where, by the finking of a Stone the motion was begun, whence it neceffarily follows, that all the parts of thefe Spheres undulated through an *Homogeneous medium* cut the Rays at right angles.

But becaufe all tranfparent *mediums* are not *Homogeneous* to one another, therefore we will next examine how this pulfe or motion will be propagated through differingly tranfparent *mediums*. And here, according to the moft acute and excellent Philofopher *Des Cartes*, I fuppofe the fign of the angle of inclination in the firft *medium* to be to the fign of refraction in the fecond, As the denfity of the firft, to the denfity of the fecond. By denfity, I mean not the denfity in refpect of gravity (with which the refractions or tranfparency of *mediums* hold no proportion) but in refpect onely to the *trajection* of the Rays of light, in which refpect they only differ in this; that the one propagates the pulfe more eafily and weakly, the other more flowly, but more ftrongly. But as for the pulfes themfelves, they will by the refraction acquire another propriety, which we fhall now endeavour to explicate.

We will fuppofe therefore in the firft Figure A C F D to be a phyfical Ray, or A B C and D E F to be two Mathematical Rays, *trajected* from a very remote point of a luminous body through an *Homogeneous* tranfparent *medium* L L L, and D A, E B, F C, to be fmall portions of the orbicular impulfes which muft therefore cut the Rays at right angles; thefe Rays meeting with the plain furface N O of a *medium* that yields an eafier *tranfitus* to the propagation of light, and falling *obliquely* on it, they will in the *medium* M M M be refracted towards the perpendicular of the furface. And becaufe this *medium* is more eafily *trajected* then the former by a third, therefore the point C of the orbicular pulfe F C will be mov'd to H four fpaces in the fame time that F the other end of it is mov'd to G three fpaces, therefore the whole refracted pulfe G H fhall be *oblique* to the refracted Rays C H K and G I; and the angle G H C fhall be an acute, and fo much the more acute by how much the greater the refraction be, then which nothing is more evident, for the fign of the inclination is to be the fign of refraction as G F to T C the diftance between the point C and the perpendicular from G on C K, which being as four to three, H C being longer then G F is longer alfo then T C, therefore the angle G H C is lefs than G T C. So that henceforth the parts of the pulfes G H and I K are mov'd afcew, or cut the Rays at *oblique* angles.

It is not my bufinefs in this place to fet down the reafons why this or that body fhould impede the Rays more, others lefs: as why Water fhould tranfmit the Rays more eafily, though more weakly than air. Onely thus

K   much

much in general I fhall hint,that I fuppofe the *medium* M M M to have lefs of the tranfparent undulating fubtile matter, and that matter to be lefs implicated by it, whereas L L L I fuppofe to contain a greater quantity of the fluid undulating fubftance,and this to be more implicated with the particles of that *medium*.

But to proceed, the fame kind of *obliquity* of the Pulfes and Rays will happen alfo when the refraction is made out of a more eafie into a more difficult *mediū*; as by the calculations of G Q & C S R which are refracted from the perpendicular. In both which calculations 'tis *obvious* to obferve, that always that part of the Ray towards which the refraction is made has the end of the *orbicular pulfe* precedent to that of the other fide. And always,the oftner the refraction is made the fame way,Or the greater the fingle refraction is, the more is this unequal progrefs. So that having found this odd propriety to be an infeparable concomitant of a refracted Ray, not ftreightned by a contrary refraction, we will next examine the refractions of the Sun-beams, as they are fuffer'd onely to pafs through a fmall paffage, *obliquely* out of a more difficult,into a more eafie *medium*.

Let us fuppofe therefore A B C in the fecond Figure to reprefent a large *Chimical Glafs-body* about two foot long, filled with very fair Water as high as A B, and inclin'd in a convenient pofture with B towards the Sun : Let us further fuppofe the top of it to be cover'd with an *opacous* body, all but the hole *a b*, through which the Sun-beams are fuffer'd to pafs into the Water,and are thereby refracted to *c d e f*,againft which part, if a Paper be expanded on the outfide, there will appear all the colours of the Rain-bow, that is, there will be generated the two principal colours, *Scarlet* and *Blue*, and all the *intermediate* ones which arife from the compofition and dilutings of thefe two, that is, *c d* fhall exhibit a *Scarlet*, which toward *d* is diluted into a *Yellow*; this is the refraction of the Ray, *i k*, which comes from the underfide of the Sun; and the Ray *e f* fhall appear of a deep *Blue*, which is gradually towards *e* diluted into a pale *Watchet-blue*. Between *d* and *e* the two *diluted* colours, *Blue* and *Yellow* are mixt and compounded into a *Green*; and this I imagine to be the reafon why *Green* is fo acceptable a colour to the eye, and that either of the two extremes are, if intenfe, rather a little offenfive, namely, the being plac'd in the middle between the two extremes, and compounded out of both thofe, *diluted* alfo, or fomewhat qualifi'd, for the *compofition*, arifing from the mixture of the two extremes *undiluted*, makes a *Purple*,which though it be a lovely colour,and pretty acceptable to the eye, yet is it nothing comparable to the ravifhing pleafure with which a curious and well tempered *Green* affects the eye. If removing the Paper, the eye be plac'd againft *c d*, it will perceive the lower fide of the Sun (or a Candle at night which is much better,becaufe it offends not the eye, and is more eafily manageable) to be of a deep *Red*, and if againft *e f* it will perceive the upper part of the luminous body to be of a deep *Blue*; and thefe colours will appear deeper and deeper, according as the Rays from the luminous body fall more *obliquely* on the furface of the Water, and thereby fuffer a greater refraction, and the

more

more diftinct, the further *c d e f* is removed from the trajecting hole.

So that upon the whole, we fhall find that the reafon of the *Phænome-na* feems to depend upon the *obliquity* of the *orbicular pulfe*, to the Lines of Radiation and in particular, that the Ray *c d* which conftitutes the *Scar-let* has its inner parts, namely thofe which are next to the middle of the luminous body, precedent to the outermoft which are contiguous to the dark and *unradiating* fkie. And that the Ray *e f* which gives a *Blue*, has its outward part, namely, that which is contiguous to the dark fkie prece-dent to the pulfe from the innermoft, which borders on the bright *area* of the luminous body.

We may obferve further, that the caufe of the *diluting* of the colours to-wards the middle, proceeds partly from the widenefs of the hole through which the Rays pafs, whereby the Rays from feveral parts of the lumi-nous body, fall upon many of the fame parts between *c* and *f* as is more manifeft by the Figure: And partly alfo from the nature of the refraction it felf, for the vividnefs or ftrength of the two terminating colours, arifing chiefly as we have feen, from the very great difference that is betwixt the outfides of thofe *oblique undulations* & the dark Rays circumambient, and that difparity betwixt the *approximate* Rays, decaying gradually: the fur-ther inward toward the middle of the luminous body they are remov'd, the more muft the colour approach to a white or an undifturbed light.

Upon the calculation of the refraction and reflection from a Ball of Water or Glafs, we have much the fame *Phænomena*, namely, an *obliquity* of the undulation in the fame manner as we have found it here. Which, be-caufe it is very much to our prefent purpofe, and affords fuch an *Inftancia crucis*, as no one that I know has hitherto taken notice of, I fhall further examine. For it does very plainly and pofitively diftinguifh, and fhew, which of the two *Hypothefes*, either the *Cartefian* or this is to be followed, by affording a generation of all the colors in the Rainbow, where accord-ing to the *Cartefian Principles* there fhould be none at all generated. And fecondly, by affording an inftance that does more clofely confine the caufe of thefe *Phænomena* of colours to this prefent *Hypothefis*.

And firft, for the *Cartefian*, we have this to object againft it, That whereas he fays (*Meteorum Cap.8.Sect.5.) Sed judicabam unicam(refractione fcilicet) ad minimu requiri,& quidem talem ut ejus effectus aliâ contrariâ (refracti-one)non deftruatur : Nam experientia docet fi fuperficies* N M & N P (*nempe refringentes*) Parallelæ forent, radios tantundem per alteram iterum erectos quantum per unam frangerentur, nullos colores depicturos ;* This Principle of his holds true indeed in a prifme where the refracting furfaces are plain, but is contradicted by the Ball or Cylinder, whether of Water or Glafs, where the refracting furfaces are Orbicular or Cylindrical. For if we ex-amine the paffage of any *Globule* or Ray of the primary *Iris*, we fhall find it to pafs out of the Ball or Cylinder again, with the fame inclination and refraction that it enter'd in withall, and that that laft refraction by means of the *intermediate* reflection fhall be the fame as if without any reflection at all the Ray had been twice refracted by two Parallel furfaces.

K 2   And

And that this is true, not onely in one, but in every Ray that goes to the conftitution of the Primary Iris; nay, in every Ray, that fuffers only two refractions, and one reflection, by the furface of the round body, we fhall prefently fee moft evident, if we repeat the *Cartefian Scheme*, mentioned in the tenth *Section* of the eighth *Chapter* of his *Meteors*, where

Schem. 6.
Fig. 3.

E F K N P in the third Figure is one of the Rays of the Primary Iris, twice refracted at F and N, and once reflected at K by the furface of the Water-ball. For, firft it is evident, that K F and K N are equal, becaufe K N being the reflected part of K F they have both the fame inclination on the furface K that is the angles F K T, and N K V made by the two Rays and the Tangent of K are equal, which is evident by the Laws of reflection; whence it will follow alfo, that K N has the fame inclination on the furface N, or the Tangent of it X N that the Ray K F has to the furface F, or the Tangent of it F Y, whence it muft neceffarily follow, that the refractions at F and N are equal, that is, K F E and K N P are equal. Now, that the furface N is by the reflection at K made parallel to the furface at F, is evident from the principles of reflection; for reflection being nothing but an inverting of the Rays, if we re-invertthe Ray K N P, and make the fame inclinations below the line T K V that it has above, it will be moft evident, that K H the inverfe of K N will be the continuation of the line F K, and that L H I the inverfe of O X is parallel to F Y. And H M the inverfe of N P is Parallel to E F for the angle K H I is equal to K N O which is equal to K F Y, and the angle K H M is equal to K N P which is equal to K F E which was to be prov'd.

So that according to the above mentioned *Cartefian* principles there fhould be generated no colour at all in a Ball of Water or Glafs by two refractions and one reflection, which does hold moft true indeed, if the furfaces be plain, as may be experimented with any kind of prifme where the two refracting furfaces are equally inclin'd to the reflecting; but in this the *Phænomena* are quite otherwife.

The caufe therefore of the generation of colour muft not be what *Des Cartes* affigns, namely, a certain *rotation* of the *Globuli ætherei*, which are the particles which he fuppofes to conftitute the *Pellucid medium*, But fomewhat elfe, perhaps what we have lately fuppofed, and fhall by and by further profecute and explain.

But, Firft I fhall crave leave to propound fome other difficulties of his, notwithftanding exceedingly ingenious *Hypothefis*, which I plainly confefs to me feem fuch; and thofe are,

Firft, if that light be (as is affirmed, *Diopt.* cap. 1. §. 8.) not fo properly a motion, as an action or propenfion to motion, I cannot conceive how the eye can come to be fenfible of the *verticity* of a *Globule*, which is generated in a drop of Rain, perhaps a mile off from it. For that *Globule* is not carry'd to the eye according to his formerly recited Principle; and if not fo, I cannot conceive how it can communicate its *rotation*, or circular motion to the line of the *Globules* between the drop and the eye. It cannot be by means of every ones turning the next before him; for if fo, then onely all the *Globules* that are in the odd places muft be turned the fame

way

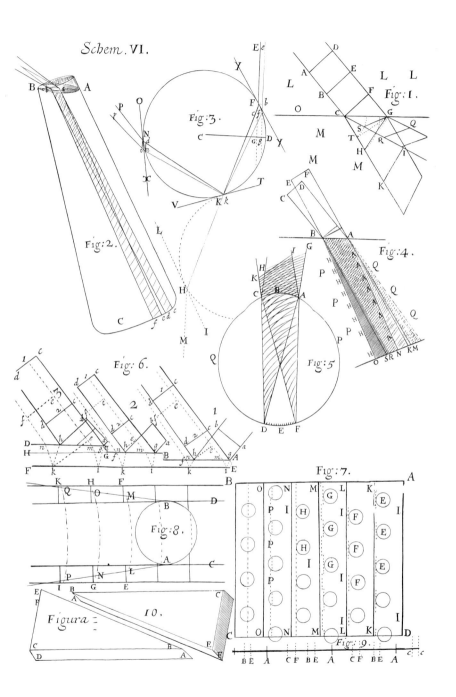

Schem. VI.

way with the firſt, namely, the 3. 5. 7. 9. 11, &c. but all the *Globules* interpoſited between them in the even places; namely, the 2.4.6.8.10.&c. muſt be the quite contrary; whence, according to the *Carteſian Hypotheſis*, there muſt be no diſtinct colour generated, but a confuſion. Next, ſince the *Carteſian Globuli* are ſuppos'd (*Principiorum Philoſoph.* Part. 3. §. 86.) to be each of them continually in motion about their centers, I cannot conceive how the eye is able to diſtinguiſh this new generated motion from their former inherent one, if I may ſo call that other wherewith they are mov'd or *turbinated,* from ſome other cauſe than refraction. And thirdly, I cannot conceive how theſe motions ſhould not happen ſome-times to oppoſe each other, and then, in ſtead of a *rotation,* there would be nothing but a direct motion generated, and conſequently no colour. And fourthly, I cannot conceive, how by the *Carteſian Hypotheſis* it is poſ-ſible to give any plauſible reaſon of the nature of the Colours generated in the thin *laminæ* of theſe our *Microſcopical Obſervations*; for in many of theſe, the refracting and reflecting ſurfaces are parallel to each other, and conſequently no *rotation* can be generated, nor is there any neceſſity of a ſhadow or termination of the bright Rays, ſuch as is ſuppos'd (*Chap. 8.* §. 5. *Et preterea obſervavi umbram quoque, ant limitationem luminis requiri:* and *Chap. 8.* §. 9.) to be neceſſary to the generation of any diſtinct co-lours; Beſides that, here is oftentimes one colour generated without any of the other appendant ones, which cannot be by the *Carteſian Hy-potheſis.*

There muſt be therefore ſome other propriety of refraction that cauſes colour. And upon the examination of the thing, I cannot conceive any one more general, inſeparable, and ſufficient, than that which I have be-fore aſſign'd. That we may therefore ſee how exactly our *Hypotheſis* agrees alſo with the *Phænomena* of the refracting round body, whether *Globe* or *Cylinder,* we ſhall next ſubjoyn our *Calculation* or *Examen* of it.

And to this end, we will calculate any two Rays: as for inſtance; let *Schem. 6.* E F be a Ray cutting the *Radius* C D (divided into 20. parts) in G 16. *Fig. 3.* parts diſtant from C, and *ef* another Ray, which cuts the ſame *Radius* in *g* 17. parts diſtant, theſe will be refracted to K and *k,* and from thence reflected to N and *n,* and from thence refracted toward P and *p*; there-fore the Arch F*f* will be 5.^{d} 5'. The Arch F K 106.^{d} 30'. the Arch *f k* 101.^{d} 2'. The line F G 6000. and *f g* 5267. therefore *h f.*733. therefore F*c* 980, almoſt. The line F K 16024. and *f k* 15436. therefore N*d* 196: and *n o* 147 almoſt, the line N n 1019 the Arch N *n* 5.^{d} 51'. therefore the Angle N *n o* is 34.^{d} 43'. therefore the Angle N *o n.* is 139.^{d} 56'. which is almoſt 50.^{d} more than a right Angle.

It is evident therefore by this *Hypotheſis,* that at the ſame time that *ef* touches *f.* E F is arrived at *c.* And by that time *ef k n* is got to *n,* E F K N is got to *d,* and when it touches N, the pulſe of the other Ray is got to *o*: and no farther, which is very ſhort of the place it ſhould have arriv'd to, to make the Ray *n p* to cut the *orbicular pulſe.* N *o* at right Angles: therefore the Angle N *o p* is an acute Angle, but the quite con-trary

trary of this will happen, if 17. and 18. be calculated in ſtead of 16. and 17. both which does moſt exactly agree with the *Phænomena* : For if the Sun, or a Candle (which is better) be placed about E *e*, and the eye about P *p*, the Rays E F *ef*. at 16. and 17. will paint the ſide of the luminous object toward *n p* *Blue*, and towards N P *Red*. But the quite contrary will happen when E F is 17. and *ef* 18. for then towards N P ſhall be a *Blue*, and towards *n p* a *Red*, exactly according to the calculation. And there appears the *Blue* of the Rainbow, where the two *Blue* ſides of the two Images unite, and there the *Red* where the two *Red* ſides unite, that is, where the two Images are juſt diſappearing; which is, when the Rays E F and N P produc'd till they meet, make an Angle of about 41. and an half; the like union is there of the two Images in the Production of the *Secundary Iris*, and the ſame cauſes, as upon calculation may appear; onely with this difference, that it is ſomewhat more faint, by reaſon of the duplicate reflection, which does always weaken the impulſe the oftner it is repeated.

Now, though the ſecond refraction made at N *n* be convenient, that is, do make the Rays glance the more, yet is it not altogether requiſite; for it is plain from the calculation, that the pulſe *d n* is ſufficiently *oblique* to the Rays K N and *k n*, as wel as the pulſe *f c* is *oblique* to the Rays F K & *f k*. And therefore if a piece of very fine Paper be held cloſe againſt N *n* and the eye look on it either through the Ball as from D, or from the other ſide, as from B. there ſhall appear a Rainbow, or colour'd line painted on it with the part toward X appearing *Red*, towards O, *Blue*; the ſame alſo ſhall happen, if the Paper be placed about K *k*, for towards T ſhall appear a *Red*, and towards V a *Blue*, which does exactly agree with this my *Hypotheſis*, as upon the calculation of the progreſs of the pulſe will moſt eaſily appear.

Nor do theſe two obſervations of the colours appearing to the eye about *p* differing from what they appear on the Paper at N contradict each other; but rather confirm and exactly agree with one another, as will be evident to him that examines the reaſons ſet down by the ingenious. *Des Cartes* in the 12. *Sect*. of the 8. *Chapter of his Meteors*, where he gives the true reaſon why the colours appear of a quite contrary order to the eye, to what they appear'd on the Paper if the eye be plac'd in ſteed of the Paper : And as in the Priſme, ſo alſo in the Water, Drop, or Globe the *Phænomena* and reaſon are much the ſame.

Having therefore ſhewn that there is ſuch a propriety in the *priſme* and water *Globule* whereby the pulſe is made *oblique* to the progreſſive, and that ſo much the more, by how much greater the refraction is, I ſhall in the next place conſider, how this conduces to the production of colours, and what kind of impreſſion it makes upon the bottom of the eye; and to this end it will be requiſite to examine this *Hypoteſis* a little more particularly.

Firſt therefore, if we conſider the manner of the progreſs of the pulſe, it will ſeem rational to conclude, that that part or end of the pulſe which precedes the other, muſt neceſſarily be ſomwhat more *obtunded*, or *impeded*
by

by the refiftance of the tranfparent *medium*, than the other part or end of it which is fubfequent, whofe way is, as it were, prepared by the other; efpecially if the adjacent *medium* be not in the fame manner enlightned or agitated. And therefore(in the fourth *Figure* of the fixth *Iconifm*)the Ray A A A H B will have its fide H H more deadned by the refiftance of the dark or quiet *medium* P P P, Whence there will be a kind of dead-nefs fuperinduc'd on the fide H H H, which will continually increafe from B, and ftrike deeper and deeper into the Ray by the line B R ; Whence all the parts of the triangle, R B H O will be of a dead *Blue* colour, and fo much the deeper, by how much the nearer they lie to the line B H H, which is moft deaded or impeded, and fo much the more *dilute*, by how much the nearer it approaches the line B R. Next on the other fide of the Ray A A N, the end A of the pulfe A H will be promoted, or made ftronger, having its paffage already prepar'd as 'twere by the other parts preceding, and fo its impreffion wil be ftronger: And becaufe of its *obliqui-ty* to the Ray, there will be propagated a kind of faint motion into Q Q the adjacent dark or quiet *medium*, which faint motion will fpread fur-ther and further into Q Q as the Ray is propagated further and further from A, namely, as far as the line M A, whence all the triangle M A N will be ting'd with a *Red*, and that *Red* will be the deeper the nearer it ap-proaches the line M A, and the *paler* or *yellower* the nearer it is the line N A. And if the Ray be continued, fo that the lines A N and B R (which are the bounds of the *Red* and *Blue diluted*) do meet and crofs each other, there will be beyond that interfection generated all kinds of *Greens*.

Now, thefe being the proprieties of every fingle refracted Ray of light, it will be eafie enough to confider what muft be the refult of very many fuch Rays collateral : As if we fuppofe infinite fuch Rays *interjacent* be-tween A K S B and A N O B, which are the terminating : For in this cafe the Ray A K S B will have its *Red* triangle intire, as lying next to the dark or quiet *medium*, but the other fide of it B S will have no *Blue*, becaufe the *medium adjacent* to it S B O, is mov'd or enlightned, and confequent-ly that light does deftroy the colour. So likewife will the Ray A N O B lofe its *Red*, becaufe the *adjacent medium* is mov'd or enlightned, but the other fide of the Ray that is *adjacent* to the dark, namely, A H O will preferve its *Blue* entire, and thefe Rays muft be fo far produc'd as till A N and B R cut each other, before there will be any *Green* produc'd. From thefe Proprieties well confider'd, may be dedue'd the reafons of all the *Phænomena* of the *prifme*, and of the *Globules* or drops of Water which conduce to the production of the Rainbow.

Next for the impreffion they make on the *Retina*, we will further ex-amine this *Hypothefis* : Suppofe therefore A B C D E F in the fifth *Figure*, to reprefent the Ball of the eye: on the *Cornea* of which A B C two Rays G A C H and K C A I (which are the terminating Rays of a lumi-nous body) falling, are by the refraction thereof collected or *converg'd* into two points at the bottom of the eye. Now, becaufe thefe termi-nating Rays, and all the *intermediate* ones which come from any part of the luminous body, are fuppos'd by fome fufficient refraction before they

enter

enter the eye, to have their pulses made *oblique* to their progreſſion, and conſequently each Ray to have potentially *ſuperinduc'd* two proprieties, or colours, *viz.* a *Red* on the one ſide, and a *Blue* on the other, which notwithſtanding are never actually manifeſt, but when this or that Ray has the one or the other ſide of it bordering on a dark or unmov'd *medium*, therefore as ſoon as theſe Rays are entred into the eye, and ſo have one ſide of each of them bordering on a dark part of the humours of the eye, they will each of them actually exhibit ſome colour; therefore A D C the production G A C H will exhibit a *Blue*, becauſe the ſide C D is *adjacent* to the dark *medium* C Q D C, but nothing of a *Red*, becauſe its ſide A D is *adjacent* to the enlightned *medium* A D F A : And all the Rays that from the points of the luminous body are collected on the parts of the *Retina* between D and F ſhall have their *Blue* ſo much the more *diluted* by how much the farther theſe points of collection are diſtant from D towards F; and the Ray A F C the production of K C A I, will exhibit a *Red*, becauſe the ſide A F is adjacent to the dark or quiet *medium* of the eye A P F A, but nothing of a *Blue*, becauſe its ſide C F is *adjacent* to the enlightned *medium* C F D C, and all the Rays from the intermediate parts of the luminous body that are collected between F and D ſhall have their *Red* ſo much the more diluted, by how much the farther they are diſtant from F towards D.

Now, becauſe by the refraction in the *Cornea*, and ſome other parts of the eye, the ſides of each Ray, which before were almoſt parallel, are made to *converge* and meet in a point at the bottom of the eye, therefore that ſide of the *pulſe* which preceded before theſe refractions, ſhall firſt touch the *Retina*, and the other ſide laſt. And therefore according as this or that ſide, or end of the pulſe ſhall be impeded, accordingly will the *impreſſions* on the *Retina* be varied; therefore by the Ray G A C H refracted by the *Cornea* to D there ſhall be on that point a ſtroke or impreſſion confus'd, whoſe weakeſt end, namely, that by the line C D ſhall precede, and the ſtronger, namely, that by the line A D ſhall follow. And by the Ray K C A I refracted to F, there ſhall be on that part a confus'd ſtroke or impreſſion, whoſe ſtrongeſt part, namely, that by the line C F ſhal precede, and whoſe weakeſt or impeded, namely, that by the line A F ſhall follow, and all the intermediate points between F and D will receive impreſſion from the *converg'd* Rays ſo much the more like the impreſſions on F and D by how much the nearer they approach that or this.

From the conſideration of the proprieties of which impreſſions, we may collect theſe ſhort definitions of Colours : That *Blue is an impreſſion on the Retina of an oblique and confus'd pulſe of light, whoſe weakeſt part precedes, and whoſe ſtrongeſt follows*. And, that *Red is an impreſſion on the Retina of an oblique and confus'd pulſe of light, whoſe ſtrongeſt part precedes, and whoſe weakeſt follows*.

Which proprieties, as they have been already manifeſted, in the Priſme and falling drops of Rain, to be the cauſes of the colours there generated, may be eaſily found to be the efficients alſo of the colours appearing in thin *laminated* tranſparent bodies; for the explication of which, all this has been premiſed.   And

And that this is fo, a little clofer examination of the *Phænomena* and the *Figure* of the body, by this *Hypothefis*, will make evident.

For firft (as we have already obferved) the *laminated* body muft be of a determinate thicknefs, that is, it muft not be thinner then fuch a determinate quantity; for I have always obferv'd, that neer the edges of thofe which are exceeding thin, the colours difappear, and the part grows white; nor muft it be thicker then another determinate quantity; for I have likewife obferv'd, that beyond fuch a thicknefs, no colours appear'd, but the Plate looked white, between which two determinate thickneffes were all the colour'd Rings; of which in fome fubftances I have found ten or twelve, in others not half fo many, which I fuppofe depends much upon the tranfparency of the *laminated* body. Thus though the confecutions are the fame in the fcumm or the fkin on the top of metals; yet in thofe confecutions the fame colour is not fo often repeated as in the confecutions in thin Glafs, or in Sope-water, or any other more tranfparent and glutinous liquor; for in thefe I have obferv'd, *Red, Yellow, Green, Blue, Purple; Red, Yellow, Green, Blue, Purple; Red, Yellow, Green, Blue, Purple; Red, Yellow,* &c. to fucceed each other, ten or twelve times, but in the other more *opacous* bodies the confecutions will not be half fo many.

And therefore fecondly, the *laminated* body muft be tranfparent, and this I argue from this, that I have not been able to produce any colour at all with an *opacous* body, though never fo thin. And this I have often try'd, by preffing a fmall *Globule* of *Mercury* between two fmooth Plates of Glafs, whereby I have reduc'd that body to a much greater thinnefs then was requifite to exhibit the colours with a tranfparent body.

Thirdly, there muft be a confiderable reflecting body adjacent to the under or further fide of the *lamina* or *plate:* for this I always found, that the greater that reflection was, the more vivid were the appearing colours.

From which Obfervations, it is moft evident, that the reflection from the under or further fide of the body is the principal caufe of the production of thefe colours; which, that it is fo, and how it conduces to that effect, I fhall further explain in the following Figure, which is here defcribed of a very great thicknefs, as if it had been view'd through the *Microfcope*; and 'tis indeed much thicker than any *Microfcope* (I have yet us'd) has been able to fhew me thofe colour'd plates of Glafs, or *Mufcovie-glafs*, which I have not without much trouble view'd with it; for though I have endeavoured to magnifie them as much as the Glaffes were capable of, yet are they fo exceeding thin, that I have not hitherto been able pofitively to determine their thicknefs. This Figure therefore I here reprefent, is wholy *Hypothetical*.

Let A B C D H F E in the fixth Figure be a *fruftum* of *Mufcovy-glafs*, thinner toward the end A E, and thicker towards D F. Let us firft fuppofe the Ray *a g h b* coming from the Sun, or fome remote luminous object to fall *obliquely* on the thinner plate B A E, part therefore is reflected back, by *c g h d,* the firft *fuperficies*; whereby the perpendicular

L

pulfe

pulfe *a b* is after reflection propagated by *c d, c d*, equally remote from each other with *a b, a b,* fo that *ag + gc,* or *b h + h d* are either of them equal to *a a,* as is alfo *c c,* but the body BAE being tranfparent, a part of the light of this Ray is refracted in the furface AB, and propagated by *g i k h* to the furface EF, whence it is reflected and refracted again by the furface AB. So that after two refractions and one reflection, there is propagated a kind of fainter Ray *e m n f,* whofe pulfe is not only weaker by reafon of the two refractions in the furface AB, but by reafon of the time fpent in pafling and repafling between the two furfaces AB and EF, *e f* which is this fainter or weaker pulfe comes behind the pulfe *c d*; fo that hereby (the furfaces AB, and EF being fo neer together, that the eye cannot *difcriminate* them from one) this confus'd or *duplicated* pulfe, whofe ftrongeft part precedes, and whofe weakeft follows, does produce on the *Retina* (or the *optick nerve* that covers the bottom of the eye) the fenfation of a *Yellow.*

And fecondly, this *Yellow* will appear fo much the deeper, by how much the further back towards the middle between *c d* and *c d* the fpurious pulfe *e f* is remov'd, as in 2 where the furface BC being further remov'd from EF, the weaker pulfe *e f* will be nearer to the middle, and will make an impreffion on the eye of a *Red.*

But thirdly, if the two reflecting furfaces be yet further remov'd afunder (as in 3 CD and EF are) then will the weaker pulfe be fo farr behind, that it will be more then half the diftance between *c d* and *c d.* And in this cafe it will rather feem to precede the following ftronger pulfe, then to follow the preceding one, and confequently a *Blue* will be generated. And when the weaker pulfe is juft in the middle beween two ftrong ones, then is a deep and lovely *Purple* generated; but when the weaker pulfe *e f* is very neer to *c d,* then is there generated a *Green,* which will be *bluer,* or *yellower,* according as the *approximate* weak pulfe does precede or follow the ftronger.

Now fourthly, if the thicker Plate chance to be cleft into two thinner Plates, as CDFE is divided into two Plates by the furface GH then from the compofition arifing from the three reflections in the furfaces CD, GH, and EF, there will be generated feveral compounded or mixt colours, which will be very differing, according as the proportion between the thickneffes of thofe two divided Plates CDHG, and GHFE are varied.

And fifthly, if thefe furfaces CD and FE are further remov'd afunder, the weaker pulfe will yet lagg behind much further, and not onely be *coincident* with the fecond, *c d,* but lagg behind that alfo, and that fo much the more, by how much the thicker the Plate be; fo that by degrees it will be *coincident* with the third *c d* backward alfo, and by degrees, as the Plate grows thicker with a fourth, and fo onward to a fifth, fixth, feventh, or eighth; fo that if there be a thin tranfparent body, that from the greateft thinnefs requifite to produce colours, does, in the manner of a Wedge, by degrees grow to the greateft thicknefs that a Plate can be of, to exhibit a colour by the reflection of Light from fuch a body, there

fhall

shall be generated several confecutions of colours, whofe order from the thin end towards the thick, fhall be *Yellow,Red,Purple, Blue,Green* ; *Yellow, Red,Purple,Blue,Green* ; *Yellow,Red,Purple,Blue,Green*; *Yellow*,&c. and thefe fo often repeated, as the weaker pulfe does lofe paces with its *Primary*, or firft pulfe, and is *coincident* with a fecond, third, fourth,fifth,fixth,&c. pulfe behind the firft. And this, as it is *coincident*, or follows from the firft *Hypothefis* I took of colours,fo upon exeriment have I found it in multitudes of inftances that feem to prove it. One thing which feems of the greateft concern in this *Hypothefis*, is to determine the greateft or leaft thicknefs requifite for thefe effects, which, though I have not been wanting in attempting, yet fo exceeding thin are thefe coloured Plates, and fo imperfect our *Microfcope*,that I have been hitherto fuccefsfull.though if my endeavours fhall anfwer my expectations,I fhall hope to gratifie the curious Reader with fome things more remov'd beyond our reach hitherto.

Thus have I,with as much brevity as I was able, endeavoured to explicate (*Hypothetically* at leaft) the caufes of the *Phænomena* I formerly recited, on the confideration of which I have been the more particular.

Firft, becaufe I think thefe I have newly given are capable of explicating all the *Phænomena* of colours, not onely of thofe appearing in the *Prifme*, Water-drop, or Rainbow, and in *laminated* or plated bodies, but of all that are in the world, whether they be fluid or folid bodies, whether in thick or thin, whether tranfparent, or feemingly opacous, as I fhall in the next Obfervation further endeavour to fhew. And fecondly, becaufe this being one of the two ornaments of all bodies difcoverable by the fight, whether looked on with, or without a *Microfcope*, it feem'd to deferve (fomewhere in this Tract, which contains a defcription of the Figure and Colour of fome minute bodies) to be fomewhat the more intimately enquir'd into.

Obferv. X. *Of* Metalline, *and other real Colours.*

HAving in the former Difcourfe, from the Fundamental caufe of Colour, made it probable, that there are but two Colours, and fhewn, that the *Phantafm* of Colour is caus'd by the fenfation of the *oblique* or uneven pulfe of Light which is capable of no more varieties than two that arife from the two fides of the *oblique* pulfe, though each of thofe be capable of infinite gradations or degrees (each of them beginning from *White*, and ending the one in the deepeft *Scarlet* or *Yellow*, the other in the deepeft *Blue*) I fhall in this *Section* fet down fome Obfervations which I have made of other colours, fuch as *Metalline* powders tinging or colour d bodies and feveral kinds of tinctures or ting'd liquors, all which, together with thofe I treated of in the former Obfervation. will, I fuppofe, comprife the feveral fubjects in which colour is obferv'd to be inherent, and the feveral manners by which it *inheres*, or is apparent

L 2   in

in them. And here I fhall endeavour to fhew by what compofition all kind of compound colours are made, and how there is no colour in the world but may be made from the various degrees of thefe two colours, together with the intermixtures of *Black* and *White*.

And this being fo, as I fhall anon fhew, it feems an evident argument to me, that all colours whatfoever, whether in fluid or folid, whether in very tranfparent or feemingly *opacous*, have the fame efficient caufe, to wit, fome kind of *refraction* whereby the Rays that proceed from fuch bodies, have their pulfe *obliquated* or confus'd in the manner I explicated in the former *Section*; that is, a *Red* is caus'd by a duplicated or confus'd pulfe, whofe ftrongeft pulfe precedes, and a weaker follows : and a *Blue* is caus'd by a confus'd pulfe, where the weaker pulfe precedes, and the ftronger follows. And according as thefe are, more or lefs, or varioufly mixt and compounded, fo are the *fenfations*, and confequently the *phantafms* of colours *diverfified*.

To proceed therefore ; I fuppofe, that all tranfparent colour'd bodies, whether fluid or folid, do confift at leaft of two parts, or two kinds of fubftances, the one of a fubftance of a fomewhat differing *refraction* from the other. That one of thefe fubftances which may be call'd the *tinging* fubftance, does confift of diftinct parts, or particles of a determinate big-nefs which are *diffeminated*, or difpers'd all over the other : That thefe particles, if the body be equally and uniformly colour'd, are evenly rang'd and difpers'd over the other contiguous body ; That where the body is deepeft ting'd, there thefe particles are rang'd thickeft; and where 'tis but faintly ting'd, they are rang'd much thinner, but uniformly. That by the mixture of another body that unites with either of thefe, which has a differing refraction from either of the other, quite differing effects will be produc'd, that is, the *confecutions* of the confus'd pulfes will be much of another kind, and confequently produce other *fenfations* and *phantafms* of colours, and from a *Red* may turn to a *Blue*, or from a *Blue* to a *Red*, &c.

Now, that this may be the better underftood, I fhall endeavour to ex-plain my meaning a little more fenfible by a *Scheme :* Suppofe we there-fore in the feventh *Figure* of the fixth *Scheme*, that A B C D reprefents a Veffel holding a ting'd liquor, let I I I I I,&c. be the clear liquor, and let the tinging body that is mixt with it be E E, &*c.* F F, &*c.* G G, &*c.* H H, &*c.* whofe particles (whether round, or fome other determinate Figure is little to our purpofe) are firft of a determinate and equal bulk. Next, they are rang'd into the form of *Quincunx*, or *Equilaterotriangu-lar* order, which that probably they are fo, and why they are fo, I fhall elfe-where endeavour to fhew. Thirdly, they are of fuch a nature, as does either more eafily or more difficultly tranfmit the Rays of light then the liquor ; if more eafily, a *Blue* is generated, and if more difficultly, a *Red* or *Scarlet*.

And firft, let us fuppofe the tinging particles to be of a fubftance that does more *impede* the Rays of light , we fhall find that the pulfe or wave of light mov'd from A D to B C, will proceed on, through the con-taining *medium* by the pulfes or waves K K, L L, M M, N N, O O; but
because

becaufe feveral of thefe Rays that go to the conftitution of thefe pulfes will be flugged or ftopped by the tinging particles E,F,G,H; therefore there fhall be a *fecundary* and weak pulfe that fhall follow the Ray, namely P P which will be the weaker: firft, becaufe it has fuffer'd many refractions in the impeding body; next, for that the Rays will be a little difpers'd or confus'd by reafon of the refraction in each of the particles, whether *round* or *angular*; and this will be more evident, if we a little more clofely examine any one particular tinging *Globule*.

Suppofe we therefore A B in the eighth *Fgure* of the fixth *Scheme*, to reprefent a tinging *Globule* or particle which has a greater refraction than the liquor in which it is contain'd : Let C D be a part of the pulfe of light which is *propagated* through the containing *medium*; this pulfe will be a little ftopt or impeded by the *Globule*, and fo by that time the pulfe is paft to E F that part of it which has been impeded by paffing through the *Globule*, will get but to L M, and fo that pulfe which has been *propagated* through the *Globule*, to wit, L M, N O, P Q, will always come behind the pulfes E F, G H, I K, &c.

Next, by reafon of the greater impediment in A B, and its *Globular* Figure, the Rays that pafs through it will be difpers'd, and very much fcatter'd. Whence C A and D B which before went *direct* and *parallel*, will after the refraction in A B, *diverge* and fpread by A P, and B Q; fo that as the Rays do meet with more and more of thefe tinging particles in their way, by fo much the more will the pulfe of light further lagg behind the clearer pulfe, or that which has fewer refractions, and thence the deeper will the colour be, and the fainter the light that is trajected through it; for not onely many Rays are reflected from the furfaces of A B, but thofe Rays that get through it are very much difordered.

By this *Hypohefis* there is no one experiment of colour that I have yet met with, but may be, I conceive, very rationably folv'd, and perhaps, had I time to examine feveral particulars requifite to the demonftration of it, I might prove it more than probable, for all the experiments about the changes and mixings of colours related in the Treatife of Colours, publifhed by the *Incomparable* Mr. *Boyle*, and multitudes of others which I have obferv'd, do fo eafily and naturally flow from thofe principles, that I am very apt to think it probable, that they own their production to no other *fecundary* caufe : As to inftance in two or three experiments. In the twentieth Experiment, this *Noble Authour* has fhewn that the deep *bluifh purple-colour* of *Violets*, may be turn'd into a *Green*, by *Alcalizate Salts*, and to a *Red* by *acid*; that is, a *Purple* confifts of two colours, a deep *Red*, and a deep *Blue*; when the *Blue* is diluted, or altered, or deftroy'd by *acid Salts*, the *Red* becomes predominant, but when the *Red* is diluted by *Alcalizate*, and the *Blue* heightned, there is generated a *Green*; for of a *Red* diluted, is made a *Yellow*, and *Yellow* and *Blue* make a *Green*.

Now, becaufe the *fpurious* pulfes which caufe a *Red* and a *Blue*, do the one follow the clear pulfe, and the other precede it, it ufually follows, that thofe *Saline* refracting bodies which do *dilute* the colour of the one, do deepen that of the other. And this will be made manifeft by almoft moſt all kinds of *Purples*, and many ſorts of *Greens*, both theſe colours conſiſting of mixt colours; for if we ſuppoſe A and A in the ninth Figure, to repreſent two pulſes of clear light, which follow each other at a convenient diſtance, A A, each of which has a *ſpurious* pulſe preceding it, as B B, which makes a *Blue*, and another following it, as C C, which makes a *Red*, the one caus'd by tinging particles that have a greater refraction, the other by others that have a leſs refracting quality then the liquor or *Menſtruum* in which theſe are diſſolv'd, whatſoever liquor does ſo alter the refraction of the one, without altering that of the other part of the ting'd liquor, muſt needs very much alter the colour of the liquor; for if the refraction of the *diſſolvent* be increas'd, and the refraction of the tinging particles not altered, then will the preceding *ſpurious* pulſe be ſhortned or ſtopt, and not out-run the clear pulſe ſo much; ſo that B B will become E E, and the *Blue* be *diluted*, whereas the other *ſpurious* pulſe which follows will be made to lagg much more, and be further behind A A than before, and C C will become *f f*, and ſo the *Yellow* or *Red* will be heightned.

A *Saline* liquor therefore, mixt with another ting'd liquor, may alter the colour of it ſeveral ways, either by altering the refraction of the liquor in which the colour ſwims: or ſecondly by varying the refraction of the coloured particles, by uniting more intimately either with ſome particular *corpuſcles* of the tinging body, or with all of them, according as it has a *congruity* to ſome more eſpecially, or to all alike: or thirdly, by uniting and interweaving it ſelf with ſome other body that is already joyn'd with the tinging particles, with which ſubſtance it may have a *congruity*, though it have very little with the particles themſelves: or fourthly, it may alter the colour of a ting'd liquor by diſ-joyning certain particles which were before united with the tinging particles, which though they were ſomewhat *congruous* to theſe particles, have yet a greater *congruity* with the newly *infus'd Saline menſtruum*. It may likewiſe alter the colour by further diſſolving the tinging ſubſtance into ſmaller and ſmaller *particles*, and ſo *diluting* the colour; or by uniting ſeveral *particles* together as in precipitations, and ſo deepning it, and ſome ſuch other ways, which many experiments and compariſons of differing trials together, might eaſily inform one of.

From theſe Principles applied, may be made out all the varieties of colours obſervable, either in liquors, or any other ting'd bodies, with great eaſe, and I hope intelligible enough, there being nothing in the *notion* of colour, or in the ſuppos'd production, but is very conceivable, and may be poſſible.

The greateſt difficulty that I find againſt this *Hypotheſis*, is, that there ſeem to be more diſtinct colours then two, that is, then Yellow and Blue. This Objection is grounded on this reaſon, that there are ſeveral Reds, which *diluted*, make not a Saffron or pale Yellow, and therefore Red, or Scarlet ſeems to be a third colour diſtinct from a deep degree of Yellow.

To which I anſwer, that Saffron affords us a deep Scarlet tincture, which may be *diluted* into as pale a Yellow as any, either by making a weak ſolution

lution of the Saffron, by infusing a small parcel of it into a great quantity of liquor, as in spirit of Wine, or else by looking through a very thin quantity of the tincture, and which may be heightn d into the loveliest Scarlet, by looking through a very thick body of this tincture, or through a thinner parcel of it, which is highly *impregnated* with the tinging body, by having had a greater quantity of the Saffron dissolv'd in a smaller parcel of the liquor.

Now, though there may be some particles of other tinging bodies that give a lovely Scarlet also, which though *diluted* never so much with liquor, or looked on through never so thin a parcel of ting'd liquor, will not yet afford a pale Yellow, but onely a kind of faint Red; yet this is no argument but that those ting'd particles may have in them the faintest degree of Yellow, though we may be unable to make them exhibit it; For that power of being *diluted* depending upon the divisibility of the ting'd body, if I am unable to make the tinging particles so thin as to exhibit that colour, it does not therefore follow, that the thing is impossible to be done; now, the tinging particles of some bodies are of such a nature, that unless there be found some way of comminuting them into less bulks then the liquor does dissolve them into, all the Rays that pass through them must necessarily receive a tincture so deep, as their appropriate refractions and bulks compar'd with the proprieties of the dissolving liquor must necessarily dispose them to empress, which may perhaps be a pretty deep Yellow, or pale Red.

And that this is not *gratis dictum*, I shall add one instance of this kind, wherein the thing is most manifest.

If you take Blue *Smalt*, you shall find, that to afford the deepest Blue, which *cæteris paribus* has the greatest particles or sands; and if you further divide, or grind those particles on a Grindstone, or *porphyry* stone, you may by *comminuting* the sands of it, *dilute* the Blue into as pale a one as you please, which you cannot do by laying the colour thin; for wheresoever any single particle is, it exhibits as deep a Blue as the whole mass. Now, there are other Blues, which though never so much ground, will not be *diluted* by grinding, because consisting of very small particles, very deeply ting'd, they cannot by grinding be actually separated into smaller particles then the operation of the fire, or some other dissolving *menstruum*, has reduc'd them to already.

Thus all kind of *Metalline* colours, whether *precipitated, sublim'd, calcin'd*, or otherwise prepar'd, are hardly chang'd by grinding, as *ultra marine* is not more *diluted*; nor is *Vermilion* or *Red-lead* made of a more faint colour by grinding; for the smallest particles of these which I have view'd with my greatest Magnifying-Glass, if they be well enlightned, appear very deeply ting'd with their peculiar colours; nor, though I have magnified and enlightned the particles exceedingly, could I in many of them perceive them to be transparent, or to be whole particles, but the smallest specks that I could find among well ground *Vermilion* and *Red-lead*, seem'd to be a Red mass, compounded of a multitude of less and less motes, which sticking together, compos'd a bulk, not one thousand thousandth part of the smallest visible sand or mote. And

And this I find generally in moſt *Metalline* colours, that though they conſiſt of parts ſo exceedingly ſmall, yet are they very deeply ting'd, they being ſo ponderous, and having ſuch a multitude of terreſtrial particles throng'd into a little room ; ſo that 'tis difficult to find any particle tranſparent or reſembling a pretious ſtone, though not impoſſible ; for I have obſerv'd divers ſuch ſhining and reſplendent colours intermixt with the particles of *Cinnaber*, both natural and artificial, before it hath been ground and broken or flaw'd into *Vermilion* : As I have alſo in *Orpiment*, *Red-lead*, and *Biſe*, which makes me ſuppoſe, that thoſe *metalline* colours are by grinding, not onely broken and ſeparated actually into ſmaller pieces, but that they are alſo flaw'd and bruſed, whence they, for the moſt part, become *opacous*, like flaw'd Cryſtal or Glaſs, &c. But for *Smalts* and *verditures*, I have been able with a *Microſcope* to perceive their particles very many of them tranſparent.

Now, that the others alſo may be tranſparent, though they do not appear ſo to the *Microſcope*, may be made probable by this Experiment : that if you take *ammel* that is almoſt *opacous*, and grind it very well on a *Porphyry*, or *Serpentine*, the ſmall particles will by reaſon of their flaws, appear perfectly *opacous* ; and that 'tis the flaws that produce this *opacouſneſs*, may be argued from this, that particles of the ſame *Ammel* much thicker if unflaw'd will appear ſomewhat tranſparent even to the eye ; and from this alſo, that the moſt tranſparent and clear Cryſtal, if heated in the fire, and then ſuddenly quenched, ſo that it be all over flaw'd, will appear *opacous* and white.

And that the particles of *Metalline* colours are tranſparent, may be argued yet further from this, that the Cryſtals, or *Vitriols* of all Metals, are tranſparent, which ſince they conſiſt of *metalline* as well as *ſaline* particles, thoſe *metalline* ones muſt be tranſparent, which is yet further confirm'd from this, that they have for the moſt part, *appropriate* colours ; ſo the *vitriol* of Gold is Yellow ; of Copper, Blue, and ſometimes Green ; of Iron, green ; of Tinn and Lead, a pale White ; of Silver, a pale Blue, &c.

And next, the *Solution* of all Metals into *menſtruums* are much the ſame with the *Vitriols*, or Cryſtals. It ſeems therefore very probable, that thoſe colours which are made by the *precipitation* of thoſe particles out of the *menſtruums* by tranſparent *precipitating* liquors ſhould be tranſparent alſo. Thus Gold *precipitates* with *oyl of Tartar*, or *ſpirit of Vrine* into a brown Yellow. Copper with ſpirit of *Vrine* into a Mucous blue, which retains its tranſparency. A ſolution of ſublimate (as the ſame Illuſtrious Authour I lately mention'd ſhews in his 40. Experiment) *precipitates* with oyl of *Tartar per deliquium*, into an Orange colour'd *precipitate* ; nor is it leſs probable, that the *calcination* of thoſe *Vitriols* by the fire, ſhould have their particles tranſparent : Thus *Saccarum Saturni*, or the *Vitriol of Lead* by *calcination* becomes a deep Orange-colour'd *minium*, which is a kind of *precipitation* by ſome Salt which proceeds from the fire ; common *Vitriol calcin'd*, yields a deep Brown Red, &c.

A third Argument, that the particles of Metals are tranſparent, is, that being *calcin'd*, and melted with Glaſs, they tinge the Glaſs with tranſpa-

rent

rent colours. Thus the *Calx* of Silver tinges the Glaſs on which it is anneal'd with a lovely Yellow, or Gold colour, &c.

And that the parts of Metals are tranſparent, may be farther argued
from the tranſparency of Leaf-gold, which held againſt the light, both
to the naked eye, and the *Microſcope*, exhibits a deep Green. And
though I have never ſeen the other Metals *laminated* ſo thin, that I was
able to perceive them tranſparent, yet, for Copper and Braſs, if we had
the ſame conveniency for *laminating* them, as we have for Gold, we might,
perhaps, through ſuch plates or leaves, find very differing degrees of Blue,
or Green; for it ſeems very probable, that thoſe Rays that rebound from
them ting'd, with a deep Yellow, or pale Red, as from Copper, or with
a pale Yellow, as from Braſs, have paſt through them; for I cannot conceive how by reflection alone thoſe Rays can receive a tincture, taking
any *Hypotheſis* extant.

So that we ſee there may a ſufficient reaſon be drawn from theſe inſtances, why thoſe colours which we are unable to *dilute* to the paleſt
Yellow, or Blue, or Green, are not therefore to be concluded not to be a
deeper degree of them; for ſuppoſing we had a great company of ſmall
Globular eſſence Bottles, or round Glaſs bubbles, about the bigneſs of a Wal-
nut, fill'd each of them with a very deep mixture of Saffron, and that
every one of them did appear of a deep Scarlet colour, and all of them
together did *exhibit* at a diſtance, a deep dy'd Scarlet body. It does not
follow, becauſe after we have come nearer to this *congeries*, or maſs, and divided it into its parts, and examining each of its parts ſeverally or apart,
we find them to have much the ſame colour with the whole maſs; it does
not, I ſay, therefore follow, that if we could break thoſe *Globules* ſmaller,
or any other ways come to ſee a ſmaller or thinner parcel of the ting'd
liquor that fill'd thoſe bubbles, that that ting'd liquor muſt always appear
Red, or of a Scarlet hue, ſince if Experiment be made, the quite contrary
will enſue; for it is capable of being *diluted* into the paleſt Yellow.

Now, that I might avoid all the Objections of this kind, by exhibiting
an Experiment that might by ocular proof convince thoſe whom other
reaſons would not prevail with, I provided me a *Priſmatical Glaſs*, made
hollow, juſt in the form of a Wedge, ſuch as is repreſented in the tenth
Figure of the ſixth *Scheme*. The two *parallelogram* ſides A B C D, A B E F,
which met at a point, were made of the cleareſt Looking-glaſs plates well
ground and poliſh'd that I could get; theſe were joyn'd with hard cement
to the *triangular* ſides, B C E, A D F, which were of Wood; the *Parallelo-*
gram baſe B C E F, likewiſe was of Wood joyn'd on to the reſt with hard
cement, and the whole *Priſmatical* Box was exactly ſtopt every where,
but onely a little hole near the baſe was left, whereby the Veſſel could be
fill'd with any liquor, or emptied again at pleaſure.

One of theſe Boxes (for I had two of them) I fill'd with a pretty deep
tincture of *Aloes*, drawn onely with fair Water, and then ſtopt the hole
with a piece of Wax, then, by holding this Wedge againſt the Light, and
looking through it, it was obvious enough to ſee the tincture of the liquor
near the edge of the Wedge where it was but very thin, to be a pale but

M *well*

well colour'd Yellow, and further and further from the edge, as the liquor grew thicker and thicker,this tincture appear'd deeper and deeper, so that near the blunt end,which was seven Inches fromthe edge and three Inches and an half thick; it was of a deep and well colour'd Red. Now, the clearer and purer this tincture be, the more lovely will the deep Scarlet be, and the fouler the tincture be, the more dirty will the Red appear; so that some dirty tinctures have afforded their deepest Red much of the colour of burnt Oker or *Spanish* brown;others as lovely a colour as *Vermilion*, and some much brighter; but several others, according as the tinctures were worse or more foul, exhibited various kinds of Reds, of very differing degrees.

The other of these Wedges, I fill'd with a most lovely tincture of Copper, drawn from the filings of it,with spirit of *Urine*, and this Wedge held as the former against the Light, afforded all manner of Blues, from the fainteft to the deepest,so that I was in good hope by these two,to have produc'd all the varieties of colours imaginable; for I thought by this means to have been able by placing the two *Parallelogram* sides together, and the edges contrary ways,to have so mov'd them to and fro one by another, as by looking through them in several places, and through severalthicknesses, I should have compounded, and confequently have seen all those colours, which by other like compositions of colours would have enfued.

But instead of meeting with what I look'd for, I met with somewhat more admirable; and that was, that I found my self utterly unable to see through them when placed both together, though they were transparent enough when afunder; and though I could see through twice the thicknes, when both of them were fill'd with the same colour'd liquors, whether both with the Yellow, or both with the Blue, yet when one was fill'd with the Yellow, the other with the Blue,and both looked through, they both appear'd dark, onely when the parts near the tops were look'd through, they exhibited Greens, and those of very great variety, as I expected,but the Purples and other colours,I could not by any means make, whether I endeavour'd to look through them both against the Sun, or whether I plac'd them against the hole of a darkned room.

But notwithstanding this mif-ghessing, I proceeded on with my trial in a dark room, and having two holes near one another, I was able, by placing my Wedges against them,to mix the ting'd Rays that past through them, and fell on a sheet of white Paper held at a convenient distance from them as I pleas'd; so that I could make the Paper appear of what colour I would,by varying the thicknesses of theWedges,and confequently the tincture of the Rays that past through the two holes, and sometimes also by varying the Paper, that is, instead of a white Paper, holding a gray, or a black piece of Paper.

Whence I experimentally found what I had before imagin'd, that all the varieties of colours imaginable are produc'd from several degrees of these two colours, namely, Yellow and Blue, or the mixture of them with light and darknes, that is, white and black. And all those almost infinite varieties which Limners and Painters are able to make by compounding

pounding thofe feveral colours they lay on their Shels or *Palads*, are nothing elfe, but fome *compofitum*, made up of fome one or more, or all of thefe four.

Now, whereas it may here again be objected, that neither can the Reds be made out of the Yellows, added together, or laid on in greater or lefs quantity, nor can the Yellows be made out of the Reds though laid never fo thin; and as for the addition of White or Black, they do nothing but either whiten or darken the colours to which they are added, and not at all make them of any other kind of colour: as for inftance, *Vermilion*, by being temper'd with White Lead, does not at all grow more Yellow, but onely there is made a whiter kind of Red. Nor does Yellow *Oker*, though laid never fo thick, produce the colour of *Vermilion*, nor though it be temper'd with Black, does it at all make a Red; nay, though it be temper'd with White, it will not afford a fainter kind of Yellow, fuch as *mafticut*, but onely a whiten'd Yellow; nor will the Blues be *diluted* or deepned after the manner I fpeak of, as *Indico* will never afford fo fine a Blue as *Vltramarine* or *Bife*; nor will it, temper'd with *Vermilion*, ever afford a Green, though each of them be never fo much temper'd with white.

To which I anfwer, that there is a great difference between *diluting* a colour and whitening of it; for *diluting* a colour, is to make the colour'd parts more thin, fo that the ting'd light, which is made by trajecting thofe ting'd bodies, does not receive fo deep a tincture; but whitening a colour is onely an intermixing of many clear reflections of light among the fame ting'd parts; deepning alfo, and darkning or blacking a colour, are very different; for deepning a colour, is to make the light pafs through a greater quantity of the fame tinging body; and darkning or blacking a colour, is onely interpofing a multitude of dark or black fpots among the fame ting'd parts, or placing the colour in a more faint light.

Firft therefore, as to the former of thefe operations, that is, diluting and deepning, moft of the colours us'd by the Limners and Painters are incapable of, to wit, *Vermilion* and *Red-lead*, and *Oker*, becaufe the ting'd parts are fo exceeding fmall, that the moft curious Grindftones we have, are not able to feparate them into parts actually divided fo fmall as the ting'd particles are; for looking on the moft curioufly ground *Vermilion*, and *Oker*, and *Red-lead*, I could perceive that even thofe fmall *corpufcles* of the bodies they left were compounded of many pieces, that is, they feem'd to be fmall pieces compounded of a multitude of leffer ting'd parts: each piece feeming almoft like a piece of Red Glafs, or ting'd Cryftal all flaw'd; fo that unlefs the Grindftone could actually divide them into fmaller pieces then thofe flaw'd particles were, which compounded that ting'd mote I could fee with my *Microfcope*, it would be impoffible to *dilute* the colour by grinding, which, becaufe the fineft we have will not reach to do in *Vermilion* or *Oker*, therefore they cannot at all, or very hardly be *diluted*.

Other colours indeed, whofe ting'd particles are fuch as may be made fmaller, by grinding their colour, may be *diluted*. Thus feveral of the

Blues may be *diluted*, as *Smalt* and *Bife* ; and *Mafticut*, which is Yellow, may be made more faint : And even *Vermilion* it felf may, by too much grinding, be brought to the colour of *Red-lead*, which is but an Orange colour, which is confeft by all to be very much upon the Yellow. Now, though perhaps fomewhat of this *diluting* of *Vermilion* by overmuch grinding may be attributed to the Grindftone, or muller, for that fome of their parts may be worn off and mixt with the colour, yet there feems not very much, for I have done it on a Serpentine-ftone with a muller made of a Pebble, and yet obferv'd the fame effect follow.

And fecondly, as to the other of thefe operations on colours, that is, the deepning of them, Limners and Painters colours are for the moft part alfo uncapable. For they being for the moft part *opacous* ; and that *opacoufnefs*, as I faid before, proceeding from the particles, being very much flaw'd, unlefs we were able to joyn and re-unite thofe flaw'd particles again into one piece, we fhall not be able to deepen the colour, which fince we are unable to do with moft of the colours which are by Painters accounted *opacous*, we are therefore unable to deepen them by adding more of the fame kind.

But becaufe all thofe *opacous* colours have two kinds of beams or Rays reflected from them, that is, Rays unting'd, which are onely reflected from the outward furface, without at all penetrating of the body, and ting'd Rays which are reflected from the inward furfaces or flaws after they have fuffer'd a two-fold refraction ; and becaufe that tranfparent liquors mixt with fuch *corpufcles*, do, for the moft part, take off the former kind of reflection ; therefore thefe colours mixt with Water or Oyl, appear much deeper than when dry, for moft part of that white reflection from the outward furface is remov'd. Nay, fome of thefe colours are very much deepned by the mixture with fome tranfparent liquor, and that becaufe they may perhaps get between thofe two flaws, and fo confequently joyn two or more of thofe flaw'd pieces together ; but this happens but in a very few.

Now, to fhew that all this is not *gratis dictum*, I fhall fet down fome Experiments which do manifeft thefe things to be probable and likely, which I have here deliver'd.

For, firft, if you take any ting'd liquor whatfoever, efpecially if it be pretty deeply ting'd, and by any means work it into a froth, the *congeries* of that froth fhall feem an *opacous* body, and appear of the fame colour, but much whiter than that of the liquor out of which it is made. For the abundance of reflections of the Rays againft thofe furfaces of the bubbles of which the froth confifts, does fo often rebound the Rays backwards, that little or no light can pafs through, and confequently the froth appears *opacous*.

Again, if to any of thefe ting'd liquors that will endure the boiling there be added a fmall quantity of fine flower (the parts of which through the *Microfcope* are plainly enough to be perceiv'd to confift of tranfparent *corpufcles*) and fuffer'd to boyl till it thicken the liquor, the mafs of the liquor will appear *opacous*, and ting'd with the fame colour, but very much whiten'd. 　　　　　　　　　　　　　　　　　　　　**Thus**

Thus, if you take a piece of tranfparent Glafs that is well colour'd, and by heating it, and then quenching it in Water, you flaw it all over, it will become *opacous*, and will exhibit the fame colour with which the piece is ting'd, but fainter and whiter.

Or, if you take a Pipe of this tranfparent Glafs, and in the flame of a Lamp melt it, and then blow it into very thin bubbles, then break thofe bubbles, and collect a good parcel of thofe *laminæ* together in a Paper, you fhall find that a fmall thicknefs of thofe Plates will conftitute an *opacous* body, and that you may fee through the mafs of Glafs before it be thus *laminated*, above four times the thicknefs: And befides, they will now afford a colour by reflection as other *opacous* (as they are call'd) colours will, but much fainter and whiter than that of the Lump or Pipe out of which they were made.

Thus alfo, if you take *Putty*, and melt it with any tranfparent colour'd Glafs, it will make it become an *opacous* colour'd lump, and to yield a paler and whiter colour than the lump by reflection.

The fame thing may be done by a preparation of *Antimony*, as has been fhewn by the Learned *Phyfician*, D^r. *C. M.* in his Excellent Obfervations and Notes on *Nery's Art of Glafs*; and by this means all tranfparent colours become *opacous*, or *ammels*. And though by being ground they lofe very much of their colour, growing much whiter by reafon of the multitude of fingle reflections from their outward furface, as I fhew'd afore, yet the fire that in the nealing or melting re-unites them, and fo renews thofe *fpurious* reflections, removes alfo thofe whitenings of the colour that proceed from them.

As for the other colours which Painters ufe, which are tranfparent, and us'd to varnifh over all other paintintings, 'tis well enough known that the laying on of them thinner or thicker, does very much *dilute* or deepen their colour.

Painters Colours therefore confifting moft of them of folid particles, fo fmall that they cannot be either re-united into thicker particles by any Art yet known, and confequently cannot be deepned; or divided into particles fo fmall as the flaw'd particles that exhibit that colour, much lefs into fmaller, and confequently cannot be *diluted*; It is neceffary that they which are to imitate all kinds of colours, fhould have as many degrees of each colour as can be procur'd.

And to this purpofe, both Limners and Painters have a very great variety both of Yellows and Blues, befides feveral other colour'd bodies that exhibit very compounded colours, fuch as Greens and Purples; and others that are compounded of feveral degrees of Yellow, or feveral degrees of Blue, fometimes unmixt, and fometimes compounded with feveral other colour'd bodies.

The Yellows, from the paleft to the deepeft Red or Scarlet, which has no intermixture of Blue, are *pale and deep Mafticut, Orpament, Englifh Oker, brown Oker, Red Lead, and Vermilion, burnt Englifh Oker, and burnt brown Oker*, which laft have a mixture of dark or dirty parts with them, *&c.*

Their

Their Blues are feveral kinds of *Smalts*, and *Verditures*, and *Bife*, and *Ultramarine*, and *Indico*, which laft has many dirty or dark parts intermixt with it.

Their compounded colour'd bodies, as *Pink*, and *Verdigrefe*,which are Greens, the one a *Popingay*, the other a *Sea-green*; then *Lac*,which is a very lovely *Purple*.

To which may be added their Black and White, which they alfo ufually call Colours, of each of which they have feveral kinds, fuch as *Bone Black*, made of *Ivory* burnt in a clofe Veffel, and *Blue Black*, made of the fmall coal of *Willow*, or fome other Wood ; and *Cullens earth*, which is a kind of brown Black, *&c*. Their ufual Whites are either artificial or natural *White Lead*, the laft of which is the beft they yet have, and with the mixing and tempering thefe colours together, are they able to make an imitation of any colour whatfoever : Their Reds or deep Yellows, they can *dilute* by mixing pale Yellows with them, and deepen their pale by mixing deeper with them ; for it is not with *Opacous* colours as it is with tranfparent, where by adding more Yellow to yellow, it is deepned, but in *opacous diluted*. They can whiten any colour by mixing White with it, and darken any colour by mixing Black, or fome dark and dirty colour. And in a word, moft of the colours, or colour'd bodies they ufe in Limning and Painting, are fuch, as though mixt with any other of their colours, they preferve their own hue, and by being in fuch very fmal parts difpers'd through the other colour'd bodies, they both, or altogether reprefent to the eye a *compofitum* of all ; the eye being unable, by reafon of their fmalnefs, to diftinguifh the peculiarly colour'd particles, but receives them as one intire *compofitum* : whereas in many of thefe, the *Microfcope* very eafily diftinguifhes each of the compounding colours diftinct, and exhibiting its own colour.

Thus have I by gently mixing *Vermilion* and *Bife* dry, produc'd a very fine Purple,or mixt colour,but looking on it with the *Microfcope*, I could eafily diftinguifh both the Red and the Blue particles, which did not at all produce the *Phantafm* of Purple.

To fumm up all therefore in a word, I have not yet found any folid colour'd body,that I have yet examin'd,perfectly *opacous* ; but thofe that are leaft tranfparent are *Metalline* and *Mineral* bodies, whofe particles generally, feeming either to be very fmall, or very much flaw'd, appear for the moft part *opacous*, though there are very few of them that I have look'd on with a *Microfcope*, that have not very plainly or circumftantially manifefted themfelves tranfparent.

And indeed, there feem to be fo few bodies in the world that are *in minimis* opacous, that I think one may make it a rational *Query*, Whether there be any body abfolutely thus *opacous* ? For I doubt not at all (and I have taken notice of very many circumftances that make me of this mind) that could we very much improve the *Microfcope*, we might be able to fee all thofe bodies very plainly tranfparent, which we now are fain onely to ghefs at by circumftances. Nay, the Object Glaffes we yet make ufe of are fuch, that they make many tranfparent bodies to the

eye,

eye, feem *opacous* through them, which if we widen the Aperture a little, and caft more light on the objects, and not charge the Glaffes fo deep, will again difclofe their tranfparency.

Now, as for all kinds of colours that are diffolvable in Water, or other liquors, there is nothing fo manifeft, as that all thofe ting'd liquors are tranfparent ; and many of them are capable of being *diluted* and compounded or mixt with other colours, and divers of them are capable of being very much chang'd and heightned, and fixt with feveral kinds of *Saline menftruums*. Others of them upon compounding, deftroy or vitiate each others colours, and *precipitate*, or otherwife very much alter each others tincture. In the true ordering and *diluting*, and deepning, and mixing, and fixing of each of which, confifts one of the greateft myfteries of the Dyers; of which particulars, becaufe our *Microfcope* affords us very little information, I fhall add nothing more at prefent ; but onely that with a very few tinctures order'd and mixt after certain ways, too long to be here fet down, I have been able to make an appearance of all the various colours imaginable, without at all ufing the help of *Salts*, or *Saline menftruums* to vary them.

As for the mutation of Colours by *Saline menftruums*, they have already been fo fully and excellently handled by the lately mention'd Incomparable *Authour*, that I can add nothing, but that of a multitude of trials that I made, I have found them exactly to agree with his Rules and Theories ; and though there may be infinite inftances, yet may they be reduc'd under a few Heads, and compris'd within a very few Rules. And generally I find, that *Saline menftruums* are moft operative upon thofe colours that are Purple, or have fome degree of Purple in them, and upon the other colours much lefs. The *fpurious* pulfes that compofe which, being (as I formerly noted) fo very neer the middle between the true ones, that a fmall variation throws them both to one fide, or both to the other, and fo confequently muft make a vaft mutation in the formerly appearing Colour.

Obferv. XI. *Of* Figures *obferv'd in fmall Sand.*

SAnd generally feems to be nothing elfe but exceeding fmall Pebbles, or at leaft fome very fmall parcels of a bigger ftone ; the whiter kind feems through the *Microfcope* to confift of fmall tranfparent pieces of fome *pellucid* body, each of them looking much like a piece of *Alum*, or *Salt Gem*; and this kind of Sand is angled for the moft part irregularly, without any certain fhape, and the *granules* of it are for the moft part flaw'd, though amongft many of them it is not difficult to find fome that are perfectly *pellucid*, like a piece of clear Cryftal, and divers likewife moft curioufly fhap'd, much after the manner of the bigger *Stiriæ* of Cryftal, or like the fmall Diamants I obferv'd in certain Flints, of which I fhall by and by relate ; which laft particular feems to argue, that this kind of Sand is not
made

made by the comminution of greater tranſparent Cryſtaline bodies, but by the *concretion* or *coagulation* of Water, or ſome other fluid body.

There are other kinds of courſer Sands, which are browner, and have their particles much bigger; theſe, view'd with a *Microſcope*, ſeem much courſer and more *opacous* ſubſtances, and moſt of them are of ſome irregularly rounded Figures; and though they ſeem not ſo *opacous* as to the naked eye, yet they ſeem very foul and cloudy, but neither do theſe want curiouſly tranſparent, no more than they do regularly figur'd and well colour'd particles, as I have often found.

There are multitudes of other kinds of Sands, which in many particulars, plainly enough diſcoverable by the *Microſcope*, differ both from theſe laſt mention'd kinds of Sands, and from one another: there ſeeming to be as great variety of Sands, as there is of Stones. And as amongſt Stones ſome are call'd precious from their excellency, ſo alſo are there Sands which deſerve the ſame Epithite for their beauty; for viewing a ſmall parcel of *Eaſt-India* Sand (which was given me by my highly honoured friend, Mr. *Daniel Colwall*) and, ſince that, another parcel, much of the ſame kind, I found ſeveral of them, both very tranſparent like precious Stones, and regularly figur'd like Cryſtal, *Corniſh* Diamants, ſome Rubies, *&c.* and alſo ting'd with very lively and deep colours, like *Rubys, Saphyrs, Emeralds,* &c. Theſe kinds of granuls I have often found alſo in *Engliſh* Sand. And 'tis eaſie to make ſuch a counterfeit Sand with deeply ting'd Glaſs, Enamels and Painters colours.

It were endleſs to deſcribe the multitudes of Figures I have met with in theſe kind of minute bodies, ſuch as *Spherical, Oval, Pyramidal, Conical, Priſmatical,* of each of which kinds I have taken notice.

But amongſt many others, I met with none more obſervable than this pretty Shell (deſcribed in the *Figure* X. of the fifth *Scheme*) which, though as it was light on by chance, deſerv'd to have been omitted (I being unable to direct any one to find the like) yet for its rarity was it not inconſiderable, eſpecially upon the account of the information it may afford us. For by it we have a very good inſtance of the curioſity of Nature in another kind of Animals which are remov'd, by reaſon of their minuteneſs, beyond the reach of our eyes; ſo that as there are ſeveral ſorts of Inſects, as Mites, and others, ſo ſmall as not yet to have had any names; (ſome of which I ſhall afterwards deſcribe) and ſmall Fiſhes, as Leeches in Vineger; and ſmal vegetables, as Moſs, and Roſe-Leave-plants; and ſmall Muſhroms, as mould: ſo are there, it ſeems, ſmall Shel-fiſh likewiſe, Nature ſhewing her curioſity in every Tribe of *Animals, Vegetables,* and *Minerals.*

I was trying ſeveral ſmall and ſingle Magnifying Glaſſes, and caſually viewing a parcel of white Sand, when I perceiv'd one of the grains exactly ſhap'd and wreath'd like a Shell, but endeavouring to diſtinguiſh it with my naked eye, it was ſo very ſmall, that I was fain again to make uſe of the Glaſs to find it; then, whileſt I thus look'd on it, with a Pin I ſeparated all the reſt of the granules of Sand, and found it afterwards to appear to the naked eye an exceeding ſmall white ſpot, no bigger than the point of a

Pin.

Pin. Afterwards I view'd it every way with a better *Microscope* and found it on both sides, and edge-ways, to resemble the Shell of a small Water-Snail with a flat spiral Shell: it had twelve wreathings, *a, b, c, d, e*, &c. all very proportionably growing one less than another toward the middle or center of the Shell, where there was a very small round white spot. I could not certainly discover whether the Shell were hollow or not, but it seem'd fill'd with somewhat, and 'tis probable that it might be *petrify'd* as other larger Shels often are, such as are mention'd in the seventeenth *Observation*.

Observ. XII. *Of* Gravel *in* Urine.

I Have often observ'd the Sand or Gravel of Urine, which seems to be a *tartareous* substance, generated out of a *Saline* and a *terrestrial* substance *crystalliz'd* together, in the form of *Tartar*, sometimes sticking to the sides of the *Urinal*, but for the most part sinking to the bottom, and there lying in the form of coarse common Sand; these, through the *Microscope*, appear to be a company of small bodies, partly transparent, and partly *opacous*, some White, some Yellow, some Red, others of more brown and duskie colours.

The Figure of them is for the most part flat, in the manner of Slats, or such like plated Stones, that is, each of them seem to be made up of several other thinner Plates, much like *Muscovie Glass*, or *English Sparr*, to the last of which, the white plated Gravel seems most likely; for they seem not onely plated like that, but their sides shap'd also into *Rhombs, Rhomboeids*, and sometimes into *Rectangles* and *squares*. Their bigness and Figure may be seen in the second *Figure* of the sixth *Plate*, which represents about a dozen of them lying upon a plate A B C D, some of which, as *a, b, c, d*, seem'd more regular than the rest, and *e*, which was a small one, sticking on the top of another, was a perfect *Rhomboeid* on the top, and had four *Rectangular* sides.

The line E which was the measure of the *Microscope*, is $\frac{1}{32}$ part of an *English* Inch, so that the greatest bredth of any of them, exceeded not $\frac{1}{128}$ part of an Inch.

Putting these into several liquors, I found *Oyl of Vitriol, spirit of Urine*, and several other *Saline menstruums* to dissolve them; and the first of these in less than a minute without *Ebullition*, Water, and several other liquors, had no sudden operation upon them. This I mention, because those liquors that dissolve them, first make them very white, not *vitiating*, but rather rectifying their Figure, and thereby make them afford a very pretty object for the *Microscope*.

How great an advantage it would be to such as are troubled with the Stone, to find some *menstruum* that might dissolve them without hurting the Bladder, is easily imagin'd, since some *injections* made of such bodies might likewise dissolve the stone; which seems much of the same nature.

It

It may therefore, perhaps, be worthy fome Phyficians enquiry, whether there may not be fomething mixt with the Urine in which the Gravel or Stone lies, which may again make it diffolve it, the firft of which feems by it's regular Figures to have been fometimes *Cryftalliz'd* out of it. For whether this *Cryftallization* be made in the manner as *Alum, Peter,*&c. are *cryftallized* out of a cooling liquor, in which, by boyling they have been diffolv'd ; or whether it be made in the manner of *Tartarum Vitriolatum,* that is, by the *Coalition* of an *acid* and a *Sulphureous* fubftance, it feems not impoffible, but that the liquor it lies in, may be again made a *diffolvent* of it. But leaving thefe inquiries to Phyficians or Chymifts, to whom it does more properly belong, I fhall proceed.

Obferv. XIII. *Of the fmall* Diamants, *or* Sparks *in* Flints.

CHancing to break a Flint ftone in pieces, I found within it a certain cavity all crufted over with a very pretty candied fubftance, fome of the parts of which, upon changing the pofture of the Stone, in refpect of the *Incident* light, exhibited a number of fmall, but very vivid reflections ; and having made ufe of my *Microfcope*, I could perceive the whole furface of that cavity to be all befet with a multitude of little *Cryftaline* or *Adamantine* bodies, fo curioufly fhap'd, that it afforded a not unpleafing object.

Having confidered thofe vivid *repercuffions* of light, I found them to be made partly from the plain external furface of thefe regularly figured bodies (which afforded the vivid reflections) and partly to be made from within the fomewhat *pellucid* body, that is, from fome furface of the body, oppofite to that fuperficies of it which was next the eye.

And becaufe thefe bodies were fo fmall, that I could not well come to make Experiments and Examinations of them, I provided me feveral fmall *ftiriæ* of Cryftals or Diamants, found in great quantities in *Cornwall*, and are therefore commonly called *Cornifh Diamants :* thefe being very *pellucid*, and growing in a hollow cavity of a Rock (as I have been feveral times informed by thofe that have obferv d them) much after the fame manner as thefe do in the Flint ; and having befides their outward furface very regularly fhap'd, retaining very near the fame Figures with fome of thofe I obferv'd in the other, became a convenient help to me for the Examination of the proprieties of thofe kinds of bodies.

And firft for the Reflections ; in thefe I found it very obfervable, That the brighteft reflections of light proceeded from within the *pellucid* body ; that is, that the Rays admitted through the *pellucid* fubftance in their getting out on the oppofite fide, were by the contiguous and ftrong reflecting furface of the Air very vividly reflected, fo that more Rays were reflected to the eye by this furface, though the Ray in entring and getting out of the Cryftal had fuffer'd a double refraction, than there were from the outward furface of the Glafs where the Ray had fuffer'd no reftraction at all.

And

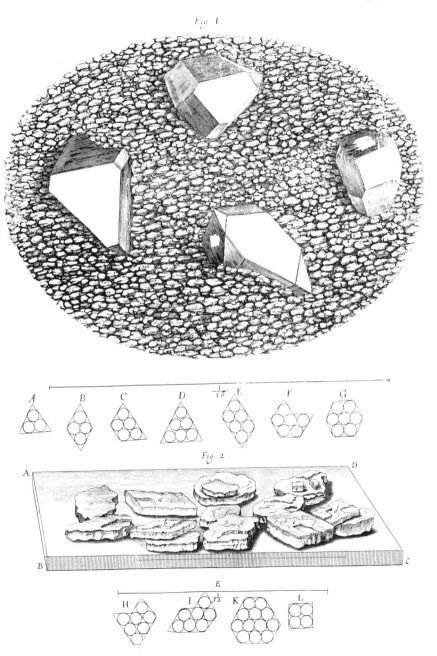

Fig. 1

A   B   C   D $\frac{1}{15}$ E   F   G

Fig. 2

A   D

B   C

E

H   I $\frac{1}{12}$ K   L

And that this was the furface of the Air that gave fo vivid a *re-percuf-fion* I try'd by this means. I funk half of a *ftiria* in Water, fo that only Water was contiguous to the under furface, and then the internal reflection was fo exceedingly faint, that it was fcarce difcernable. Again, I try'd to alter this vivid reflection by keeping off the Air, with a body not fluid, and that was by rubbing and holding my finger very hard againft the under furface, fo as in many places the pulp of my finger did touch the Glafs, without any *interjacent* air between; then obferving the reflection, I found, that wherefoever my finger or fkin toucht the furface, from that part there was no reflection, but in the little furrows or creafes of my fkin, where there remain'd little fmall lines of air, from them was return'd a very vivid reflection as before. I try'd further, by making the furface of very pure Quickfilver to be contiguous to the under furface of this *pellucid* body, and then the reflection from that was fo exceedingly more vivid than from the air, as the reflection from air was than the reflection from the Water; from all which trials I plainly faw, that the ftrong reflecting air was the caufe of this *Phænomenon*.

And this agrees very well with the *Hypothefis* of light and *Pellucid* bodies which I have mention'd in the defcription of *Mufcovy-glafs*; for we there fuppofe Glafs to be a *medium*, which does lefs refift the pulfe of light, and confequently, that moft of the Rays incident on it enter into it, and are refracted towards the *perpendicular*; whereas the air I fuppofe to be a body that does more refift it, and confequently more are *re-percufs'd* then do enter it: the fame kind of trials have I made, with *Cryftalline Glafs*, with drops of fluid bodies, and feveral other ways, which do all feem to agree very exactly with this *Theory*. So that from this Principle well eftablifh'd, we may deduce feverall Corollaries not unworthy obfervation.

And the firft is, that it plainly appears by this, that the production of the Rainbow is as much to be afcribed to the reflection of the concave furface of the air, as to the refraction of the *Globular* drops: this will be evidently manifeft by thefe Experiments, if you *foliate* that part of a Glafs-ball that is to reflect an *Iris*, as in the *Cartefian* Experiment, above mention'd, the reflections will be abundantly more ftrong, and the colours more vivid: and if that part of the furface be touch'd with Water, fcarce affords any fenfible colour at all.

Next we learn, that the great reafon why *pellucid* bodies beaten fmall are white, is from the multitude of reflections, not from the particles of the body, but from the *contiguous* furface of the air. And this is evidently manifefted, by filling the *Interftitia* of thofe powder'd bodies with Water, whereby their whitenefs prefently difappears. From the fame reafon proceeds the whitenefs of many kinds of Sands, which in the *Microfcope* appear to be made up of a multitude of little *pellucid* bodies, whofe brighteft reflections may by the *Microfcope* be plainly perceiv'd to come from their internal furfaces; and much of the whitenefs of it may be deftroy'd by the affufion of fair Water to be contiguous to thofe furfaces.

The whitenefs alfo of froth, is for the moft part to be afcribed to the

reflection

reflection of the light from the surface of the air within the Bubbles, and very little to the reflection from the surface of the Water it self: for this last reflection does not return a quarter so many Rays, as that which is made from the surface of the air, as I have certainly found by a multitude of Observations and Experiments.

The whiteness of *Linnen, Paper, Silk,* &c. proceeds much from the same reason, as the *Microscope* will easily discover; for the Paper is made up of an abundance of *pellucid* bodies, which afford a very plentifull reflection from within, that is, from the concave surface of the air contiguous to its component particles; wherefore by the affusion of Water, Oyl, Tallow, Turpentine, *&c.* all those reflections are made more faint, and the beams of light are suffer'd to traject & run through the Paper more freely.

Hence further we may learn the reason of the whiteness of many bodies, and by what means they may be in part made *pellucid*: As white Marble for instance, for this body is composed of a *pellucid* body exceedingly flaw'd, that is, there are abundance of thin, and very fine cracks or chinks amongst the multitude of particles of the body, that contain in them small parcels of air, which do so *re-percuss* and drive back the penetrating beams, that they cannot enter very deep within that body, which the *Microscope* does plainly inform us to be made up of a *Congeries* of *pellucid* particles. And I further found it somewhat more evidently by some attempts I made towards the making transparent Marble, for by heating the Stone a little, and soaking it in Oyl, Turpentine, Oyl of Turpentine, *&c*, I found that I was able to see much deeper into the body of Marble then before; and one trial, which was not with an unctuous substance, succeeded better than the rest, of which, when I have a better opportunity, I shall make further trial.

This also gives us a probable reason of the so much admired *Phænomena* of the *Oculus Mundi,* an *Oval* stone, which commonly looks like white Alabaster, but being laid a certain time in Water, it grows *pellucid,* and transparent, and being suffer'd to lie again dry, it by degrees loses that transparency, and becomes white as before. For the Stone being of a hollow spongie nature, has in the first and last of these appearances, all those pores fill'd with the obtunding and reflecting air; whereas in the second, all those pores are fill'd with a *medium* that has much the same refraction with the particles of the Stone, and therefore those two being *contiguous,* make, as 'twere, one *continued medium,* of which more is said in the 15. *Observation.*

There are a multitude of other *Phænomena,* that are produc'd from this same Principle, which as it has not been taken notice of by any yet that I know, so I think, upon more diligent observation, will it not be found the least considerable. But I have here onely time to hint *Hypotheses,* and not to prosecute them so fully as I could wish; many of them having a vast extent in the production of a multitude of *Phænomena,* which have been by others, either not attempted to be explain'd, or else attributed to some other cause than what I have assign'd, and perhaps than the right; and therefore I shall leave this to the prosecution of such as have more leisure:

onely

onely before I leave it, I muft not pretermit to hint, that by this Principle, multitudes of the *Phænomena* of the air,as about *Mifts, Clouds, Meteors, Haloes,*&c. are moft plainly and (perhaps) truly explicable; multitudes alfo of the *Phænomena* in colour'd bodies, as liquors, *&c.* are deducible from it.

And from this I fhall proceed to a fecond confiderable *Phænomenon*
which thefe Diamants exhibit, and that is the regularity of their *Figure*,
which is a propriety not lefs general than the former ; It comprifing within its extent, all kinds of *Metals*, all kinds of *Minerals*,moft *Precious ftones*,
all kinds of *Salts*,multitudes of *Earths*,and almoft all kinds of *fluid bodies*.
And this is another propiety, which, though a little fuperficially taken
notice of by fome, has not, that I know, been fo much as attempted to
be explicated by any.

This propriety of bodies,as I think it the moft worthy, and next in order to be confider'd after the contemplation of the *Globular Figure*, fo
have I long had a defire as wel as a determination to have profecuted it if I
had had an opportunity,having long fince propos'd to my felf the method
of my enquiry therein, it containing all the allurements that I think any
enquiry is capable of: For,firft I take it to proceed from the moft fimple
principle that any kind of form can come from, next the *Globular*, which
was therefore the firft I fet upon, and what I have therein perform'd, I
leave the Judicious Reader to determine. For as that form proceeded
from a propiety of fluid bodies, which I have call'd *Congruity*, or *Incongruity*; fo I think, had I time and opportunity, I could make probable,
that all thefe regular Figures that are fo confpicuoufly *various* and *curious*, and do fo adorn and beautifie fuch multitudes of bodies, as I have
above hinted,arife onely from three or four feveral pofitions or poftures
of *Globular* particles,and thofe the moft plain,obvious, and neceffary con-
junctions of fuch figur'd particles that are poffible, fo that fuppofing fuch
and fuch plain and obvious caufes concurring the *coagulating particles*
muft neceffarily compofe a body of fuch a determinate regular Figure,
and no other ; and this with as much neceffity and obvioufnefs as a fluid
body encompaft with a *Heterogeneous* fluid muft be protruded into a
Spherule or *Globe*. And this I have *ad oculum* demonftated with a company of bullets,and fome few other very fimple bodies ; fo that there was
not any regular Figure,which I have hitherto met withall, of any of thofe
bodies that I have above named, that I could not with the compofition of
bullets or globules, and one or two other bodies, imitate, even almoft
by fhaking them together. And thus for inftance may we find that the
Globular bullets will of themfelves,if put on an inclining plain,fo that they
may run together, naturally run into a *triangular* order, compofing all
the variety of figures that can be imagin'd to be made out of *æquilateral
triangles* ; and fuch will you find,upon trial,all the furfaces of *Alum* to be
compos'd of: For three bullets lying on a plain, as clofe to one another as
they can compofe an *æquilatero-triangular* form, as in A in the 7.*Scheme.*
If a fourth be joyn'd to them on either fide as clofely as it can, they four
compofe the moft regular Rhombus confifting of two *æquilateral triangles*,
as

as B. If a fifth be joyn'd to them on either fide in as clofe a pofition as it can, which is the propriety of the *Texture*, it makes a *Trapezium*, or four-fided Figure, two of whofe angles are 120. and two 60. degrees, as C. If a fixth be added, as before, either it makes an *æquilateral triangle*, as D, or a Rhomboeid, as E, or an *Hex-angular Figure*, as F, which is com-pos'd of two *primary Rhombes*. If a feventh be added, it makes either an *æquilatero-hexagonal* Figure, as G, or fome kind of fix-fided *Figure*, as H, or I. And though there be never fo many placed together, they may be rang'd into fome of thefe lately mentioned Figures, all the angles of which will be either 60. degrees, or 120. as the figure K. which is an *æquiangular hexagonal* Figure is compounded of 12. *Globules*, or may be of 25, or 27, or 36, or 42, *&c.* and by thefe kinds of texture, or pofition of globular bodies, may you find out all the variety of regular fhapes, into which the fmooth furfaces of *Alum* are form'd, as upon ex-amination any one may eafily find ; nor does it hold only in fuperficies, but in folidity alfo, for it's obvious that a fourth *Globule* laid upon the third in this texture, compofes a regular *Tetrahedron*, which is a very ufual Figure of the *Cryftals* of *Alum*. And (to haften) there is no one Figure into which *Alum* is obferv'd to be cryftallized, but may by this texture of *Globules* be imitated, and by no other.

I could inftance alfo in the Figure of *Sea-falt*, and *Sal-gem*, that it is com-pos'd of a texture of *Globules*, placed in a *cubical* form, as L, and that all the Figures ofthofe Salts may be imitated by this texture of *Globules*, and by no other whatfoever. And that the forms of *Vitriol* and of *Salt-Peter*, as alfo of *Cryftal*, *Hore-froft*, &c. are compounded of thefe two textures, but modulated by certain proprieties: But I have not here time to in-fift upon, as I have not neither to fhew by what means *Globules* come to be thus context, and what thofe *Globules* are, and many other particulars requifite to a full and intelligible explication of this propriety of bodies. Nor have I hitherto found indeed an opportunity of profecuting the in-quiry fo farr as I defign'd ; nor do I know when I may, it requiring abun-dance of time, and a great deal of affiftance to go through with what I defign'd ; the model of which was this :

Firft, to get as exact and full a collection as I could, of all the differing kinds of Geometrical figur'd bodies, fome three or four feveral bodies of each kind.

Secondly, with them to get as exact a Hiftory as poffibly I could learn of their places of Generation or finding, and to enquire after as many circumftances that tended to the Illuftrating of this Enquiry, as poffibly I could obferve.

Thirdly, to make as many trials as upon experience I could find re-quifite, in Diffolutions and Coagulations of feveral cryftallizing Salts ; for the needfull inftruction and information in this Enquiry.

Fourthly, to make feveral trials on divers other bodies, as Metals, Minerals, and Stones, by diffolving them in feveral *Menftruums*, and cryftalizing them, to fee what Figures would arife from thofe feveral *Compofitums*.

Fifthly,

Fitfthly, to make Compofitions and Coagulations of feveral Salts together into the fame mafs, to obferve of what Figure the product of them would be; and in all, to note as many circumftances as I fhould judge conducive to my Enquiry.

Sixthly, to enquire the clofenefs or rarity of the texture of thefe bodies, by examining their gravity, and their refraction, &c.

Seventhly, to enquire particularly what operations the fire has upon feveral kinds of Salts, what changes it caufes in their Figures, Textures, or Energies.

Eighthly, to examine their manner of diffolution, or acting upon thofe bodies diffoluble in them; The texture of thofe bodies before and after the procefs. And this for the Hiftory.

Next for the Solution, To have examin'd by what, and how many means, fuch and fuch Figures, actions and effects could be produc'd poffibly.

And laftly, from all circumftances well weigh'd, I fhould have endeavoured to have fhewn which of them was moft likely, and (if the informations by thefe Enquiries would have born it) to have demonftrated which of them it muft be, and was.

But to proceed, As I believe it next to the Globular the moft fimple; fo do I, in the fecond place, judge it not lefs pleafant; for that which makes an Enquiry pleafant, are, firft a noble *Inventum* that promifes to crown the fuccefsfull endeavour; and fuch muft certainly the knowledge of the efficient and concurrent caufes of all thefe curious Geometrical Figures be, which has made the Philofophers hitherto to conclude nature in thefe things to play the Geometrician, according to that faying of *Plato*, Ὁ Θεὸς γεωμετρεῖ. Or next, a great variety of matter in the Enquiry; and here we meet with nothing lefs than the *Mathematicks* of nature, having every day a new Figure to contemplate, or a variation of the fame in another body,

Which do afford us a third thing, which will yet more fweeten the Enquiry, and that is, a multitude of information; we are not fo much to grope in the dark, as in moft other Enquiries, where the *Inventum* is great; for having fuch a multitude of inftances to compare, and fuch eafie ways of generating, or compounding and of deftroying the form, as in the *Solution* and *Cryftallization* of Salts, we cannot but learn plentifull information to proceed by. And this will further appear from the univerfality of the Principle which Nature has made ufe of almoft in all inanimate bodies. And therefore, as the contemplation of them all conduces to the knowledg of any one; fo from a Scientifical knowledge of any one does follow the fame of all, and every one.

And fourthly, for the ufefulnefs of this knowledge, when acquir'd, certainly none can doubt, that confiders that it caries us a ftep forward into the Labirinth of Nature, in the right way towards the end we propofe our felves in all Philofophical Enquiries. So that knowing what is the form of Inanimate or Mineral bodies, we fhall be the better able to proceed in our next Enquiry after the forms of Vegetative

tive bodies; and laſt of all, of Animate ones, that ſeeming to be the higheſt ſtep of natural knowledge that the mind of man is capable of.

Obſerv. XIV. *Of ſeveral kindes of* frozen *Figures.*

I Have very often in a Morning, when there has been a great *hoar-froſt,* with an indifferently magnifying *Microſcope,* obſerv'd the ſmall *Stiriæ,* or Cryſtalline beard, which then uſually covers the face of moſt bodies that lie open to the cold air, and found them to be generally *Hexangular priſmatical* bodies, much like the long Cryſtals of *Salt-peter,* ſave onely that the ends of them were differing: for whereas thoſe of *Nitre* are for the moſt part *pyramidal,* being terminated either in a point or edge; theſe of Froſt were hollow, and the cavity in ſome ſeem'd pretty deep, and this cavity was the more plainly to be ſeen, becauſe uſually one or other of the ſix *parallelogram* ſides was wanting, or at leaſt much ſhorter then the reſt.

But this was onely the Figure of the *Bearded hoar-froſt;* and as for the particles of other kinds of *hoar-froſts,* they ſeem'd for the moſt part irregular, or of no certain Figure. Nay, the parts of thoſe curious branchings, or *vortices,* that uſually in cold weather tarniſh the ſurface of Glaſs, appear through the *Microſcope* very rude and unſhapen, as do moſt other kinds of frozen *Figures,* which to the naked eye ſeem exceeding neat and curious, ſuch as the Figures of *Snow,* frozen *Urine, Hail,* ſeveral *Figures* frozen in common Water,*&c.* Some Obſervations of each of which I ſhall hereunto annex, becauſe if well conſider'd and exami'nd, they may, perhaps, prove very inſtructive for the finding out of what I have endeavoured in the preceding Obſervation to ſhew, to be (next the *Globular Figure* which is caus'd by *congruity,* as I hope I have made probable in the ſixth *Obſervation*) the moſt ſimple and plain operation of Nature, of which, notwithſtanding we are yet ignorant.

I.

Several Obſervables in the ſix-branched *Figures form'd on the ſurface of* Urine *by freezing.*

Schem. 8.
Fig. 1. 1 The Figures were all frozen almoſt even with the ſurface of the *Urine* in the Veſſel, but the bigger ſtems were a little *prominent* above that ſurface, and the parts of thoſe ſtems which were neareſt the center (*a*) were biggeſt above the ſurface.

2 I have obſerv'd ſeveral kinds of theſe Figures, ſome ſmaller, no bigger then a Two-pence, others ſo bigg, that I have by meaſure found one of its ſtems or branches above four foot long; and of theſe, ſome were pretty round, having all their branches pretty neer alike; other of them were more extended towards one ſide, as uſually thoſe very large ones
were

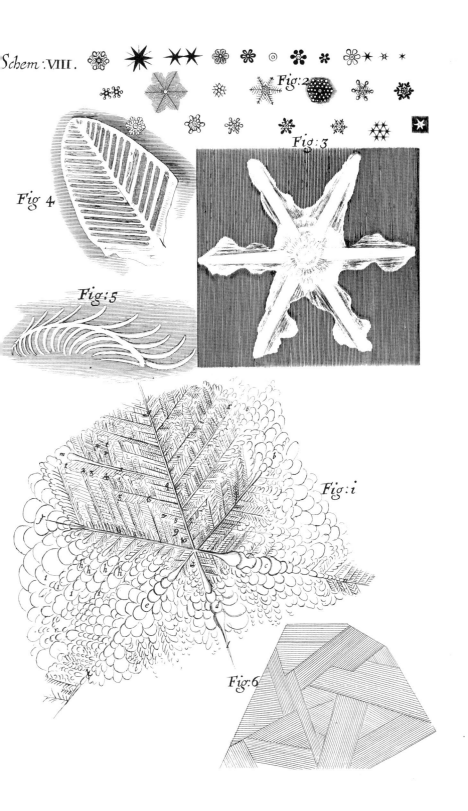

Schem:VIII.

Fig:2

Fig:3

Fig 4

Fig:5

Fig:1

Fig:6

were, which I have obferv'd in Ditches which have been full of foul water.

3 None of all thefe Figures I have yet taken notice of, had any regular pofition in refpect of one another, or of the fides of the Veffel ; nor did I find any of them equally to exactnefs extended every way from the center *a*.

4 Where ever there was a center, the branchings from it, *a b, a c, a d, a e, a f, a g,* were never fewer, or more then fix, which ufually concurr'd, or met one another very neer in the fame point or center, *a* ; though oftentimes not exactly ; and were enclin'd to each other by an angle, of very neer fixty degrees, I fay, very neer, becaufe, though having endeavoured to meafure them the moft acurately I was able, with the largeft Compaffes I had, I could not find any fenfible variation from that meafure, yet the whole fix-branched Figure feeming to compofe a folid angle, they muft neceffarily be fomewhat lefs.

5 The middle lines or ftems of thefe branches, *a b, a c, a d, a e, a f, a g,* feem'd fomewhat whiter, and a little higher than any of the *intermediate* branchings of thefe Figures ; and the center *a,* was the moft *prominent* part of the whole Figure, feeming the *apex* of a folid angle or *pyramid,* each of the fix plains being a little enclin'd below the furface of the *Vrin.*

6 The lateral branchings iffuing out of the great ones, fuch as *o p, m q,* &c. were each of them inclin'd to the great ones, by the fame angle of about fixty degrees, as the great ones were one to another, and always the bigger branchings were *prominent* above the lefs, and the lefs above the leaft, by proportionate *gradations.*

7 The *lateral* branches fhooting out of the great ones, went all of them from the center, and each of them was parallel to that great branch, next to which it lay ; fo that as all the branches on one fide were parallel to one another, fo were they all of them to the *approximate* great branch, as *p o, q r,* as they were parallel to each other, and fhot from the center, fo were they parallel alfo to the great branch *a b.*

8 Some of the ftems of the fix branches proceeded ftraight, and of a thicknefs that gradually grew fharper towards the end, as *a g.*

9 Others of the ftems of thofe branches grew bigger and knotty towards the middle, and the branches alfo as well as ftems, from Cylinders grew into Plates, in a moft admirable and curious order, fo exceeding regular and delicate, as nothing could be more, as is vifible in *a b, a c, a d, a e, a f,* but towards the end of fome of thefe ftems, they began again to grow fmaller and to recover their former branchings, as about *k* and *n.*

10 Many of the *lateral* branches had *collateral* branches (if I may fo call them) as *q m* had many fuch as *f t,* and moft of thofe again *fubcollateral,* as *v w,* and thefe again had others lefs, which one may call *laterofubcollateral,* and thefe again others, and they others, *&c.* in greater Figures.

11 The branchings of the main Stems joyn'd not together by any regular line, nor did one fide of the one lie over the other fide of the other, but the fmall *collateral* and *fubcollateral* branches did lie at top of one

another

another according to a certain order or method, which I always obſerv'd to be this.

12 That ſide of a *collateral* or *ſubcollateral*, &c. branch, lay over the ſide of the *approximate* (as the feathers in the wing of a Bird) whoſe branchings proceeded parallel to the laſt biggeſt ſtem from which it ſprung, and not to the biggeſt ſtem of all, unleſs that were a ſecond ſtem backwards.

13 This rule that held in the branchings of the *Sexangular Figure* held alſo in the branchings of any other great or ſmall ſtem, though it did not proceed from a center.

14 The exactneſs and curioſity of the figuration of theſe branches, was in every particular ſo tranſcendent, that I judge it almoſt impoſſible for humane art to imitate.

15 Taſting ſeveral cleer pieces of this *Ice*, I could not find any *Urinous* taſte in them, but thoſe few I taſted, ſeem'd as *inſipid* as water.

16 A figuration ſomewhat like this, though indeed in ſome particulars much more curious, I have ſeveral times obſerv'd in *regulus martis ſtellatus*, but with this difference, that all the ſtems and branchings are bended in a moſt excellent and regular order, whereas in *Ice* the ſtems and branchings are ſtreight, but in all other particulars it agrees with this, and ſeems indeed nothing but one of theſe ſtars, or branched Figures frozen on *Urine*, diſtorted, or wreathed a little, with a certain proportion : *Lead* alſo that has *Arſenick*, and ſome other things mixt with it, I have found to have its ſurface, when ſuffer'd to cool, figured ſomewhat like the branchings of *Urine*, but much ſmaller.

17 But there is a *Vegetable* which does exceedingly imitate theſe branches, and that is, *Fearn*, where the main ſtem may be obſerv'd to ſhoot out branches, and the ſtems of each of theſe *lateral* branches, to ſend forth *collateral*, and thoſe *ſubcollateral*, and thoſe *latero ſubcollateral*, &c. and all thoſe much after the ſame order with the branchings, diviſions, and ſubdiviſions in the branchings of theſe Figures in frozen *Urine* ; ſo that if the Figures of both be well conſider'd, one would gheſs that there were not much greater need of a *ſeminal principle* for the production of *Fearn*, then for the production of the branches of *Urine*, or the *Stella martis*, there ſeeming to be as much form and beauty in the one as in the other.

And indeed, this Plant of *Fearn*, if all particulars be well conſider'd, will ſeem of as ſimple, and uncompounded a form as any *Vegetable*, next to *Mould* or *Muſhromes*, and would next after the invention of the forms of thoſe, deſerve to be enquir'd into ; for notwithſtanding ſeveral have affirm'd it to have ſeed, and to be propagated thereby ; yet, though I have made very diligent enquiry after that particular, I cannot find that there is any part of it that can be imagin'd to be more ſeminal then another : But this onely here by the by :

For the freezing Figures in *Urine*, I found it requiſite,

Firſt, that the Superficies be not diſturbed with any wind, or other commotion of the air, or the like.

Secondly,

Secondly, that it be not too long expofed, fo as that the whole bulk be frozen, for oftentimes, in fuch cafes, by reafon of the fwelling the of *Ice*, or from fome other caufe, the curious branched Figures difappear.

Thirdly, an artificial freezing with *Snow* and *Salt*, apply'd to the out-fide of the containing Veffel, fucceeds not well, unlefs there be a very little quantity in the Veffel.

Fourthly, If you take any cleer and fmooth Glafs, and wetting all the infide of it with *Urine*, you expofe it to a very fharp freezing, you will find it cover'd with a very regular and curious Figure.

II.

Obfervables in figur'd Snow.

Expofing a piece of black Cloth, or a black Hatt to the falling *Snow*, *Schem. 8.* I have often with great pleafure, obferv'd fuch an infinite variety of cu- *Fig. 2.* rioufly figur'd *Snow*, that it would be as impoffible to draw the Figure and fhape of every one of them, as to imitate exactly the curious and Geometrical *Mechanifme* of Nature in any one. Some coorfe draughts, fuch as the coldnefs of the weather, and the ill provifions, I had by me for fuch a purpofe, would permit me to make, I have here added in the Second *Figure* of the Eighth *Scheme*.

In all which I obferv'd, that if they were of any regular Figures, they were always branched out with fix principal branches, all of equal length, fhape and make, from the center, being each of them inclin'd to either of the next branches on either fide of it, by an angle of fixty degrees.

Now, as all thefe ftems were for the moft part in one flake exactly of the fame make, fo were they in differing Figures of very differing ones; fo that in a very little time I have obferv'd above an hundred feveral cizes and fhapes of thefe ftarry flakes.

The branches alfo out of each ftem of any one of thefe flakes, were ex-actly alike in the fame flake; fo that of whatever Figure one of the branches were, the other five were fure to be of the fame, very exactly, that is, if the branchings of the one were fmall *Perallelipipeds* or Plates, the branchings of the other five were of the fame; and generally, the branchings were very conformable to the rules and method obferv'd be-fore, in the Figures on *Urine*, that is, the branchings from each fide of the ftems were parallel to the next ftem on that fide, and if the ftems were plated, the branches alfo were the fame; if the ftems were very long, the branches alfo were fo, &c.

Obferving fome of thefe figur'd flakes with a *Microfcope*, I found them not to appear fo curious and exactly figur'd as one would have imagin'd, but like Artificial Figures, the bigger they were magnify'd, the more ir-regularites appear'd in them; but this irregularity feem'd afcribable to the thawing and breaking of the flake by the fall, and not at all to the defect of the *plaftick* virtue of Nature, whofe curiofity in the formation of moft of thefe kind of regular Figures, fuch as thofe of *Salt, Minerals,* &c.

appears

appears by the help of the *Microscope*, to be very many degrees smaller then the most acute eye is able to perceive without it. And though one of these six-branched Stars appear'd here below much of the shape described in the Third *Figure* of the Eighth *Scheme*; yet I am very apt to think, that could we have a sight of one of them through a *Microscope* as they are generated in the Clouds before their Figures are vitiated by external accidents, they would exhibit abundance of curiosity and neatness there also, though never so much magnify'd: For since I have observ'd the Figures of *Salts* and *Minerals* to be some of them so exceeding small, that I have scarcely been able to perceive them with the *Microscope*, and yet have they been regular, and since (as far as I have yet examin'd it) there seems to be but one and the same cause that produces both these effects, I think it not irrational to suppose that these pretty figur'd Stars of *Snow*, when at first generated might be also very regular and exact.

III.

Several kinds of Figures in Water frozen.

Putting fair Water into a large capacious Vessel of *Glass*, and exposing it to the cold, I observ'd after a little time, several broad, flat, and thin *laminæ*, or plates of *Ice*, crossing the bulk of the water and one another very irregularly, onely most of them seem'd to turn one of their edges towards that side of the Glass which was next it, and seem'd to grow, as 'twere from the inside of the Vessel inwards towards the middle, almost like so many blades of *Fern*. Having taken several of these plates out of water on the blade of a Knife, I observ'd them figur'd much after the manner of *Herring bones*, or *Fern blades*, that is, there was one bigger stem in the middle like the back-bone, and out of it, on either side, were a multitude of small *stiriæ*, or *icicles*, like the smaller bones, or the smaller branches in *Fern*, each of these branches on the one side, were parallel to all the rest on the same side, and all of them seem'd to make an angle with the stem, towards the top, of sixty degrees, and towards the bottom or root of this stem, of 120. See the fourth *Figure* of the 8. *Plate*.

I observ'd likewise several very pretty varieties of Figures in Water, frozen on the top of a broad flat Marble-stone, expos'd to the cold with a little Water on it, some like feathers, some of other shapes, many of them were very much of the shape exprest in the fifth Figure of the 8. *Scheme*, which is extremely differing from any of the other Figures.

I observ'd likewise, that the shootings of *Ice* on the top of Water, beginning to freez, were in streight *prismatical* bodies much like those of *roch-peter*, that they crost each other usually without any kind of order or rule, that they were always a little higher then the surface of the Water that lay between them; that by degrees those *interjacent* spaces would be fill'd with *Ice* also, which usually would be as high as the surface of the rest.

In flakes of *Ice* that had been frozen on the top of Water to any considerable

Figur : 1

Figur : 2.

Fig : 3.

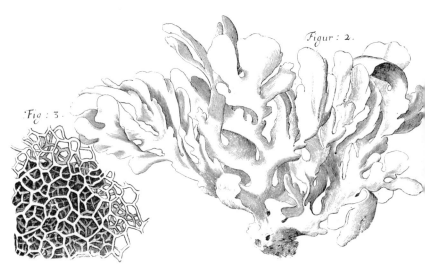

fiderable thicknefs, I obferv'd that both the upper and the under fides of it were curioufly quill'd, furrow'd, or grain'd, as it were, which when the Sun fhone on the Plate, was exceeding eafily to be perceiv'd to be much after the fhape of the lines in the 6. *Figure* of the 8. *Scheme*, that is, they confifted of feveral ftreight ends of parallel Plates, which were of divers lengths and angles to one another without any certain order.

The caufe of all which regular Figures (and of hundreds of others, namely of *Salts, Minerals, Metals*, &c. which I could have here inferted, would it not have been too long) feems to be deducible from the fame Principles, which I have (in the 13. *Obfervation*) hinted only, having not yet had time to compleat a *Theory* of them. But indeed (which I there alfo hinted) I judge it the fecond ftep by which the *Pyramid* of natural knowledge (which is the knowledge of the form of bodies) is to be afcended : And whofoever will climb it, muft be well furnifh'd with that which the Noble *Verulam* calls *Scalam Intellectus* ; he muft have fcaling Ladders, otherwife the fteps are fo large and high, there will be no getting up them, and confequently little hopes of attaining any higher ftation, fuch as to the knowledge of the moft fimple principle of Vegetation manifefted in Mould and Mufhromes, which, as I elfewhere endeavoured to fhew, feems to be the third ftep; for it feems to me, that the Intellect of man is like his body, deftitute of wings, and cannot move from a lower to a higher and more fublime ftation of knowledg, otherwife then ftep by ftep, nay even there where the way is prepar'd and already made paffible ; as in the *Elements of Geometry*, or the like, where it is fain to climb a whole *feries* of Propofitions by degrees, before it attains the knowledge of one *Probleme*. But if the afcent be high, difficult and above its reach, it muft have recourfe to a *novum organum*, fome new engine and contrivance, fome new kind of *Algebra*, or *Analytick Art* before it can furmount it.

Obferv. X V. *Of* Kettering-ftone, *and of the pores of* Inanimate *bodies*.

THis Stone which is brought from *Kettering* in *Northampton-fhire*, and digg'd out of a Quarry, as I am inform'd, has a grain altogether admirable, nor have I ever feen or heard of any other ftone that has the like. It is made up of an innumerable company of fmall bodies, not all of the fame cize or fhape, but for the moft part, not much differing from a Globular form, nor exceed they one another in Diameter above three or four times ; they appear to the eye, like the Cobb or Ovary of a *Herring*, or fome fmaller fifhes, but for the moft part, the particles feem fomewhat lefs, and not fo uniform; but their variation from a perfect globular ball, feems to be only by the preffure of the *contiguous* bals which have a little depreft and protruded thofe toucht fides inward, and forc'd

the

Schem. 9. Fig. 1.

the other fides as much outwards beyond the limits of a Globe; juft as it would happen, if a heap of exactly round Balls of foft Clay were heap'd upon one another; or, as I have often feen a heap of fmall Globules of *Quickfilver*, reduc'd to that form by rubbing it much in a glaz'd Veffel, with fome flimy or fluggifh liquor, fuch as Spittle, when though the top of the upper Globules be very neer fpherical, yet thofe that are preft upon others, exactly imitate the forms of thefe lately mention'd grains.

Where thefe grains touch each other, they are fo firmly united or fettled together, that they feldom part without breaking a hole in one or th'other of them, fuch as *a, a, a, b, c, c,* &c. Some of which fractions, as *a, a, a, a,* where the touch has been but light, break no more then the outward cruft, or firft fhell of the ftone, which is of a white colour, a little dafh'd with a brownifh Yellow, and is very thin, like the fhell of an Egg : and I have feen fome of thofe grains perfectly refemble fome kind of Eggs, both in colour and fhape : But where the union of the *contiguous granules* has been more firm, there the divulfion has made a greater Chafm, as at *b, b, b,* in fo much that I have obferv'd fome of them quite broken in two, as at *c, c, c,* which has difcovered to me a further refemblance they have to Eggs, they having an appearance of a white and yelk, by two differing fubftances that envelope and encompafs each other.

That which we may call the white was pretty whitifh neer the yelk, but more dufkie towards the fhell; fome of them I could plainly perceive to be fhot or radiated like a *Pyrites* or *fire-ftone*; the yelk in fome I faw hollow, in others fill'd with a dufkie brown and porous fubftance like a kind of pith.

The fmall pores, or *interftitia e e e e* betwixt the Globules, I plainly faw, and found by other trials to be every way pervious to air and water, for I could blow through a piece of this ftone of a confiderable thicknefs, as eafily as I have blown through a Cane, which minded me of the pores which *Des Cartes* allow his *materia fubtilis* between the *æthereal* globules.

The object, through the *Microfcope*, appears like a *Congeries* or heap of Pibbles, fuch as I have often feen caft up on the fhore, by the working of the Sea after a great ftorm, or like (in fhape, though not colour) a company of fmall Globules of Quickfilver, look'd on with a *Microfcope*, when reduc'd into that form by the way lately mentioned. And perhaps, this laft may give fome hint at the manner of the formation of the former : For fuppofing fome *Lapidefcent* fubftance to be generated, or fome way brought (either by fome commixture of bodies in the Sea it felf, or protruded in, perhaps, out of fome *fubterraneous* caverns) to the bottom of the Sea, and there remaining in the form of a liquor like Quickfilver, *heterogeneous* to the ambient *Saline* fluid, it may by the working and tumblings of the Sea to and fro be jumbled and comminuted into fuch Globules as may afterwards be hardned into Flints, the lying of which one upon another, when in the Sea, being not very hard, by reafon of the weight of the incompaffing fluid, may caufe the undermoft to be a little, though not much, varied from a globular Figure. But this only by the by.　　　　　　　　　　　　　　　　　　　　After

After what manner this *Kettering-stone* should be generated I cannot learn, having never been there to view the place, and observe the circumstances; but it seems to me from the structure of it to be generated from some substance once more fluid, and afterwards by degrees growing harder, almost after the same manner as I supposed the generation of Flints to be made.

But whatever were the cause of its curious texture, we may learn this information from it; that even in those things which we account vile, rude, and coorse, Nature has not been wanting to shew abundance of curiosity and excellent Mechanisme.

We may here find a Stone by help of a *Microscope*, to be made up of abundance of small Balls, which do but just touch each other, and yet there being so many contacts, they make a firm hard mass, or a Stone much harder then Free-stone.

Next, though we can by a *Microscope* discern so curious a shape in the particles, yet to the naked eye there scarce appears any such thing; which may afford us a good argument to think, that even in those bodies also, whose *texture* we are not able to discern, though help'd with *Microscopes*, there may be yet *latent* so curious a *Schematisme*, that it may abundantly satisfie the curious searcher, who shall be so happy as to find some way to discover it.

Next, we here find a Stone, though to the naked eye a very close one, yet every way perforated with innumerable pores, which are nothing else but the *interstitia*, between those multitudes of minute globular particles, that compose the bulk it self; and these pores are not only discover'd by the *Microscope*, but by this contrivance.

I took a pretty large piece of this stone, and covering it all over with cement, save only at two opposite parts, I found my self able, by blowing in at one end that was left open, to blow my spittle, with which I had wet the other end, into abundance of bubbles, which argued these pores to be open and pervious through the whole stone, which affords us a very pretty instance of the porousness of some seemingly close bodies, of which kind I shall anon have occasion to subjoyn many more, tending to prove the same thing.

I must not here omit to take notice, that in this body there is not a *vegetative* faculty that should so contrive this structure for any peculiar use of *Vegetation* or growth, whereas in the other instances of vegetable porous bodies, there is an *anima*, or *forma informans*, that does contrive all the Structures and *Mechanismes* of the constituting body, to make them subservient and usefull to the great Work or Function they are to perform. And so I ghess the pores in Wood, and other vegetables, in bones, and other Animal substances, to be as so many channels, provided by the Great and Alwise Creator, for the conveyance of appropriated juyces to particular parts. And therefore, that this may tend, or be pervious all towards one part, and may have impediments, as valves or the like, to any other; but in this body we have very little reason to suspect there should be any such design, for it is equally pervious every way, not onely forward,

ward, but backwards, and fide-ways, and feems indeed much rather to be *Homogeneous* or fimilar to thofe pores, which we may with great probability believe to be the channels of *pellucid* bodies, not directed, or more open any one way, then any other, being equally pervious every way. And, according as thefe pores are more or greater in refpect of the *interftitial* bodies, the more tranfparent are the fo conftituted concretes; and the fmaller thofe pores are, the weaker is the *Impulfe* of light communicated through them, though the more quick be the progrefs.

Upon this Occafion, I hope it will not be altogether unfeafonable, if I propound my conjectures and *Hypothefis* about the *medium* and conveyance of light.

I fuppofe then, that the greateft part of the *Interftitia* of the world, that lies between the bodies of the Sun and Starrs, and the Planets, and the Earth, to be an exceeding fluid body, very apt and ready to be mov'd, and to communicate the motion of any one part to any other part, though never fo far diftant: Nor do I much concern my felf, to determine what the Figure of the particles of this exceedingly fubtile fluid *medium* muft be; nor whether it have any interftitiated pores or vacuities, it being fufficient to folve all the *Phænomena* to fuppofe it an exceedingly fluid, or the moft fluid body in the world, and as yet impoffible to determine the other difficulties.

That being fo exceeding fluid a body, it eafily gives paffage to all other bodies to move to and fro in it.

That it neither receives from any of its parts, or from other bodies; nor communicates to any of its parts, or to any other body, any impulfe, or motion in a direct line, that is not of a determinate quicknefs. And that when the motion is of fuch determinate fwiftnefs, it both receives, and communicates, or propagates an impulfe or motion to any imaginable diftance in ftreight lines, with an unimaginable celerity and vigour.

That all kind of folid bodies confift of pretty maffie particles in refpect of the particles of this fluid *medium*, which in many places do fo touch each other, that none of this fluid *medium* interpofes much after the fame mannner (to ufe a grofs fimilitude) as a heap of great ftones compafs one great *congeries* or mafs in the midft of the water.

That all fluid bodies which we may call *tangible*, are nothing but fome more fubtile parts of thofe particles, that ferve to conftiture all *tangible* bodies.

That the water, and fuch other fluid bodies, are nothing but a *congeries* of particles agitated or made fluid by it in the fame manner as the particles of *Salt* are agitated or made fluid by a parcel of water, in which they are diffolv'd, and fubfiding to the bottom of it, conftitute a fluid body, much more maffie and denfe, and lefs fluid then the pure water it felf.

That the air on the other fide is a certain company of particles of quite another kind, that is, fuch as are very much fmaller, and more eafiely moveable by the motion of this fluid *medium*; much like thofe very fubtile parts of *Cochenel*, and other very deep tinging bodies, where by a very
fmall

{mall parcel of matter is able to tinge and diffuſe it ſelf over a very great quantity of the fluid diſſolvent ; or ſomewhat after that manner, as ſmoak, and ſuch like minute bodies, or ſteams, are obſerv'd to tinge a very great quantity of air ; onely this laſt ſimilitude is deficient in one propriety, and that is a perpetuity or continuance in that ſtate of com-mixture with the air, but the former does more neerly approach to the nature and manner of the air's being diſſolv'd by this fluid or *Æther*. And this Similitude will further hold in theſe proprieties ; that as thoſe tinctures may be increaſed by certain bodies,ſo may they be precipitated by others ; as I ſhall afterwards ſhew it to be very probable, that the like accidents happen even to the Air it ſelf.

Further, as theſe ſolutions and tinctures do alter the nature of theſe fluid bodies,as to their aptneſs to propagate a motion or impulſe through them, even ſo does the particles of the Air, Water, and other fluid bo-dies, and of Glaſs, Cryſtal, *&c.* which are commixt with this bulk of the *Æther*, alter the motion of the propagated pulſe of light ; that is, where theſe more bulkie particles are more plentifull, and conſequently a leſſer quantity of the *Æther* between them to be mov'd,there the motion muſt neceſſarily be the ſwifter,though not ſo robuſt, which will produce thoſe effects, which I have (I hope) with ſome probability, aſcribed to it in the digreſſion about Colours, at the end of the *Obſervations* on *Muſ-covy-glaſs*.

Now, that other Stones, and thoſe which have the cloſeſt and hardeſt textures, and ſeem (as far as we are able to diſcover with our eyes, though help'd with the beſt *Microſcopes*) freeſt from pores, are yet not-withſtanding repleniſh'd with them ; an Inſtance or two will, I ſuppoſe, make more probable.

A very ſolid and unflaw'd piece of cleer white *Marble*, if it be well poliſh'd and glaz'd, has ſo curiouſly ſmooth a ſurface, that the beſt and moſt poliſh'd ſurface of any wrought-glaſs, ſeems not to the naked eye, nor through a *Microſcope*, to be more ſmooth, and leſs porous. And yet, that this hard cloſe body is repleniſh'd with abundance of pores, I think theſe following Experiments will ſufficiently prove.

The firſt is, That if you take ſuch a piece, and for a pretty while boyl it in Turpentine and Oyl of Turpentine, you ſhall find that the ſtone will be all imbu'd with it ; and whereas before it look'd more white,but more opacous, now it will look more greaſie, but be much more tranſparent, and if you let it lie but a little while, and then break off a part of it, you ſhall find the unctuous body to have penetrated it to ſuch a determinate depth every way within the ſurface. This may be yet eaſier try'd with a piece of the ſame *Marble*, a little warm'd in the fire,and then a little Pitch or Tarr melted on the top of it ; for theſe black bodies, by their inſinu-ating themſelves into the inviſible pores of the ſtone, ting it with ſo black a hue, that there can be no further doubt of the truth of this aſſertion, that it abounds with ſmall imperceptible pores.

Now, that other bodies will alſo ſink into the pores of *Marble*, beſides *unctuous*, I have try'd, and found, that a very Blue tincture made in
ſpirit

ſpirit of Urine would very readily and eaſily ſink into it, as would alſo ſeveral tinctures drawn with *ſpirit of Wine*.

Nor is *Marble* the only ſeemingly cloſe ſtone, which by other kinds of Experiments may be found porous; for I have by this kind of Experiment on divers other ſtones found much the ſame effect, and in ſome, indeed much more notable. Other ſtones I have found ſo porous, that with the *Microſcope* I could perceive ſeveral ſmall winding holes, much like Worm-holes, as I have noted in ſome kind of *Purbeck-ſtone*, by looking on the ſurface of a piece newly flaw'd off; for if otherwiſe, the ſurface has been long expos'd to the Air, or has been ſcraped with any tool, thoſe ſmall caverns are fill'd with duſt, and diſappear.

And to confirm this *Conjecture*, yet further, I ſhall here inſert an excellent account, given into the *Royal Society* by that Eminently Learned Phyſician, Doctor *Goddard*, of an Experiment, not leſs inſtructive then curious and accurate, made by himſelf on a very hard and ſeemingly cloſe ſtone call'd *Oculus Mundi*, as I find it preſerv'd in the Records of that Honourable Society.

A ſmall ſtone of the kind, call'd by ſome Authours, *Oculus Mundi*, being dry and cloudy, weigh'd $5\frac{209}{256}$ *Grains*.

The ſame put under water for a night, and ſomewhat more, became tranſparent, and the ſuperficies being wiped dry, weighed $6\frac{3}{256}$ *Grains*.

The difference between theſe two weights, $0\frac{50}{256}$ of a *Grain*.

The ſame Stone kept out of water one Day and becoming cloudy again weighed, $5\frac{225}{256}$ *Graines*.

Which was more then the firſt weight, $0\frac{16}{256}$ of a *Grain*.

The ſame being kept two Days longer weighed, $5\frac{202}{256}$ *Graines*. Which was leſs then at firſt, $0\frac{7}{256}$ of a *Grain*.

Being kept dry ſomething longer it did not grow ſenſibly lighter.

Being put under water for a night and becoming again tranſparent and wiped dry, the weight was, $6\frac{3}{256}$ *Grains*, the ſame with the firſt after putting in water, and more then the laſt weight after keeping of it dry, $0\frac{57}{256}$ of a *Grain*.

Another Stone of the ſame kind being variegated with milky *white* and *gray* like ſome ſorts of *Agates*, while it lay under water, was alwaies invironed with little Bubbles, ſuch as appear in

water

water a little before boyling, next the fides of the Veffel.

There were alfo fome the like Bubbles on the Surface of the water juft over it, as if either fome exhalations came out of it, or that it did excite fome fermentation in the parts of the water contiguous to it.

There was little fenfible difference in the tranfparency of this Stone, before the putting under water, and after : To be fure the milky-*white* parts continued as before, but more difference in weight then in the former. For whereas before the putting into the water the weight was $18\frac{27}{128}$ *Graines*. After it had lyen in about four and twenty hours the weight was $20\frac{27}{128}$ *Graines*, fo the difference was, $1\frac{8}{128}$ *Graines*.

The fame Stone was infufed in the water fcalding hot, and fo continued for a while after it was cold, but got no more weight then upon infufing in the cold, neither was there any fenfible Difference in the weight both times.

In which Experiment, there are three Obfervables, that feem very manifeftly to prove the poroufnefs of thefe feemingly clofe bodies: the firft is their acquiring a tranfparency, and lofing their whitenefs after fteeping in water, which will feem the more ftrongly to argue it, if what I have already faid about the making tranfparent, or clarifying of fome bodies, as the white powder of beaten Glafs, and the froth of fome glutinous tranfparent liquor be well confider'd ; for thereby it will feem rational to think that this tranfparency arifes from the infinuation of the water (which has much the fame refraction with fuch ftony particles, as may be difcoverd by Sand view'd with a *Microfcope*) into thofe pores which were formerly repleat with air (that has a very differing refraction, and confequently is very reflective) which feems to be confirm'd by the fecond Obfervable, namely, the increafe of weight after fteeping, and decreafe upon drying. And thirdly, feem'd yet more fenfibly confirm'd by the multitude of bubbles in the laft Experiment.

We find alfo moft Acid Salts very readily to diffolve and feparate the parts of this body one from another ; which is yet a further Argument to confirm the poroufnefs of bodies, and will ferve as fuch, to fhew that even Glafs alfo has an abundance of pores in it, fince there are feveral liquors, that with long ftaying in a Glafs, will fo *Corrode* and eat into it, as at laft, to make it pervious to the liquor it contain'd, of which I have feen very many Inftances.

Since therefore we find by other proofs, that many of thofe bodies
which

which we think the moſt ſolid ones, and appear ſo to our ſight, have not-withſtanding abundance of thoſe groſſer kind of pores, which will ad-mit ſeveral kinds of liquors into them, why ſhould we not believe that Glaſs, and all other tranſparent bodies abound with them, ſince we have many other arguments, beſides the propagation of light, which ſeem to argue for it ?

And whereas it may be objected, that the propagation of light is no argument that there are thoſe atomical pores in glaſs, ſince there are _Hy-potheſes_ plauſible enough to ſolve thoſe _Phænomena_, by ſuppoſing the pulſe onely to be communicated through the tranſparent body.

To this I anſwer, that that _Hypotheſis_ which the induſtrious _Moreanus_ has publiſh'd about the ſlower motion of the end of a Ray in a denſer _medium_, then in a more rare and thin, ſeems altogether unſufficient to ſolve abundance of _Phænomena_, of which this is not the leaſt conſiderable, that it is impoſſible from that ſuppoſition, that any colours ſhould be gene-rated from the refraction of the Rays ; for ſince by that _Hypotheſis_ the _undulating pulſe_ is always carried perpendicular, or at right angles with the Ray or Line of direction, it follows, that the ſtroke of the pulſe of light, after it has been once or twice refracted (through a Priſme, for ex-ample) muſt affect the eye with the ſame kind of ſtroke as if it had not been refracted at all. Nor will it be enough for a Defendant of that _Hy-potheſis_, to ſay, that perhaps it is becauſe the refractions have made the Rays more weak, for if ſo, then two refractions in the two parallel ſides of a _Quadrangular Priſme_ would produce colours, but we have no ſuch _Phænomena_ produc'd.

There are ſeveral Arguments that I could bring to evince that there are in all tranſparent bodies ſuch atomical pores. And that there is ſuch a fluid body as I am arguing for, which is the _medium_, or Inſtrument, by which the pulſe of Light is convey'd from the _lucid body_ to the en-lightn'd. But that it being a digreſſion from the Obſervations I was re-cording, about the Pores of _Kettering Stone_, it would be too much ſuch, if I ſhould protract it too long ; and therefore I ſhall proceed to the next _Obſervation_.

Obſerv. XVI. _Of_ Charcoal, _or burnt_ Vegetables.

CHarcoal, or a Vegetable burnt black, affords an object no leſs pleaſant than inſtructive ; for if you take a ſmall round Charcoal, and break it ſhort with your fingers, you may perceive it to break with a very ſmooth and ſleek ſurface, almoſt like the ſurface of black ſealing Wax ; this ſurface, if it be look'd on with an ordinary _Microſcope_, does manifeſt abundance of thoſe pores which are alſo viſible to the eye in many kinds of _Wood_, rang'd round the pith, both a in kind of circular order, and a radiant one. Of theſe there are a multitude in the ſubſtance of the Coal, every where almoſt perforating and drilling it from end to end ; by

means

means of which, be the Coal never fo long, you may eafily blow through it; and this you may prefently find, by wetting one end of it with Spittle, and blowing at the other.

But this is not all, for befides thofe many great and confpicuous irregular fpots or pores, if a better *Microfcope* be made ufe of, there will appear an infinite company of exceedingly fmall, and very regular pores, fo thick and fo orderly fet, and fo clofe to one another, that they leave very little room or fpace between them to be fill'd with a folid body, for the apparent *interftitia*, or feparating fides of thefe pores feem fo thin in fome places, that the texture of a Honey-comb cannot be more porous. Though this be not every where fo, the intercurrent partitions in fome places being very much thicker in proportion to the holes.

Moft of thefe fmall pores feem'd to be pretty round, and were rang'd in rows that radiated from the pith to the bark; they all of them feem'd to be continued open pores, running the whole length of the Stick; and that they were all perforated, I try'd by breaking off a very thin fliver of the Coal crofs-ways, and then with my *Microfcope*, diligently furveying them againft the light, for by that means I was able to fee quite through them.

Thefe pores were fo exceeding fmall and thick, that in a line of them, $\frac{1}{18}$ part of an Inch long, I found by numbring them no lefs then 150. fmall pores; and therefore in a line of them an Inch long, muft be no lefs then 2700. pores, and in a circular *area* of an Inch diameter, muft be about 5725350. of the like pores; fo that a Stick of an Inch Diameter, may containe no lefs then feven hundred and twenty five thonfand, befides 5 Millions of pores, which would, I doubt not, feem even incredible, were not every one left to believe his own eyes. Nay, having fince examin'd *Cocus, black and green Ebony, Lignum Vitæ*, &c. I found, that all thefe Woods have their pores, abundantly fmaller then thofe of foft light Wood; in fo much, that thofe of *Guajacum* feem'd not above an eighth part of the bignefs of the pores of Beech, but then the *Interftitia* were thicker; fo prodigioufly curious are the contrivances, pipes, or fluces by which the *Succus nutritius*, or Juyce of a Vegetable is convey'd from place to place.

This *Obfervation* feems to afford us the true reafon of feveral *Phænomena* of Coals; as

Firft, why they look black; and for this we need go no further then the *fcheme*, for certainly, a body that has fo many pores in it as this is difcover'd to have, from each of which no light is reflected, muft neceffarily look black, efpecially, when the pores are fomewhat bigger in proportion to the intervals then they are cut in the *Scheme*, black being nothing elfe but a privation of Light, or a want of reflection; and wherefover this reflecting quality is deficient, there does that part look black, whether it be from a poroufnefs of the body, as in thisInftance, or in a deadning and dulling quality, fuch as I have obferv'd in the *Scoria* of Lead, Tin, Silver, Copper, &c.

Next, we may alfo as plainly fee the reafon of its fhining quality, and
that

that is from the even breaking off of the ſtick, the ſolid *interſtitia* having a regular termination or ſurface, and having a pretty ſtrong reflecting quality, the many ſmall reflections become united to the naked eye, and make a very pretty ſhining ſurface.

Thirdly, the reaſon of its hardneſs and brittleneſs ſeems evident, for ſince all the watery or liquid ſubſtance that moiſtn'd and toughn'd thoſe *Interſtitia* of the more ſolid parts, are evaporated and remov'd, that which is left hehind becomes of the nature almoſt of a ſtone, which will not at all, or very little, bend without a *divulſon* or *ſolution* of its *continuity*.

It is not my deſign at preſent, to examine the uſe and *Mechaniſme* of theſe parts of Wood, that being more proper to another Enquiry ; but rather to hint, that from this Experiment we may learn,

Firſt, what is the cauſe of the blackneſs of many burnt bodies, which we may find to be nothing elſe but this ; that the heat of the fire agitating and rarifying the wateriſh, tranſparent, and volatile water that is contain'd in them, by the continuation of that action, does ſo totally expel and drive away all that which before fill d the pores, and was diſpers'd alſo through the ſolid maſs of it, and thereby caus'd an univerſal kind of tranſparency, that it not onely leaves all the pores empty, but all the *Interſtitia* alſo ſo dry and *opacous*, and perhaps alſo yet further perforated, that that light onely is reflected back which falls upon the very outward edges of the pores, all they that enter into the pores of the body, never returning, but being loſt in it.

Now, that the Charring or coaling of a body is nothing elſe, may be eaſily believ'd by one that ſhall conſider the means of its production, which may be done after this, or any ſuch manner. The body to be charr'd or coal'd, may be put into a *Crucible*, Pot, or any other Veſſel that will endure to be made red-hot in the Fire without breaking, and then cover'd over with Sand, ſo as no part of it be ſuffer'd to be open to the Air, then ſet into a good Fire, and there kept till the Sand has continu'd red hot for a quarter, half, an hour or two, or more, according to the nature and bigneſs of the body to be coal'd or charr'd, then taking it out of the Fire, and letting it ſtand till it be quite cold, the body may be taken out of the Sand well charr'd and cleans'd of its wateriſh parts ; but in the taking of it out, care muſt be had that the Sand be very neer cold, for elſe, when it comes into the free air, it will take fire, and readily burn away.

This may be done alſo in any cloſe Veſſel of Glaſs, as a *Retort*, or the like, and the ſeveral fluid ſubſtances that come over may be receiv'd in a fit *Recipient*, which will yet further countenance this *Hypotheſis* : And their manner of charring Wood in great quantity comes much to the ſame thing, namely, an application of a great heat to the body, and preſerving it from the free acceſs of the devouring air ; this may be eaſily learn'd from the Hiſtory of Charring of Coal, moſt excellently deſcrib'd and publiſh'd by that moſt accompliſh'd Gentleman, Mr. *John Evelin*, in the 100, 101, 103, pages of his *Sylva*, to which I ſhall therefore refer the curious Reader that deſires a full information of it.

Next

Next, we may learn what part of the Wood it is that is the *combuſtible* matter; for ſince we ſhall find that none, or very little of thoſe fluid ſubſtances that are driven over into the Receiver are *combuſtible*, and that moſt of that which is left behind is ſo, it follows, that the ſolid *interſtitia* of the Wood are the *combuſtible* matter. Further, the reaſon why uncharr'd Wood burns with a greater flame then that which is charr'd, is as evident, becauſe thoſe wateriſh or volatil parts iſſuing out of the fired Wood, every way, not onely ſhatter and open the body, the better for the fire to enter, but iſſuing out in vapours or wind, they become like ſo many little *æolipiles*, or Bellows, whereby they blow and agitate the fir'd part, and conduce to the more ſpeedy and violent conſumption or diſſolution of the body.

Thirdly, from the Experiment of charring of Coals (whereby we ſee that notwithſtanding the great heat, and the duration of it, the ſolid parts of the Wood remain, whileſt they are preſerv'd from the free acceſs of the air undiſſipated) we may learn, that which has not, that I know of, been publiſh'd or hinted, nay, not ſo much as thought of, by any; and that in ſhort is this.

Firſt, *that the Air* in which we live, move, and breath, and which encompaſſes very many, and cheriſhes moſt bodies it encompaſſes, that this Air is the *menſtruum*, or univerſal diſſolvent of all *ſulphureous* bodies.

Secondly, *that this action* it performs not, till the body be firſt ſufficiently heated, as we find requiſite alſo to the diſſolution of many other bodies by ſeveral other *menſtruums*.

Thirdly, *that this action* of diſſolution, produces or generates a very great heat, and that which we call Fire; and this is common alſo to many diſſolutions of other bodies, made by *menſtruums*, of which I could give multitudes of Inſtances.

Fourthly, *that this action* is perform'd with ſo great a violence, and does ſo minutely act, and rapidly agitate the ſmalleſt parts of the *combuſtible* matter, that it produces in the *diaphanous medium* of the Air, the action or pulſe of light, which what it is, I have elſe-where already ſhewn.

Fifthly, *that the diſſolution* of ſulphureous bodies is made by a ſubſtance inherent, and mixt with the Air, that is like, if not the very ſame, with that which is fixt in *Salt-peter*, which by multitudes of Experiments that may be made with *Saltpeter*, will, I think, moſt evidently be demonſtrated.

Sixthly, *that in this diſſolution* of bodies by the Air, a certain part is united and mixt, or diſſolv'd and turn'd into the Air, and made to fly up and down with it in the ſame manner as a *metalline* or other body diſſolv'd into any *menſtruums*, does follow the motions and progreſſes of that *menſtruum* till it be precipitated.

Seventhly, That as there is one part that is diſſoluble by the Air, ſo are there other parts with which the parts of the Air mixing and uniting, do make a *Coagulum*, or *precipitation*, as one may call it, which cauſes it to be ſeparated from the Air, but this *precipitate* is ſo light, and in ſo ſmall and rarify'd or porous cluſters, that it is very volatil, and is eaſily carry'd up by the motion of the Air, though afterwards, when the heat and
agitation

agitation that kept it rarify'd ceafes, it eafily condenfes, and commixt with other indiffoluble parts, it fticks and adheres to the next bodies it meets withall; and this is a certain *Salt* that may be extracted out of *Soot.*

Eighthly, that many indiffoluble parts being very apt and prompt to be rarify'd, and fo, whileft they continue in that heat and agitation, are lighter then the Ambient Air, are thereby thruft and carry'd upwards with great violence, and by that means carry along with them, not onely that *Saline concrete* I mention'd before, but many terreftrial, or indiffoluble and irrarefiable parts, nay, many parts alfo which are diffoluble, but are not fuffer'd to ftay long enough in a fufficient heat to make them prompt and apt for that action. And therefore we find in *Soot*, not onely a part, that being continued longer in a competent heat, will be diffolv'd by the Air, or take fire and burn; but a part alfo which is fixt, terreftrial, and irrarefiable.

Ninthly, that as there are thefe feveral parts that will rarifie and fly, or be driven up by the heat, fo are there many others, that as they are indiffoluble by the *aerial menftruum,* fo are they of fuch fluggifh and grofs parts, that they are not eafily rarify'd by heat, and therefore cannot be rais'd by it; the volatility or fixtnefs of a body feeming to confift only in this, that the one is of a texture, or has component parts that will be eafily rarify'd into the form of Air, and the other, that it has fuch as will not, without much ado, be brought to fuch a conftitution; and this is that part which remains behind in a white body call'd Afhes, which contains a fubftance, or *Salt*, which Chymifts call *Alkali:* what the particular natures of each of thefe bodies are, I fhall not here examine, intending it in another place, but fhall rather add that this *Hypothefis* does fo exactly agree with all *Phænomena* of Fire, and fo genuinely explicate each particular circumftance that I have hitherto obferv'd, that it is more then probable, that this caufe which I have affign'd is the true adequate, real, and onely caufe of thofe *Phænomena*; And therefore I fhall proceed a little further, to fhew the nature and ufe of the Air.

Tenthly, therefore the diffolving parts of the Air are but few, that is, it feems of the nature of thofe *Saline menftruums,* or fpirits, that have very much flegme mixt with the fpirits, and therefore a fmall parcel of it is quickly glutted, and will diffolve no more; and therefore unlefs fome frefh part of this *menftruum* be apply'd to the body to be diffolv'd, the action ceafes, and the body leaves to be diffolv'd and to fhine, which is the Indication of it, though plac'd or kept in the greateft heat; whereas *Salt-peter* is a *menftruum,* when melted and red-hot, that abounds more with thofe Diffolvent particles, and therefore as a fmall quantity of it will diffolve a great fulphureous body, fo will the diffolution be very quick and violent.

Therefore in the *Eleventh* place, it is obfervable, that, as in other folutions, if a copious and quick fupply of frefh *menftruum,* though but weak, be poured on, or applied to the diffoluble body, it quickly confumes it: So this *menftruum* of the Air, if by Bellows, or any other fuch contrivance, it be copioufly apply'd to the fhining body, is found to

diffolve

diſſolve it as ſoon, and as violently as the more ſtrong *menſtruum* of melted *Nitre*.

Therefore twelfthly, it ſeems reaſonable to think that there is no ſuch thing as an Element of Fire that ſhould attract or draw up the flame, or towards which the flame ſhould endeavour to aſcend out of a deſire or appetite of uniting with that as its *Homogeneal* primitive and generating Element ; but that that ſhining tranſient body which we call *Flame*, is nothing elſe but a mixture of Air, and volatil ſulphureous parts of diſſoluble or combuſtible bodies, which are acting upon each other whil'ſt they aſcend, that is, flame ſeems to be a mixture of Air, and the combuſtible volatil parts of any body, which parts the encompaſſing Air does diſſolve or work upon, which action, as it does intend the heat of the *aerial* parts of the diſſolvent, ſo does it thereby further rarifie thoſe parts that are acting, or that are very neer them, whereby they growing much lighter then the heavie parts of that *Menſtruum* that are more remote, are thereby protruded and driven upward ; and this may be eaſily obſerv'd alſo in diſſolutions made by any other *menſtruum*, eſpecially ſuch as either create heat or bubbles. Now, this action of the *Menſtuum*, or *Air*, on the diſſoluble parts, is made with ſuch violence, or is ſuch, that it imparts ſuch a motion or pulſe to the *diaphanous* parts of the Air, as I have elſewhere ſhewn is requiſite to produce light.

This *Hypotheſis* I have endeavoured to raiſe from an Infinite of Obſervations and Experiments, the proceſs of which would be much too long to be here inſerted, and will perhaps another time afford matter copious enough for a much larger Diſcourſe, the Air being a Subject which (though all the world has hitherto liv'd and breath'd in, and been unconverſant about) has yet been ſo little truly examin'd or explain'd, that a diligent enquirer will be able to find but very little information from what has been (till of late) written of it : But being once well underſtood, it will, I doubt not, inable a man to render an intelligible, nay probable, if not the true reaſon of all the *Phænomena* of Fire, which, as it has been found by Writers and Philoſophers of all Ages a matter of no ſmall difficulty, as may be ſufficiently underſtood by their ſtrange *Hypotheſes*, and unintelligible Solutions of ſome few *Phænomena* of it ; ſo will it prove a matter of no ſmall concern and uſe in humane affairs, as I ſhall elſewhere endeavour to manifeſt when I come to ſhew the uſe of the Air in reſpiration, and for the preſervation of the life, nay, for the conſervation and reſtauration of the health and natural conſtitution of mankind as well as all other aereal *animals*, as alſo the uſes of this principle or propriety of the Air in chymical, mechanical, and other operations. In this place I have onely time to hint an *Hypotheſis*, which, if God permit me life and opportunity, I may elſewhere proſecute, improve and publiſh. In the mean time, before I finiſh this Diſcourſe, I muſt not forget to acquaint the Reader, that having had the liberty granted me of making ſome trials on a piece of *Lignum foſſile* ſhewn to the Royal Society, by the eminently Ingenious and Learned Phyſician, Doctor *Ent*, who receiv'd it for a Preſent from the famous *Ingenioſo Cavalliero de Pozzi*, it being one of the faireſt

and

and beſt pieces of *Lignum foſſile* he had ſeen; Having (I ſay) taken a ſmall piece of this Wood, and examin'd it, I found it to burn in the open Air almoſt like other Wood, and inſteed of a reſinous ſmoak or fume, it yielded a very bituminous one, ſmelling much of that kind of ſent: But that which I chiefly took notice of, was, that cutting off a ſmall piece of it, about the bigneſs of my Thumb, and charring it in a *Crucible* with Sand, after the manner I above preſcrib'd, I found it infinitely to abound with the ſmaller ſort of pores, ſo extreme thick, and ſo regularly perforating the ſubſtance of it long-ways, that breaking it off a-croſs, I found it to look very like an Honey-comb; but as for any of the ſecond, or bigger kind of pores, I could not find that it had any; ſo that it ſeems, whatever were the cauſe of its production, it was not without thoſe ſmall kind of pores which we have onely hitherto found in Vegetable bodies: and comparing them with the pores which I have found in the Charcoals that I by this means made of ſeveral other kinds of Wood, I find it reſemble none ſo much as thoſe of Firr, to which it is not much unlike in grain alſo, and ſeveral other proprieties.

And therefore, what ever is by ſome, who have written of it, and particularly by *Franciſco Stelluto*, who wrote a Treatiſe in *Italian* of that Subject, which was Printed at *Rome*, 1637. affirm'd that it is a certain kind of Clay or Earth, which in tract of time is turn'd into Wood, I rather ſuſpect the quite contrary, that it was at firſt certain great Trees of Fir or Pine, which by ſome Earthquake, or other caſualty, came to be buried under the Earth, and was there, after a long time's reſidence(according to the ſeveral natures of the encompaſſing adjacent parts)either rotted and turn'd into a kind of Clay, or *petrify'd* and turn'd into a kind of Stone, or elſe had its pores fill'd with certain Mineral juices, which being ſtayd in them, and in tract of time coagulated, appear'd, upon cleaving out, like ſmall Metaline Wires, or elſe from ſome flames or ſcorching forms that are the occaſion oftentimes, and uſually accompany Earthquakes, might be blaſted and turn'd into Coal, or elſe from certain *ſubterraneous* fires which are affirm'd by that Authour to abound much about thoſe parts (namely, in a Province of *Italy*, call'd *Umbria*, now the *Dutchie* of *Spoletto*, in the Territory of *Todi*, anciently call'd *Tudor*; and between the two Villages of *Colleſecco* and *Roſaro* not far diſtant from the high-way leading to *Rome*, where it is found in greater quantity then elſewhere)are by reaſon of their being encompaſſed with Earth, and ſo kept cloſe from the diſſolving Air, charr'd and converted into Coal. It would be too long a work to deſcribe the ſeveral kinds of pores which I met withall, and by this means diſcovered in ſeveral other Vegetable bodies; nor is it my preſent deſign to expatiate upon Inſtances of the ſame kind, but rather to give a Specimen of as many kinds as I have had opportunity as yet of obſerving, reſerving the proſecution and enlarging on particulars till a more fit opportunity; and in proſecution of this deſign, I ſhall here add:

Obſerv,

Fig: 1

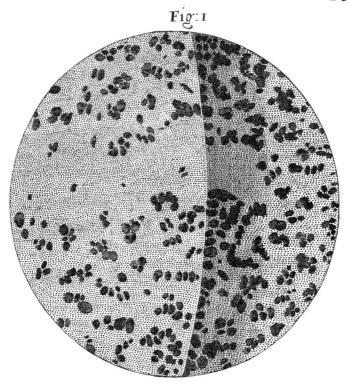

Fig: 2:

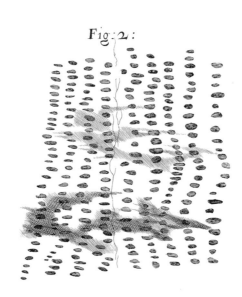

Obſerv. XVII. *Of* Petrify'd wood, *and other* Petrify'd bodies.

OF this ſort of ſubſtance, I obſerv'd ſeveral pieces of very differing kinds, both for their outward ſhape, colour, grain, *texture*, hardneſs, &c. ſome being brown and rediſh, others gray, like a Hone; others black, and Flint-like: ſome ſoft, like a Slate or Whetſtone, others as hard as a Flint, and as brittle. That which I more particular examin'd, was a piece about the bigneſs of a mans hand, which ſeem'd to have been a part of ſome large tree, that by rottenneſs had been broken off from it before it began to be *petrify'd*.

And indeed, all that I have yet ſeen, ſeem to have been rotten Wood before the petrifaction was begun; and not long ſince, examining and viewing a huge great *Oak*, that ſeem'd with meer age to be rotten as it ſtood, I was very much confirm'd in this opinion; for I found, that the grain, colour, and ſhape of the Wood, was exactly like this *petrify'd* ſubſtance; and with a *Microſcope*, I found, that all thoſe *Microſcopical* pores, which in ſappy or firm and ſound Wood are fill'd with the natural or innate juices of thoſe Vegetables, in this they were all empty, like thoſe of *Vegetables charr'd*; but with this difference, that they ſeem'd much larger then I have ſeen any in *Char-coals*; nay, even then thoſe of Coals made of great blocks of Timber, which are commonly call'd *Old-coals*.

The reaſon of which difference may probably be, that the charring of Vegetables, being an operation quickly perform'd, and whileſt the Wood is ſappy, the more ſolid parts may more eaſily ſhrink together, and contract the pores or *interſtitia* between them, then in the rotten Wood, where that natural juice ſeems onely to be waſh'd away by *adventitious* or unnatural moiſture; and ſo though the natural juice be waſted from between the firm parts, yet thoſe parts are kept aſunder by the *adventitious* moyſtures, and ſo by degrees ſettled in thoſe poſtures.

And this I likewiſe found in the *petrify'd* Wood, that the pores were ſomewat bigger then thoſe of *Charcoal*, each pore being neer upon half as bigg again, but they did not bear that diſproportion which is expreſt in the tenth *Scheme*, between the ſmall ſpecks or pores in the firſt Figure (which repreſenteth the pores of Coal or Wood charr'd) and the black ſpots of the ſecond Figure (which repreſent the like *Microſcopical* pores in the *petrify'd* Wood) for theſe laſt were drawn by a *Microſcope* that magnify'd the object above ſix times more in Diameter then the *Microſcope* by which thoſe pores of Coal were obſerv'd.

Now, though they were a little bigger, yet did they keep the exact figure and order of the pores of Coals and of rotten Wood, which laſt alſo were much of the ſame cize.

The other Obſervations on this *petrify'd* ſubſtance, that a while ſince, by the appointment of the *Royal Society*, I made, and preſented to them an account of, were theſe that follow, which had the honour done them by by the moſt accompliſh'd Mr. *Evelin*, my highly honour'd friend, to be inſerted and publiſhed among thoſe excellent Obſervations wherewith his *Sylva* is repleniſh'd, and would therefore have been here omitted, had not the Figure of them, as they appear'd through the *Microſcope* been before that engraven.

This *Petrify'd* ſubſtance reſembled Wood, in that

Firſt, all the parts of it ſeem'd not at all *diſlocated*, or alter'd from their natural Poſition, whil'ſt they were Wood, but the whole piece retain'd the exact ſhape of Wood, having many of the conſpicuous pores of wood ſtill remaining pores, and ſhewing a manifeſt difference viſible enough between the grain of the Wood and that of the bark, eſpecially when any ſide of it was cut ſmooth and polite; for then it appear'd to have a very lovely grain, like that of ſome curious cloſe Wood.

Next (it reſembled Wood) in that all the ſmaller and (if I may ſo call thoſe which are onely viſible with a good magnifying Glaſs) *Microſcopical* pores of it appear (both when the ſubſtance is cut and poliſh'd *tranſverſly* and *parallel* to the pores of it) perfectly like the *Microſcopical* pores of ſeveral kinds of Wood, eſpecially like and equal to thoſe of ſeveral ſorts of rotten Wood which I have ſince obſerv'd, retaining both the ſhape, poſition and magnitude of ſuch pores. It was differing from Wood :

Firſt, in *weight*, being to common water as $3\frac{1}{4}$ to 1. whereas there are few of our *Engliſh* Woods, that when very dry are found to be full as heavie as water.

Secondly, in *hardneſs*, being very neer as hard as a Flint; and in ſome places of it alſo reſembling the grain of a Flint : and, like it, it would very readily cut Glaſs, and would not without difficulty, eſpecially in ſome parts of it, be ſcratch'd by a black hard Flint : It would alſo as readily ſtrike fire againſt a Steel, or againſt a Flint, as any common Flint.

Thirdly, in the *cloſeneſs* of it, for though all the *Microſcopical* pores of this *petrify'd* ſubſtance were very conſpicuous in one poſition, yet by altering that poſition of the poliſh'd ſurface to the light, it was alſo manifeſt, that thoſe pores appear'd darker then the reſt of the body, onely becauſe they were fill'd up with a more duſkie ſubſtance, and not becauſe they were hollow.

Fourthly, in its *incombuſtibleneſs*, in that it would not burn in the fire; nay, though I kept it a good while red-hot in the flame of a Lamp, made very *intenſe* by the blaſt of a ſmall Pipe, and a large Charcoal, yet it ſeem'd not at all to have diminiſh'd its extenſion; but only I found it to have chang'd its colour, and to appear of a more dark and duſkie brown colour; nor could I perceive that thoſe parts which ſeem'd to have been Wood at firſt, were any thing waſted, but the parts appear'd as ſolid and cloſe as before. It was further obſervable alſo, that as it did not conſume like Wood, ſo neither did it crack and flie like a Flint, or ſuch like hard Stone, nor was it long before it appear'd red-hot.

Fifthly, in its *diſſolubleneſs*; for putting ſome drops of diſtill'd *Vinegar* upon the Stone, I found it preſently to yield very many Bubbles, juſt like thoſe which may be obſerv'd in ſpirit of *Vinegar* when it corrodes *corals*, though

though perhaps many of thofe fmall Bubbles might proceed from fome fmall parcels of Air which were driven out of the pores of this *petrify'd* fubftance by the infinuating liquid *menftruum*.

Sixthly, in its *rigidnefs* and *friability*, being not at all flexible but brittle like a Flint, infomuch that I could with one knock of a Hammer break off a piece of it, and with a few more, reduce that into a pretty fine powder.

Seventhly, it feem'd alfo very differing from Wood to the *touch, feeling* more cold then Wood ufually does, and much like other clofe ftones and Minerals.

The Reafons of all which *Phænomena* feem to be,

That this *petrify'd* Wood having lain in fome place where it was well foak'd with *petrifying* water (that is. fuch a water as is well *impregnated* with ftony and earthy particles) did by degrees feparate, either by ftraining and *filtration*, or perhaps, by *precipitation, cohefion* or *coagulation*, abundance of ftony particles from the permeating water, which ftony particles, being by means of the fluid *vehicle* convey d, not onely into the *Microfcopical* pores, and fo perfectly ftoping them up, but alfo into the pores or *interftitia*, which may. perhaps, be even in the texture or *Schematifme* of that part of the Wood, which, through the *Microfcope*, appears moft folid, do thereby fo augment the weight of the Wood, as to make it above three times heavier then water, and perhaps, fix times as heavie as it was when Wood.

Next, they thereby fo lock up and fetter the parts of the Wood, that the fire cannot eafily make them flie away but the action of the fire upon them is onely able to *Char* thofe parts, as it were, like a piece of Wood, if it be clos'd very faft up in Clay, and kept a good while red-hot in the fire, will by the heat of the fire be charr'd and not confum'd, which may, perhaps, alfo be fomewhat of the caufe, why the *petrify'd* fubftance appear'd of a dark brown colour after it had been burnt.

By this *intrufion* of the *petrifying* particles, this fubftance alfo becomes hard and *friable*; for the fmaller pores of the Wood being perfectly wedg'd, and ftuft up with thofe ftony particles, the fmall parts of the Wood have no places or pores into which they may flide upon bending, and confequently little or no flexion or yielding at all can be caus'd in fuch a fubftance.

The remaining particles likewife of the Wood among the ftony particles, may keep them from cracking and flying when put into the fire, as they are very apt to do in a Flint.

Nor is Wood the onely fubftance that may by this kind of *tranfmutation* be chang'd into ftone; for I my felf have feen and examin'd very many kinds of fubftances, and among very credible Authours, we may meet with Hiftories of fuch *Metamorphofes* wrought almoft on all kind of fubftances, both *Vegetable* and *Animal*, which Hiftories, it is not my bufinefs at prefent, either to relate, or *epitomife*, but only to fet down fome Obfervation I lately made on feveral kind of *petrify'd* Shels, found about *Keinfham*, which lies within four or five miles of *Briftol*, which are commonly call'd *Serpentine-ftones*, Exami-

Examining several of these very curiously figur'd bodies (which are commonly thought to be Stones form'd by some extraordinary *Plastick virtue latent* in the Earth it self) I took notice of these particulars:

First, that these figured bodies, or stones, were of very differing substances, as to hardness: some of Clay, some Marle, some soft Stone, almost of the hardness of those soft stones which Masons call Fire-stone, others as hard as Portland stone, others as hard as Marble, and some as hard a a Flint or Cryftal.

Next, they were of very differing substances as to transparency and colour; some white, some almost black, some brown, some Metalline, or like Marchasites; some transparent like white Marble, others like flaw'd Cryftal,some gray, some of divers colours; some radiated like these long *petrify'd drops*, which are commonly found at the *Peak*, and in other *subterraneous caverns*, which have a kind of pith in the middle.

Thirdly, that they were very different as to the manner of their outward figuration; for some of them seem'd to have been the substance that had fill'd the Shell of some kind of Shel-fish; others, to have been the substance that had contain'd or enwrapp'd one of those Shels,on both which,the perfect impression either of the inside or outside of such Shells seem'd to be left, but for the most part, those impressions seem'd to be made by an imperfect or broken Shell, the great end or mouth of the Shell being always wanting, and oftentimes the little end, and sometimes half, and in some there were impressions, just as if there had been holes broken in the figurating, imprinting or moulding Shell; some of them seem d to be made by such a Shell very much brused or flaw'd, insomuch that one would verily have thought that very figur'd stone had been broken or brused whilst a gelly, as 'twere, and so hardned, but within in the grain of the stone, there appear'd not the least sign of any such bruse or breaking, but onely on the very uttermost superficies.

Fourthly, they were very different,as to their outward covering some having the perfect Shell, both in figure, colour, and substance, sticking on upon its surface, and adhering to it, but might very easily be separated from it, and like other common *Cockle* or *Scolop-shels*, which some of them most accurately resembled,were very dissoluble in common *Vinegar*, others of them,especially those *Serpentine*, or *Helical stones* were cover'd or retained the shining or Pearl-colour'd substance of the inside of a Shel, which substance, on some parts of them, was exceeding thin, and might very easily be rubbed off; on other parts it was pretty thick, and retained a white coat, or flaky substance on the top, just like the outsides of such Shells; some of them had very large pieces of the Shell very plainly sticking on to them, which were easily to be broken or flaked off by degrees: they likewise, some of them retain'd all along the surface of them very pretty kind of *sutures*, such as are observ'd in the skulls of several kinds of living creatures, which *sutures* were most curiously shap'd in the manner of leaves, and every one of them in the same Shell, exactly one like another, which I was able to discover plainly enough with my naked eye, but more perfectly and distinctly with my *Microscope*; all
these

thefe *futures*, by breaking fome of thefe ftones,I found to be the *termini*, or boundings of certain *diaphragms*, or partitions,which feem'd to divide the cavity of the Shell into a multitude of very proportionate and regular *cells* or *caverns*, thefe *Diaphragms*, in many of them, I found very perfect and compleat, of a very diftinct fubftance from that which fill'd the cavities, and exactly of the fame kind with that which covered the outfide, being for the moft part whitifh, or *mother-of-pearl* colour'd.

As for the cavities between thofe *Diaphragms*, I found fome of them fill'd with Marle, and others with feveral kinds of ftones, others, for the moft part hollow, onely the whole cavity was ufually covered over with a kind of *tartareous petrify'd* fubftance, which ftuck about the fides, and was there fhot into very curious regular Figures, juft as *Tartar*, or other diffolv'd Salts are obferv'd to ftick and *cryftallize* about the fides of the containing Veffels; or like thofe little *Diamants* which I before obferved to have covered the vaulted cavity of a Flint; others had thefe cavities all lin'd with a kind of *metalline* or *marchafite-like* fubftance, which with a *Microfcope* I could as plainly fee moft curioufly and regularly figured, as I had done thofe in a Flint.

From all which, and feveral other particulars which I obferv'd, I cannot but think, that all thefe, and moft other kinds of ftony bodies which are found thus ftrangely figured,do owe their formation and figuration, not to any kind of *Plaftick virtue* inherent in the earth, but to the Shells of certain Shel-fifhes, which, either by fome Deluge, Inundation, Earthquake, or fome fuch other means, came to be thrown to that place, and there to be fill'd with fome kind of Mudd or Clay, or *petrifying* Water, or fome other fubftance, which in tract of time has been fettled together and hardned in thofe fhelly moulds into thofe fhaped fubftances we now find them; that the great and thin end of thefe Shells by that Earthquake, or what ever other extraordinay caufe it was that brought them thither, was broken off; and that many others were otherwife broken, bruifed and disfigured; that thefe Shells which are thus *fpirallied* and feparated with *Diaphragmes*,were fome kind of *Nautili* or *Porcelane fhells*; and that others were fhells of *Cockles*,*Mufcles*,*Periwincles*,*Scolops*,&c. of various forts; that thefe Shells in many, from the particular nature of the containing or enclos'd Earth, or fome other caufe, have in tract of time rotted and mouldred away, and onely left their impreffions, both on the containing and contained fubftances; and fo left them pretty loofe one within another, fo that they may be eafily feparated by a knock or two of a Hammer. That others of thefe Shells, according to the nature of the fubftances adjacent to them, have, by a long continuance in that pofture, been *petrify'd* and turn'd into the nature of ftone, juft as I even now obferv'd feveral forts of Wood into to be. That oftentimes the Shell may be found with one kind of fubftance within, and quite another without, having, perhaps, been fill'd in one place, and afterwards tranflated to another, which I have very frequently obferv'd in *Cockle*, *Mufcle*, *Periwincle*, and other fhells, which I have found by the Sea fide. Nay, further that fome parts of the fame Shell may be fill'd in one place, and
fome

fome other caverns in another, and others in a third, or a fourth, or a fifth place, for fo many differing fubftances have I found in one of thefe *petrify'd* Shells, and perhaps all thefe differing from the encompaffing earth or ftone ; the means how all which varieties may be caus'd, I think, will not be difficult to conceive, to any one that has taken notice of thofe Shells, which are commonly found on the Sea fhore : And he that fhall throughly examine feveral kinds of fuch curioufly form'd ftones, will (I am very apt to think) find reafon to fuppofe their generation or formation to be afcribable to fome fuch accidents as I have mention'd, and not to any *Plaftick virtue :* For it feems to me quite contrary to the infinite prudence of Nature, which is obfervable in all its works and productions, to defign every thing to a determinate end, and for the attaining of that end, makes ufe of fuch ways as are (as farr as the knowledge of man has yet been able to reach) altogether confonant, and moft agreeable to man's reafon, and of no way or means that does contradict, or is contrary to humane Ratiocination ; whence it has a long time been a general obfervation and *maxime,* that *Nature does nothing in vain* ; It feems, I fay, contrary to that great Wifdom of Nature, that thefe prettily fhap'd bodies fhould have all thofe curious Figures and contrivances (which many of them are adorn'd and contriv'd with) generated or wrought by a *Plaftick virtue,* for no higher end then onely to exhibit fuch a form ; which he that fhall throughly confider all the circumftances of fuch kind of Figur'd bodies, will, I think, have great reafon to believe, though, I confefs, one cannot prefently be able to find out what Nature's defigns are. It were therefore very defirable, that a good collection of fuch kind of figur'd ftones were collected ; and as many particulars, circumftances, and informations collected with them as could be obtained, that from fuch a Hiftory of Obfervations well rang'd, examin'd and digefted, the true original or production of all thofe kinds of ftones might be perfectly and furely known ; fuch as are *Thunderftones, Lapides Stellares, Lapides Judaici,* and multitudes of other, whereof mention is made in *Aldrovandus Wormius,* and other Writers of Minerals.

Obferv. XVIII. *Of the* Schematifme *or* Texture *of* Cork, *and of the Cells and Pores of fome other fuch frothy Bodies.*

I Took a good clear piece of Cork, and with a Pen-knife fharpen'd as keen as a Razor, I cut a piece of it off, and thereby left the furface of it exceeding fmooth, then examining it very diligently with a *Microfcope,* me thought I could perceive it to appear a little porous ; but I could not fo plainly diftinguifh them, as to be fure that they were pores, much lefs what Figure they were of : But judging from the lightnefs and yielding quality of the Cork, that certainly the texture could not be fo

curious,

curious, but that poſſibly, if I could uſe ſome further diligence, I might find it to be diſcernable with a *Microſcope*, I with the ſame ſharp Pen-knife, cut off from the former ſmooth ſurface an exceeding thin piece of it, and placing it on a black objeſt Plate, becauſe it was it ſelf a white body, and caſting the light on it with a deep *plano-convex Glaſs*, I could exceeding plainly perceive it to be all perforated and porous, much like a Honey-comb, but that the pores of it were not regular; yet it was not unlike a Honey-comb in theſe particulars.

Firſt, in that it had a very little ſolid ſubſtance, in compariſon of the empty cavity that was contain'd between, as does more manifeſtly appear by the Figure A and B of the X I. *Scheme*, for the *Interſtitia*, or walls (as I may ſo call them) or partitions of thoſe pores were neer as thin in proportion to their pores, as thoſe thin films of Wax in a Honey-comb (which encloſe and conſtitute the *ſexangular cells*) are to theirs.

Next, in that theſe pores, or cells, were not very deep, but conſiſted of a great many little Boxes, ſeparated out of one continued long pore, by certain *Diaphragms*, as is viſible by the Figure B, which repreſents a ſight of thoſe pores ſplit the long-ways.

I no ſooner diſcern'd theſe (which were indeed the firſt *microſcopical* pores I ever ſaw, and perhaps, that were ever ſeen, for I had not met with any Writer or Perſon, that had made any mention of them before this) but me thought I had with the diſcovery of them, preſently hinted to me the true and intelligible reaſon of all the *Phænomena* of Cork; As,

Firſt, if I enquir'd why it was ſo exceeding light a body? my *Micro-ſcope* could preſently inform me that here was the ſame reaſon evident that there is found for the lightneſs of froth, an empty Honey-comb, Wool, a Spunge, a Pumice-ſtone, or the like; namely, a very ſmall quantity of a ſolid body, extended into exceeding large dimenſions.

Next, it ſeem'd nothing more difficult to give an intelligible reaſon, why Cork is a body ſo very unapt to ſuck and drink in Water, and con-ſequently preſerves it ſelf, floating on the top of Water, though left on it never ſo long : and why it is able to ſtop and hold air in a Bottle, though it be there very much condens'd and conſequently preſſes very ſtrongly to get a paſſage out, without ſuffering the leaſt bubble to paſs through its ſubſtance. For, as to the firſt, ſince our *Microſcope* informs us that the ſubſtance of Cork is altogether fill'd with Air, and that that Air is per-feſtly encloſed in little Boxes or Cells diſtinſt from one another. It ſeems very plain, why neither the Water, nor any other Air can eaſily inſinu-ate it ſelf into them, ſince there is already within them an *intus exiſtens*, and conſequently, why the pieces of Cork become ſo good floats for Nets, and ſtopples for Viols, or other cloſe Veſſels.

And thirdly, if we enquire why Cork has ſuch a ſpringineſs and ſwel-ling nature whem compreſs'd? and how it comes to ſuffer ſo great a com-preſſion, or ſeeming penetration of dimenſions, ſo as to be made a ſub-ſtance as heavie again and more, bulk for bulk, as it was before compreſ-ſion, and yet ſuffer'd to return, is found to extend it ſelf again into the ſame ſpace? Our *Microſcope* will eaſily inform us, that the whole maſs

conſiſts

confifts of an infinite company of fmall Boxes or Bladders of Air, which is a fubftance of a fpringy nature, and that will fuffer a confiderable con-denfation (as I have feveral times found by divers trials, by which I have moft evidently condens'd it into lefs then a twentieth part of its ufual di-menfions neer the Earth, and that with no other ftrength then that of my hands without any kind of forcing Engine,fuch as Racks,Leavers,Wheels, Pullies, or the like, but this onely by and by) and befides, it feems very probable that thofe very films or fides of the pores,have in them a fpring-ing quality, as almoft all other kind of Vegetable fubftances have, fo as to help to reftore themfelves to their former pofition.

And could we fo eafily and certainly difcover the *Schematifme* and *Texture* even of thefe films,and of feveral other bodies,as we can thefe of Cork; there feems no probable reafon to the contrary, but that we might as readily render the true reafon of all their *Phænomena*; as namely,what were the caufe of the fpringinefs, and toughnefs of fome, both as to their flexibility and reftitution. What, of the friability or brittlenefs of fome others, and the like; but till fuch time as our *Microfcope*, or fome other means,enable us to difcover the true *Schematifm* and *Texture* of all kinds of bodies, we muft grope, as it were, in the dark, and onely ghefs at the true,reafons of things by fimilitudes and comparifons.

But, to return to our Obfervation. I told feveral lines of thefe pores, and found that there were ufually about threefcore of thefe fmall Cells placed end-ways in the eighteenth part of an Inch in length,whence I concluded there muft be neer eleven hundred of them, or fomewhat more then a thoufand in the length of an Inch, and therefore in a fquare Inch above a Million, or 1166400. and in a Cubick Inch,above twelve hundred Millions, or 1259712000. a thing almoft incredible, did not our *Microfcope* affure us of it by ocular demonftration; nay, did it not difco-ver to us the pores of a body, which were they *diaphragm'd*,like thofe of Cork, would afford us in one Cubick Inch, more then ten times as many little Cells, as is evident in feveral charr'd Vegetables; fo prodigioufly curious are the works of Nature, that even thefe confpicuous pores of bodies, which feem to be the channels or pipes through which the *Succus nutritius*, or natural juices of Vegetables are convey'd, and feem to cor-refpond to the veins, arteries and other Veffels in fenfible creatures, that thefe pores I fay, which feem to be the Veffels of nutrition to the vafteft body in the World, are yet fo exceeding fmall, that the *Atoms* which *Epi-curus* fancy'd would go neer too prove too bigg to enter them, much more to conftitute a fluid body in them.And how infinitely fmaller then muft be the Veffels of a Mite, or the pores of one of thofe little Vegetables I have difcovered to grow on the back-fide of a Rofe-leaf, and fhall anon more fully defcribe, whofe bulk is many millions of times lefs then the bulk of the fmall fhrub it grows on; and even that fhrub, many millions of times lefs in bulk then feveral trees (that have heretofore grown in *England,* and are this day flourifhing in other hotter Climates, as we are very cre-dibly inform'd) if at leaft the pores of this fmall Vegetable fhould keep any fuch proportion to the body of it , as we have.found thefe pores

Fig:I.

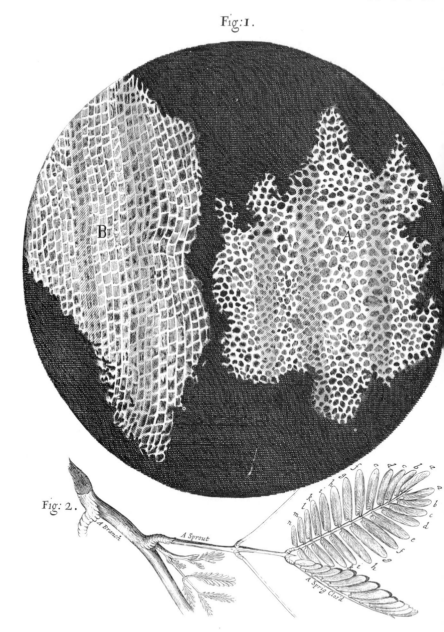

B

A

Fig: 2.

A Branch

A Sprout

A Sprig Clos'd

of other Vegetables to do to their bulk. But of thefe pores I have faid more elfewhere.

To proceed then, Cork feems to be by the tranfverfe conftitution of the pores, a kind of *Fungus* or Mufhrome, for the pores lie like fo many Rays tending from the center, or pith of the tree, outwards; fo that if you cut off a piece from a board of Cork tranfverfly, to the flat of it, you will, as it were, fplit the pores, and they will appear juft as they are exprefs'd in the Figure B of the XI. *Scheme*. But if you fhave off a very thin piece from this board, parallel to the plain of it, you will cut all the pores tranfverfly, and they will appear almoft as they are exprefs'd in the Figure A, fave onely the folid *Interftitia* will not appear fo thick as they are there reprefented.

So that Cork feems to fuck its nourifhment from the fubjacent bark of the Tree immediately, and to be a kind of excrefcence, or a fubftance diftinct from the fubftances of the entire Tree, fomething *analogus* to the Mufhrome, or Mofs on other Trees, or to the hairs on Animals. And having enquir'd into the Hiftory of Cork, I find it reckoned as an excrefcency of the bark of a certain Tree, which is diftinct from the two barks that lie within it, which are common alfo to other trees; That 'tis fome time before the Cork that covers the young and tender fprouts comes to be difcernable; That it cracks, flaws, and cleaves into many great chaps, the bark underneath remaining entire; That it may be feparated and remov'd from the Tree, and yet the two under-barks (fuch as are alfo common to that with other Trees) not at all injur'd, but rather helped and freed from an external injury. Thus *Jonftonus* in *Dendrologia*, fpeaking *de Subere*, fays, *Arbor eft procera, Lignum eft robuftum, dempte cortice in aquis non fluitat, Cortice in orbem detracto juvatur, crafcefcens enim præftringit & ftrangulat, intra triennium iterum repletur: Caudex ubi adolefcit craffus, cortex fuperior denfus carnofus, duos digitos craffus, fcaber, rimofus, & qui nifi detrahatur dehifcit, alioque fubnafcente expellitur, inte-rior qui fubeft novellus ita rubet ut arbor minio picta videatur.* Which Hiftories, if well confider'd, and the tree, fubftance, and manner of grow-ing, if well examin'd, would, I am very apt to believe, much confirm this my conjecture about the origination of Cork.

Nor is this kind of Texture peculiar to Cork onely; for upon exami-nation with my *Microfcope*, I have found that the pith of an Elder, or al-moft any other Tree, the inner pulp or pith of the Cany hollow ftalks of feveral other Vegetables: as of Fennel, Carrets, Daucus, Bur-docks, Teafels, Fearn, fome kinds of Reeds, &c. have much fuch a kind of *Schematifme*, as I have lately fhewn that of Cork, fave onely that here the pores are rang'd the long-ways, or the fame ways with the length of the Cane, whereas in Cork they are tranfverfe.

The pith alfo that fills that part of the ftalk of a Feather that is above the Quil, has much fuch a kind of texture, fave onely that which way fo-ever I fet this light fubftance, the pores feem'd to be cut tranfverfly; fo that I ghefs this pith which fills the Feather, not to confift of abundance of long pores feparated with Diaphragms, as Cork does, but to be a kind
of

of folid or hardned froth, or a *congeries* of very fmall bubbles confolidated in that form, into a pretty ftiff as well as tough concrete, and that each Cavern, Bubble, or Cell, is diftinctly feparate from any of the reft, without any kind of hole in the encompaffing films, fo that I could no more blow through a piece of this kinde of fubftance, then I could through a piece of Cork, or the found pith of an Elder.

But though I could not with my *Microfcope*, nor with my breath, nor any other way I have yet try'd, difcover a paffage out of one of thofe cavities into another, yet I cannot thence conclude, that therefore there are none fuch, by which the *Succus nutritius*, or appropriate juices of Vegetables, may pafs through them; for, in feveral of thofe Vegetables, whil'ft green, I have with my *Microfcope*, plainly enough difcover'd thefe Cells or Poles fill'd with juices, and by degrees fweating them out : as I have alfo obferved in green Wood all thofe long *Microfcopical* pores which appear in Charcoal perfectly empty of any thing but Air.

Now, though I have with great diligence endeavoured to find whether there be any fuch thing in thofe *Microfcopical* pores of Wood or Piths, as the *Valves* in the heart, veins, and other paffages of Animals, that open and give paffage to the contain'd fluid juices one way, and fhut themfelves, and impede the paffage of fuch liquors back again, yet have I not hitherto been able to fay any thing pofitive in it; though, me thinks, it feems very probable, that Nature has in thefe paffages, as well as in thofe of Animal bodies, very many appropriated Inftruments and contrivances, whereby to bring her defigns and end to pafs, which 'tis not improbable, but that fome diligent Obferver, if help'd with better *Microfcopes*, may in time detect.

And that this may be fo, feems with great probability to be argued from the ftrange *Phænomena* of fenfitive Plants, wherein Nature feems to perform feveral Animal actions with the fame *Schematifm* or *Originization* that is common to all Vegetables, as may appear by fome no lefs inftructive then curious Obfervations that were made by divers Eminent Members of the *Royal Society* on fome of thefe kind of Plants, whereof an account was delivered in to them by the moft Ingenious and Excellent *Phyfician*, Doctor *Clark*, which, having that liberty granted me by that moft Illuftrious Society, I have hereunto adjoyn'd.

Obfervations on the Humble *and* Senfible Plants *in* M Chiffin's *Garden in Saint* James's *Park, made* Auguft *the* 9[th,] 1661. *Prefent, the* Lord *Brouncker,* Sr. *Robert Moray,* Dr. *Wilkins,* Mr. *Evelin,* Dr. *Henfhaw, and* Dr. *Clark.*

There are four Plants, two of which are little fhrub Plants, with a little fhort ftock, about an Inch above the ground, from whence are fpread feveral fticky branches, round, ftreight, and fmooth, smooth in the diſtances between the Sprouts, but juſt under the Sprouts there are two ſharp thorny prickles, broad in the letting on, as in the Bramble, one juſt under the Sprout, the other on the oppoſite ſide of the branch.

The diſtances betwixt the Sprouts are uſually ſomething more then an Inch, and many upon a Branch, according to its length, and they grew ſo, that if the lower Sprout be on the left ſide of the Branch, the next above is on the right, and ſo to the end, not ſprouting by pairs. See Schem.11. Fig 2.

At the end of each Sprout are generally four ſprigs, two at the Extremity, and one on each ſide, juſt under it. At the firſt ſprouting of theſe from the Branch to the Sprig where the leaves grow, they are full of little ſhort white hairs, which wear off as the leaves grow, and then they are ſmooth as the Branch.

Upon each of theſe ſprigs, are, for the moſt part, eleven pair of leaves, neatly ſet into the uppermoſt part of the little ſprig, exactly one againſt another, as it were in little *articulations*, ſuch as Anatomiſts call *Enarthroſis*, where the round head of a Bone is received into another fitted for its motion ; and ſtanding very fitly to ſhut themſelves and touch, the pairs juſt above them cloſing ſomewhat upon them, as in the ſhut ſprig ; ſo is the little round *Pedunculus* of this leaf fitted into a little cavity of the ſprig, viſible to the eye in a ſprig new pluck'd, or in a ſprig withered on the Branch, from which the leaves eaſily fall by touching.

The leaf being almoſt an oblong ſquare, and ſet into the *Pedunculus*, at one of the lower corners, receiveth from that not onely a *Spine*, as I may call it, which, paſſing through the leaf, divides it ſo length-ways that the outer-ſide is broader then the inner next the ſprig, but little *fibres* paſſing obliquely towards the oppoſite broader ſide, ſeem to make it here a little muſcular, and fitted to move the whole leaf, which, together with the whole ſprig, are ſet full with little ſhort whitiſh hairs.

One

One of thefe Plants, whofe branch feem'd to be older and more grown then the other, onely the tender Sprouts of it, after the leaves are fhut, fall and hang down ; of the other, the whole branches fall to the ground, if the Sun fhine very warm, upon the firft taking off the Glafs, which I therefore call the *humble Plant*.

The other two, which do never fall, nor do any of their branches flagg and hang down, fhut not their leaves, but upon fomewhat a hard ftroke ; the ftalks feem to grow up from a root, and appear more *herbaceous*, they are round and fmooth, without any prickle, the Sprouts from them have feveral pairs of fprigs, with much lefs leaves then the other on them, and have on each fprig generally feventeen pair.

Upon touching any of the fprigs with leaves on, all the leaves on that fprig contracting themfelves by pairs, joyned their upper fuperficies clofe together.

Upon the dropping a drop of *Aqua fortis* on the fprig betwixt the leaves, $f f$ all the leaves above fhut prefently, thofe below by pairs fucceffively after, and by the lower leaves of the other branches, $l l$, $k k$, &c. and fo every pair fucceffively, with fome little diftance of time betwixt, to the top of each fprig, and fo they continu'd fhut all the time we were there. But I returning the next day, and feveral days fince, found all the leaves dilated again on two of the fprigs ; but from $f f$, where the *Aqua fortis* had dropped upwards, dead and withered ; but thofe below on the fame fprig, green, and clofing upon the touch, and are fo at this day, *Auguft* 14.

With a pair of Scifters, as fuddenly as it could be done, one of the leaves $b b$ was clipped off in the middle, upon which that pair, and the pair above, clofed prefently, after a little interval, $d d$, then $e e$, and fo the reft of the pairs, to the bottom of the fprig, and then the motion began in the lower pairs, $l l$, on the other fprigs, and fo fhut them by pairs upwards, though not with fuch diftinct diftances.

Under

Under a pretty large branch with its fprigs on, there lying a large Shell betwixt two and three Inches below it, there was rubbed on a ftrong fented oyl, after a little time all the leaves on that fprig were fhut, and fo they continued all the time of our ftay there, but at my returne the next day, I found the pofition of the Shell alter'd, and the leaves expanded as before, and clofing upon the touch.

Upon the application of the Sun-beams by a Burning-glafs, the more *humble Plant* fell, the other fhut their leaves.

We could not fo apply the fmoak of *Sulpher*, as to have any vifible effect from that, at two or three times trial; but on another trial, the fmoak touching the leaves, it fucceeded.

The *humble Plant* fell upon taking off the Glafs wherewith it was covered.

Cutting off one of the little Sprouts, two or three drops of liquor were thruft out of the part from whence that was cut, very cleer, and pellucid, of a bright greenifh colour, tafting at firft a little bitterifh, but after leaving a licorifh-like tafte in my mouth.

Since, going two or three times when it was cold, I took the Glaffes from the more *humble Plant*, and it did not fall as formerly, but fhut its leaves onely. But coming afterwards, when the Sun fhone very warm, as foon as it was taken off, it fell as before.

Since I pluck'd off another fprig, whofe leaves were all fhut, and had been fo fome time, thinking to obferve the liquor fhould come from that I had broken off, but finding none, though with preffing, to come, I, as dexteroufly as I could, pull'd off one whofe leaves were expanded, and then had upon the fhutting of the leaves, a little of the mention'd liquor, from the end of the fprig I had broken from the Plant. And this twice fucceffively, as often almoft as I durft rob the Plant.

But my curiofity carrying me yet further, I cut off one of the harder branches of the ftronger Plant, and there came of the liquor,

liquor, both from that I had cut, and that I had cut it from, without preſſure.

Which made me think, that the motion of this Plant upon touching, might be from this, that there being a conſtant *inter-courſe* betwixt every part of this Plant and its root, either by a *circulation* of this liquor, or a conſtant preſſing of the ſubtiler parts of it to every extremity of the Plant. Upon every preſſure, from whatſoever it proceeds, greater then that which keeps it up, the ſubtile parts of this liquor are thruſt downwards, towards its *articulations* of the leaves, where, not having room preſently to get into the ſprig, the little round *pedunculus*, from whence the *Spine* and thoſe oblique *Fibres* I mention'd riſe, being dilated, the *Spine* and *Fibres* (being continued from it) muſt be contracted and ſhortned, and ſo draw the leaf upwards to joyn with its fellow in the ſame condition with it ſelf, where, being cloſed, they are held together by the implications of the little whitiſh hair, as well as by the ſtill retreating liquor, which diſtending the *Fibres* that are continued lower to the branch and root, ſhorten them above ; and when the liquor is ſo much forced from the Sprout, whoſe *Fibres* are yet tender, and not able to ſupport themſelves, but by that tenſneſs which the liquor filling their *interſtices* gives them, the Sprout hangs and flags.

But, perhaps, he that had the ability and leiſure to give you the exact *Anatomy* of this pretty Plant, to ſhew you its *Fibres,* and viſible *Canales,* through which this fine liquor circulateth, or is moved, and had the faculty of better and more copiouſly expreſſing his Obſervations and conceptions, ſuch a one would eaſily from the motion of this liquor, ſolve all the *Phænomena,* and would not fear to affirm, that it is no obſcure ſenſation this Plant hath. But I have ſaid too much, I humbly ſubmit, and am ready to ſtand corrected.

I have not yet made ſo full and ſatisfactory Obſervations as I deſire on this Plant, which ſeems to be a Subject that will afford abundance of information.

information. But as farr as I have had opportunity to examine it, I have difcovered with my *Microfcope* very curious ftructures and contrivances; but defigning much more accurate examinations and trials, both with my *Microfcope*, and otherwife, as foon as the feafon will permit, I fhall not till then add any thing of what I have already taken notice of; but as farr as I have yet obferv'd, I judge the motion of it to proceed from caufes very differing from thofe by which Gut-ftrings, or Lute-ftrings, the beard of a wilde *Oat*, or the beard of the Seeds of *Geranium*, *Mofcatum*, or *Musk-grafs* and other of kinds of *Cranes-bill*, move themfelves. Of which I fhall add more in the fubfequent Obfervations on thofe bodies.

Obferv. XIX. *Of a* Plant *growing in the blighted or yellow fpecks of* Damask-rofe-leaves, Bramble-leaves, *and fome other kind of leaves.*

I Have for feveral years together, in the Moneths of *June*, *July*, *Auguft*, and *September* (when any of the green leaves of *Rofes* begin to dry and grow yellow) obferv'd many of them, efpecially the leaves of the old fhrubs of *Damask-Rofes*, all befpecked with yellow ftains, and the under-fides juft againft them, to have little yellow hillocks of a gummous fub-ftance, and feveral of them to have fmall black fpots in the midft of thofe yellow ones, which, to the naked eye, appear'd no bigger then the point of a Pin, or the fmalleft black fpot or tittle of Ink one is able to make with a very fharp pointed Pen.

Examining thefe with a *Microfcope*, I was able plainly to diftinguifh, up and down the furface, feveral fmall yellow knobs, of a kind of yellowifh red gummy fubftance, out of which I perceiv'd there fprung multitudes of little cafes or black bodies like Seed-cods, and thofe of them that were quite without the hillock of Gumm, difclos'd themfelves to grow out of it with a fmall Straw-colour'd and tranfparent ftem, the which feed and ftem appear'd very like thofe of common Mofs (which I elfe-where defcribe) but that they were abundantly lefs, many hundreds of them being not able to equalize one fingle feed Cod of Mofs.

I have often doubted whether they were the feed Cods of fome little Plant, or fome kind of fmall Buds, or the Eggs of fome very fmall Infect, they appear'd of a dark brownifh red, fome almoft quite black, and of a Figure much refembling the feed-cod of Mofs, but their ftalks on which they grew were of a very fine tranfparent fubftance, almoft like the ftalk of mould, but that they feem'd fomewhat more yellow.

That which makes me to fuppofe them to be Vegetables, is for that I perceiv'd many of thofe hillocks bare or deftitute, as if thofe bodies lay yet conceal'd, as G. In others of them, they were juft fpringing out of their gummy hillocks, which all feem'd to fhoot directly outwards, as at A. In others, as at B, I found them juft gotten out, with very little or no ftalk,

and

and the Cods of an indifferent cize;but in others,as C, I found them begin to have little short stalks, or stems; in others, as D, those stems were grown bigger, and larger; and in others, as at E, F, H, I, K, L, &c. those stems and Cods were grown a great deal bigger, and the stalks were more bulky about the root, and very much taper'd towards the top, as at F and L is most visible.

I did not find that any of them had any seed in them, or that any of them were hollow, but as they grew bigger and bigger, I found those heads or Cods begin to turn their tops towards their roots, in the same manner as I had observ'd that of Moss to do; so that in all likelihood, Nature did intend in that posture, what she does in the like seed-cods of greater bulk, that is, that the seed, when ripe, should be shaken out and dispersed at the end of it, as we find in Columbine Cods, and the like.

The whole Oval O O O O in the second *Figure* of the 12. *Scheme* represents a small part of a Rose leaf, about the bigness of the little Oval in the hillock, C, marked with the Figure X. in which I have not particularly observ'd all the other forms of the surface of the Rose-leaf, as being little to my present purpose.

Now, if these Cods have a seed in them so proportion'd to the Cod, as those of *Pinks*, and *Carnations*, and *Columbines*, and the like, how unimaginably small must each of those seeds necessarily be, for the whole length of one of the largest of those Cods was not $\frac{1}{500}$ part of an Inch; some not above $\frac{1}{1000}$, and therefore certainly, very many thousand of them would be unable to make a bulk that should be visible to the naked eye; and if each of these contain the Rudiments of a young Plant of the same kind, what must we say of the pores and constituent parts of that?

The generation of this Plant seems in part,ascribable to a kind of *Mildew* or *Blight*,whereby the parts of the leaves grow scabby, or putrify'd, as it were, so as that the moisture breaks out in little scabs or spots, which, as I said before, look like little knobs of a red gummous substance.

From this putrify'd scabb breaks out this little Vegetable; which may be somewhat like a *Mould* or *Moss*; and may have its *equivocal* generation much after the same manner as I have supposed *Moss* or *Mould* to have, and to be a more simple and uncompounded kind of vegetation, which is set a moving by the *putrifactive* and *fermentative* heat, joyn'd with that of the ambient aerial, when (by the putrifaction and decay of some other parts of the vegetable, that for a while staid its progress) it is unfetter'd and left at liberty to move in its former course, but by reason of its *regulators*, moves and acts after quite another manner then it did when a *coagent* in the more compounded *machine* of the more perfect Vegetable.

And from this very same Principle, I imagine the *Misleto* of Oaks, Thorns, Appletrees, and other Trees, to have its original: It seldom or never growing on any of those Trees,till they begin to wax decrepid,and decay with age, and are pester'd with many other infirmities.

Hither also may be referr'd those multitudes and varieties of *Mushroms*, such as that,call'd *Jews-ears*, all sorts of *gray* and *green* Mosses, &c. which

infest

infeſt all kind of Trees,ſhrubs,and the like,eſpecially when they come to any bigneſs. And this we ſee to be very much the method of Nature throughout its operations, *putrifactive Vegetables* very often producing a Vegetable of a much leſs compounded nature, and of a much inferiour tribe ; and *putrefactive* animal ſubſtances degenerating into ſome kind of animal production of a much inferiour rank,and of a more ſimple nature.

Thus we find the humours and ſubſtances of the body,upon *putrifacti-on*,to produce ſtrange kinds of moving Vermine : the *putrifaction* of the ſlimes and juices of the Stomack and Guts, produce Worms almoſt like Earth-worms,the Wheals in childrens hands produce a little Worm,call'd a *Wheal-worm* : The bloud and milk, and other humours, produce other kinds of Worms, at leaſt, if we may believe what is deliver'd to us by very famous Authors ; though, I confeſs, I have not yet been able to diſcover ſuch my ſelf.

And whereas it may ſeem ſtrange that *Vinegar, Meal,* muſty *Casks* , &c. are obſerv'd to breed their differing kinds of Inſects, or living creatures, whereas they being Vegetable ſubſtances, ſeem to be of an inferiour kind, and ſo unable to produce a creature more noble, or of a more compounded nature then they themſelves are of, and ſo without ſome concurrent ſeminal principle, may be thought utterly unfit for ſuch an operation ; I muſt add, that we cannot preſently poſitively ſay, there are no animal ſubſtances, either mediately, as by the ſoil or fatning of the Plant from whence they ſprung,or more immediately,by thereal mixture or compoſition of ſuch ſubſtances, join'd with them ; or perchance ſome kind of Inſect, in ſuch places where ſuch kind of *putrifying* or *fermenting* bodies are, may, by a certain inſtinct of nature, eject ſome ſort of ſeminal principle, which cooperating with various kinds of *putrifying* ſubſtances, may produce various kinds of Inſects,or Animate bodies : For we find in moſt ſorts of thoſe lower degrees of Animate bodies, that the *putrifying* ſubſtances on which theſe Eggs, Seeds, or ſeminal principles are caſt by the Inſect, become, as it were, the *Matrices* or Wombs that conduce very much to their generation, and may perchance alſo to their variation and alteration, much after the ſame manner. as, by ſtrange and unnatural copulations, ſeveral new kinds of Animals are produc'd, as *Mules,* and the like, which are uſually call d Monſtrous, becauſe a little unuſual, though many of them have all their principal parts as perfectly ſhap'd and adapted for their peculiar uſes, as any of the moſt perfect Animals. If therefore the *putrifying* body,on which any kind of ſeminal or vital principle chances to be caſt become ſomewhat more then meerly a nurſing and foſtering helper in the generation and production of any kind of Animate body, the more neer it approaches the true nature of a Womb, the more power will it have on the by-blow it incloſes. But of this ſomewhat more in the deſcription of the *Water-gnat.* Perhaps ſome more accurate Enquiries and Obſervations about theſe matters might bring the Queſtion to ſome certainty, which would be of no ſmall concern in Natural Philoſophy.

But that *putrifying* animal ſubſtances may produce animals of an inferior kind, I fee not any fo very great a difficulty, but that one may, without much abfurdity, admit : For as there may be multitudes of contrivances that go to the making up of one compleat Animate body ; fo, That fome of thofe *coadjutors*, in the perfect exiftence and life of it, may be vitiated, and the life of the whole deftroyed, and yet feveral of the conftituting contrivances remain intire, I cannot think it beyond imagination or poffibility ; no more then that a like accidental procefs, as I have elfwhere hinted, may alfo be fuppofed to explicate the method of Nature in the *Metamorphofis* of Plants. And though the difference between a Plant and an Animal be very great, yet I have not hitherto met with any fo *cogent* an Argument, as to make me pofitive in affirming thefe two to be altogether *Heterogeneous*, and of quite differing kinds of Nature : And befides, as there are many *Zoophyts*, and fenfitive Plants (divers of which I have feen, which are of a middle nature, and feem to be Natures tranfition from one degree to another, which may be obferv'd in all her other paffages, wherein fhe is very feldom obferv'd to leap from one ftep to another) fo have we, in fome Authors, Inftances of Plants turning into Animals, and Animals into Plants, and the like ; and fome other very ftrange (becaufe unheeded) proceedings of Nature ; fomething of which kind may be met with, in the defcription of the *Water-Gnat*, though it be not altogether fo direct to the prefent purpofe.

But to refer this Difcourfe of Animals to their proper places, I fhall add, that though one fhould fuppofe, or it fhould be prov'd by Obfervations, that feveral of thefe kinds of Plants are accidentally produc'd by a cafual *putrifaction*, I fee not any great reafon to queftion, but that, notwithftanding its own production was as 'twere cafual, yet it may germinate and produce feed, and by it propagate its own, that is, a new Species. For we do not know, but that the Omnipotent and All-wife Creator might as directly defign the ftructure of fuch a Vegetable, or fuch an Animal to be produc'd out of fuch or fuch a *putrifaction* or change of this or that body, towards the conftitution or ftructure of which, he knew it neceffary, or thought it fit to make it an ingredient ; as that the digeftion or moderate heating of an Egg, either by the Female, or the Sun, or the heat of the Fire, or the like, fhould produce this or that Bird ; or that *Putrifactive* and warm fteams fhould, out of the blowings, as they call them, that is, the Eggs of a Flie, produce a living Magot, and that, by degrees, be turn'd into an *Aurelia*, and that, by a longer and a proportion'd heat, be *tranfmuted* into a Fly. Nor need we therefore to fuppofe it the more imperfect in its kind, then the more compounded Vegetable or Animal of which it is a part ; for he might as compleatly furnifh it with all kinds of contrivances neceffary for its own exiftence, and the propagation of its own Species, and yet make it a part of a more compounded body : as a Clock-maker might make a Set of Chimes to be a part of a Clock, and yet, when the watch part or ftriking part are taken away, and the hindrances of its motion remov'd, this chiming part may go as accurately, and ftrike its tune as exactly, as if it were ftill a part of the compounded *Automaton.* So, though the original caufe, or

feminal

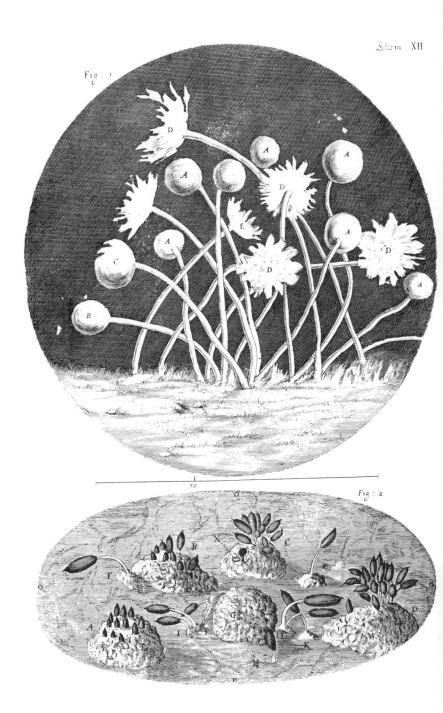

Fig: 1

Fig: 2

feminal principle from which this minute Plant on Rofe leaves did fpring were, before the corruption caus'd by the Mill-dew, a component part of the leaf on which it grew, and did ferve as a *coagent* in the producti-on and conftitution of it, yet might it be fo confummate, as to produce a feed which might have a power of propagating the fame fpecies:the works of the Creator feeming of fuch an excellency,that though they are unable to help to the perfecting of the more compounded exiftence of the greater Plant or Animal,they may have notwithftanding an ability of acting fingly upon their own internal principle, fo as to produce a Vegetable body, though of a lefs compounded nature, and to proceed fo farr in the me-thod of other Vegetables, as to bear flowers and feeds, which may be ca-pabale of propagating the like. So that the little cafes which appear to grow on the top of the flender ftalks, may, for ought I know, though I fhould fuppofe them to fpring from the perverting of the ufual courfe of the parent Vegetable, contain a feed, which, being fcatter'd on other leaves of the fame Plant, may produce a Plant of much the fame kind.

Nor are Damafk-Rofe leaves the onely leaves that produce thefe kinds of Vegetable fproutings; for I have obferv'd them alfo in feveral other kinds of Rofe leaves, and on the leaves of feveral forts of Briers, and on Bramble leaves they are oftentimes to be found in very great clufters; fo that I have found in one clufter,three,four, or five hundred of them, making a very confpicuous black fpot or fcab on the back fide of the leaf.

Obferv. X X. *Of* blue Mould, *and of the firft Principles of Ve-getation arifing from* Putrefaction.

THe Blue and White and feveral kinds of hairy mouldy fpots, which are obfervable upon divers kinds of *putrify'd* bodies, whether Ani-mal fubftances,or Vegetable,fuch as the fkin, raw or drefs'd, flefh,bloud, humours, milk, green Cheefe, &c. or rotten fappy Wood, or Herbs, Leaves,Barks, Roots, &c. of Plants, are all of them nothing elfe but fe-veral kinds of fmall and varioufly figur'd Mufhroms, which, from conve-nient materials in thofe *putrifying* bodies, are, by the concurrent heat of the Air, excited to a certain kind of vegetation, which will not be un-worthy our more ferious fpeculation and examination. as I fhall by and by fhew. But, firft, I muft premife a fhort defcription of this *Specimen,* which I have added of this Tribe, in the firft Figure of the X I I. *Scheme,* which is nothing elfe but the appearance of a fmall white fpot of hairy mould,multitudes of which I found to befpeck & whiten over the red co-vers of a fmall book,which, it feems, were of Sheeps-fkin,that being more apt to gather mould, even in a dry and clean room, then other leathers. Thefe fpots appear'd,through a good *Microfcope,*to be a very pretty fhap'd Vegetative body, which, from almoft the fame part of the Leather, fhot
out

out multitudes of fmall long cylindrical and tranfparent ftalks, not exactly ftreight, but a little bended with the weight of a round and white knob
that grew on the top of each of them; many of thefe knobs I obferv'd
to be very round, and of a fmooth furface, fuch as A A, &c. others
fmooth likewife, but a little oblong, as B; feveral of them a little broken,
or cloven with chops at the top, as C; others flitter'd as 'twere, or flown
all to pieces, as D D. The whole fubftance of thefe pretty bodies was
of a very tender conftitution, much like the fubftance of the fofter kind
of common white Mufhroms, for by touching them with a Pin, I found
them to be brufed and torn; they feem'd each of them to have a diftinct root of their own; for though they grew neer together in a clufter,
yet I could perceive each ftem to rife out of a diftinct part or pore of the
Leather; fome of thefe were fmall and fhort, as feeming to have been but
newly fprung up, of thefe the balls were for the moft part round, others
were bigger, and taller, as being perhaps of a longer growth, and of thefe,
for the moft part, the heads were broken, and fome much wafted, as E;
what thefe heads contain'd I could not perceive; whether they were
knobs and flowers, or feed cafes, I am not able to fay, but they feem'd
moft likely to be of the fame nature with thofe that grow on Mufhroms,
which they did, fome of them, not a little refemble.

Both their fmell and tafte, which are active enough to make a fenfible
impreffion upon thofe organs, are unpleafant and noifome.

I could not find that they would fo quickly be deftroy'd by the actual
flame of a Candle, as at firft fight of them I conceived they would be, but
they remain'd intire after I had paft that part of the Leather on which
they ftuck three or four times through the flame of a Candle; fo that, it
feems they are not very apt to take fire, no more then the common white
Mufhroms are when they are fappy.

There are a multitude of other fhapes, of which thefe *Microfcopical*
Mufhroms are figur'd, which would have been a long Work to have defcribed, and would not have fuited fo well with my defign in this Treatife,
onely, amongft the reft, I muft not forget to take notice of one that was a
little like to, or refembled, a Spunge, confifting of a multitude of little
Ramifications almoft as that body does, which indeed feems to be a kind
of Water-Mufhrom, of a very pretty texture, as I elfe-where manifeft.
And a fecond, which I muft not omit, becaufe often mingled, and neer
adjoining to thefe I have defcrib'd, and this appear'd much like a Thicket
of bufhes, or brambles, very much branch'd, and extended, fome of them,
to a great length, in proportion to their Diameter, like creeping brambles.

The manner of the growth and formation of this kind of Vegetable, is
the third head of Enquiry, which, had I time, I fhould follow: the figure
and method of Generation in this concrete feeming to me, next after
the Enquiry into the formation, figuration, or chryftalization of Salts, to
be the moft fimple, plain, and eafie; and it feems to be a *medium*
through which he muft neceffarily pafs, that would with any likelihood
inveftigate the *forma informans* of Vegetables: for as I think that he fhall
find it a very difficult tafk, who undertakes to difcover the form of Saline

line cryſtallizations, without the conſideration and preſcience of the nature and reaſon of a Globular form, and as difficult to explicate this configuration of Muſhroms, without the previous conſideration of the form of Salts; ſo will the enquiry into the forms of Vegetables be no leſs, if not much more difficult, without the fore-knowledge of the forms of Muſhroms, theſe ſeveral Enquiries having no leſs dependance one upon another then any ſelect number of Propoſitions in Mathematical Elements may be made to have.

Nor do I imagine that the ſkips from the one to another will be found very great, if beginning from fluidity, or body without any form, we deſcend gradually, till we arrive at the higheſt form of a bruite Animals Soul, making the ſteps or foundations of our Enquiry, *Fluidity, Orbiculation, Fixation, Angulization,* or *Cryſtallization Germination* or *Ebullition, Vegetation, Plant animation, Animation, Senſation, Imagination.*

Now, that we may the better proceed in our Enquiry, It will be requiſite to conſider:

Firſt, that Mould and Muſhroms require no ſeminal property. but the former may be produc'd at any time from any kind of *putrifying* Animal, or Vegetable Subſtance, as Fleſh, &c. kept moiſt and warm, and the latter, if what *Mathiolus* relates be true, of making them by Art, are as much within our command, of which Matter take the *Epitomie* which Mr. *Parkinſon* has deliver'd in his *Herbal,* in his Chapter of *Muſhroms,* becauſe I have not *Mathiolus* now by me: *Unto theſe Muſhroms* (ſaith he) *may alſo be adjoyn'd thoſe which are made of Art (whereof* Mathiolus *makes mention) that grow naturally among certain ſtones in* Naples, *and that the ſtones being digg'd up, and carried to* Rome, *and other places, where they ſet them in their Wine Cellars, covering them with a little Earth, and ſprinkling a little warm water thereon, would within four days produce Muſhroms fit to be eaten, at what time one will: As alſo that Muſhroms may be made to grow at the foot of a wilde* Poplar Tree, *within four days after, warm water wherein ſome leaves have been diſſolv'd ſhall be pour'd into the Root (which muſt be ſlit) and the ſtock above ground.*

Next, that as Muſhroms may be generated without ſeed, ſo does it not appear that they have any ſuch thing as ſeed in any part of them; for having conſidered ſeveral kinds of them, I could never find any thing in them that I could with any probability gheſs to be the ſeed of it, ſo that it does not as yet appear (that I know of) that Muſhroms may be generated from a ſeed, but they rather ſeem to depend merely upon a convenient conſtitution of the matter out of which they are made, and a concurrence of either natural or artificial heat.

Thirdly, that by ſeveral bodies (as Salts and Metals both in Water and in the air, and by ſeveral kinds of ſublimations in the Air) actuated and guided with a congruous heat, there may be produc'd ſeveral kinds of bodies as curiouſly, if not of a more compos'd Figure; ſeveral kinds of riſing or Ebulliating Figures ſeem to manifeſt; as witneſs the ſhooting in the Rectification of ſpirits of *Urine, Hart-horn, Bloud,* &c. witneſs alſo the curious branches of evaporated diſſolutions, ſome of them againſt
the

the fides of the containing Jar: others ftanding up, or growing an end, out of the bottom, of which I have taken notice of a very great variety. But above all the reft, it is a very pretty kind of Germination which is afforded us in the Silver Tree, the manner of making which with Mercury and Silver, is well known to the Chymifts, in which there is an Ebullition or Germination, very much like this of Mufhroms, if I have been rightly inform'd of it.

Fourthly, I have very often taken notice of, and alfo obferv'd with a *Microfcope*, certain excrefcencies or Ebullitions in the fnuff of a Candle, which, partly from the fticking of the fmoaky particles as they are carryed upwards by the current of the rarify'd Air and flame, and partly alfo from a kind of Germination or Ebullition of fome actuated unctuous parts which creep along and filter through fome fmall ftring of the Week, are formed into pretty round and uniform heads, very much refembling the form of hooded Mufhroms, which, being by any means expos'd to the frefh Air, or that air which encompaffes the flame, they are prefently lick'd up and devour'd by it, and vanifh.

The reafon of which *Phænomenon* feems to me, to be no other then this:

That when a convenient thread of the Week is fo bent out by the fides of the fnuff that are about half an Inch or more, remov'd above the bottom, or loweft part of the flame, and that this part be wholly included in the flame; the Oyl (for the reafon of filtration, which I have elfewhere rendred) being continualy driven up the fnuff, is driven likewife into this ragged bended-end, and this being remov'd a good diftance, as half an Inch or more, above the bottom of the flame, the parts of the air that paffes by it, are already, almoft fatiated with the diffolution of the boiling unctuous fteams that iffued out below, and therefore are not onely glutted, that is, can diffolve no more then what they are already acting upon, but they carry up with them abundance of unctuous and footy particles, which meeting with that rag of the Week, that is plentifully fill'd with Oyl, and onely fpends it as faft as it evaporates, and not at all by diffolution or burning, by means of thefe fteamy parts of the filterated Oyl iffuing out at the fides of this ragg, and being inclos'd with an air that is already fatiated and cannot prey upon them nor burn them, the afcending footy particles are ftay'd about it and fix'd, fo as that about the end of that ragg or filament of the fnuff, whence the greateft part of the fteams iffue, there is conglobated or fix'd a round and pretty uniform cap, much refembling the head of a Mufhrom, which, if it be of any great bignefs, you may obferve that its underfide will be bigger then that which is above the ragg or ftem of it; for the Oyl that is brought into it by filtration, being by the bulk of the cap a little fhelter'd from the heat of the flame, does by that means iffue as much out from beneath from the ftalk or downwards, as it does upwards, and by reafon of the great accefs of the adventitious fmoak from beneath, it increafes moft that way. That this may be the true reafon of this *Phænomenon*, I could produce many Arguments and Experiments to make it probable: As,

Firft, that the *Filtration* carries the Oyl to the top of the Week, at leaft as high as thefe raggs, is vifible to one that will obferve the fnuff of a burning Candle with a *Microfcope*, where he may fee an Ebullition or bubbling of the Oyl, as high as the fnuff looks black.

Next, that it does fteam away more then burn; I could tell you of the dim burning of a Candle, the longer the fnuff be which arifes from the abundance of vapours out of the higher parts of it.

And, thirdly, that in the middle of the flame of the Candle, neer the top of the fnuff, the fire or diffolving principle is nothing neer fo ftrong, as neer the bottom and out edges of the flame, which may be obferv d by the burning afunder of a thread, that will firft break in thofe parts that the edges of the flame touch, and not in the middle.

And I could add feveral Obfervables that I have taken notice of in the flame of a Lamp actuated with Bellows, and very many others that confirm me in my opinion, but that it is not fo much to my prefent purpofe, which is onely to confider this concreet in the fnuff of a Candle, fo farr as it has any refemblance of a Mufhrom, to the confideration of which, that I may return, I fay, we may alfo obferve:

In the firft place, that the droppings or trillings of Lapidefcent waters in Vaults under ground, feem to conftitute a kind of *petrify'd* body, form'd almoft like fome kind of Mufhroms inverted, in fo much that I have feen fome knobb'd a little at the lower end, though for the moft part, indeed they are otherwife fhap'd, and taper'd towards the end; the generation of which feems to be from no other reafon but this, that the water by foaking through the earth and Lime (for I ghefs that fubftance to add much to it *petrifying* quality) does fo impregnate it felf with ftony particles, that hanging in drops in the roof of the Vault, by reafon that the foaking of the water is but flow, it becomes expos'd to the Air, and thereby the outward part of the drop by degrees grows hard, by reafon that the water gradually evaporating the ftony particles neer the outfides of the drop begin to touch, and by degrees, to dry and grow clofer together, and at length conftitute a cruft or fhell about the drop; and this foaking by degrees, being more and more fupply'd, the drop grows longer and longer, and the fides harden thicker and thicker into a Quill or Cane, and at length, that hollow or pith becomes almoft ftop'd up, and folid: afterwards the foaking of the *petrifying* water, finding no longer a paffage through the middle, burfts out, and trickles down the outfide, and as the water evaporates, leaves new fuperinduc'd fhells, which more and more fwell the bulk of thofe Iceicles; and becaufe of the great fupply from the Vault, of *petrifying* water, thofe bodies grow bigger and bigger next to the Vault, and taper or fharpen towards the point; for the accefs from the arch of the Vault being but very flow, and confequently the water being fpread very thinly over the furface of the Iceicle, the water begins to fettle before it can reach to the bottom, or corner end of it; whence, if you break one of thefe, you would almoft imagine it a ftick of Wood *petrify'd*, it having fo pretty a refemblance of pith and grain, and if you look on the outfide of a piece, or of one whole, you would think no lefs, both from its vegetable roundnefs and

T

tapering

tapering form; but whereas all Vegetables are obferv'd to fhoot and grow perpendicularly upwards, this does fhoot or propend directly downwards.

By which laft Obfervables, we fee that there may be a very pretty body fhap'd and concreeted by Mechanical principles, without the leaft fhew or probability of any other feminal *formatrix*.

And fince we find that the great reafon of the *Phænomena* of this pretty *petrifaction*, are to be reduc'd from the gravity of a fluid and pretty volatil body impregnated with ftony particles, why may not the *Phænomena* of Ebullition or Germination be in part poffibly enough deduc'd from the levity of an impregnated liquor, which therefore perpendicularly afcending by degrees, evaporates and leaves the more folid and fix'd parts behind in the form of a Mufhrom, which is yet further diverfify'd and fpecificated by the forms of the parts that impregnated the liquor, and compofe or help to conftitute the Mufhrom.

That the foremention'd Figures of growing Salts, and the Silver Tree, are from this principle, I could very eafily manifeft; but that I have not now a convenient opportunity of following it, nor have I made a fufficient number of Experiments and Obfervations to propound, explicate, and prove fo ufefull a *Theory* as this of Mufhroms: for, though the contrary principle to that of *petrify'd* Iceicles may be in part a caufe; yet I cannot but think, that there is fomewhat a more complicated caufe, though yet Mechanical, and poffible to be explain'd.

We therefore have further to enquire of it, what makes it to be fuch a liquor, and to afcend, whether the heat of the Sun and Air, or whether that of *firmentiation* and *putrifaction*, or both together; as alfo whether there be not a third or fourth; whether a Saline principle be not a confiderable agent in this bufinefs alfo as well as heat; whether alfo a fixation, precipitation or fettling of certain parts out of the aerial Mufhrom may not be alfo a confiderable coadjutor in the bufinefs. Since we find that many pretty beards or *ftiriæ* of the particles of Silver may be precipitated upon a piece of Brafs put into a *folution* of Silver very much diluted with fair water, which look not unlike a kind of mould or hoar upon that piece of metal; and the hoar froft looks like a kind of mould; and whether there may not be feveral others that do concurr to the production of a Mufhrom, having not yet had fufficient time to profecute according to my defires, I muft referr this to a better opportunity of my own, or leave and recommend it to the more diligent enquiry and examination of fuch as can be mafters both of leifure and conveniencies for fuch an Enquiry.

And in the mean time, I muft conclude, that as far as I have been able to look into the nature of this Primary kind of life and vegetation, I cannot find the leaft probable argument to perfwade me there is any other concurrent caufe then fuch as is purely Mechanical, and that the effects or productions are as neceffary upon the concurrence of thofe caufes as that a Ship, when the Sails are hoift up, and the Rudder is fet to fuch a pofition, fhould, when the Wind blows, be mov'd in fuch a way or courfe

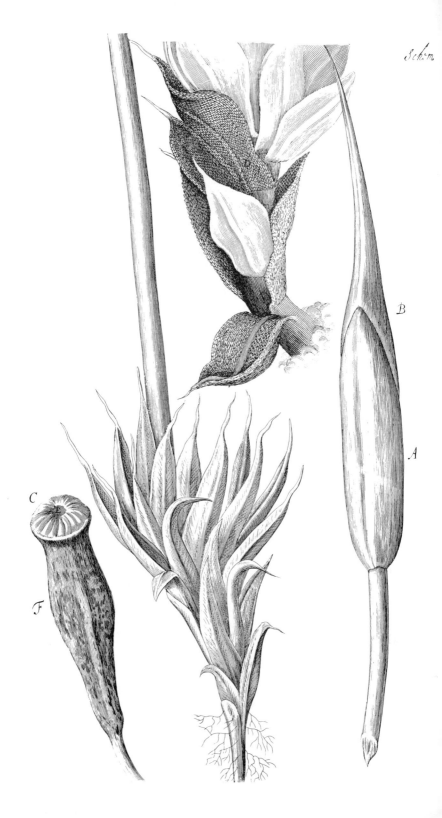

to that or t'other place; Or, as that the brufed Watch, which I mention in the defcription of Mofs, fhould, when thofe parts which hindred its motion were fallen away, begin to move, but after quite another manner then it did before.

Obferv. XXI. *Of Mofs, and feveral other fmall vegetative Subftances.*

MOfs is a Plant, that the wifeft of Kings thought neither unworthy his fpeculation, nor his Pen, and though amongft Plants it be in bulk one of the fmalleft, yet it is not the leaft confiderable: For, as to its fhape, it may compare for the beauty of it with any Plant that grows, and bears a much bigger breadth; it has a root almoft like a feedy Parfnep, furnifh'd with fmall ftrings and fuckers, which are all of them finely branch'd, like thofe of the roots of much bigger Vegetables; out of this fprings the ftem or body of the Plant, which is fomewhat *Quadrangular,* rather then *Cylindrical,* moft curiously *fluted* or ftrung with fmall creafes, which run, for the moft part, *parallel* the whole ftem; on the fides of this are clofe and thick fet, a multitude of fair, large, well-fhap'd leaves, fome of them of a rounder, others of a longer fhape, according as they are younger or older when pluck'd; as I ghefs by this, that thofe Plants that had the ftalks growing from the top of them, had their leaves of a much longer fhape, all the furface of each fide of which, is curiously cover'd with a multitude of little oblong tranfparent bodies, in the manner as you fee it exprefs'd in the leaf B, in the XIII. *Scheme.*

This Plant, when young and fpringing up, does much refemble a Houfleek, having thick leaves, almoft like that, and feems to be fomwhat of kin to it in other particulars; alfo from the top of the leaves, there fhoots out a fmall white and tranfparent hair, or thorn: This ftem, in time, come to fhoot out into a long, round and even ftalk, which by cutting tranfverfly, when dry, I manifeftly found to be a ftiff, hard, and hollow Cane, or Reed, without any kind of knot, or ftop, from its bottom, where the leaves encompafs'd it, to the top, on which there grows a large feed cafe, A, cover'd with a thin, and more whitifh fkin, B, terminated in a long thorny top, which at firft covers all the Cafe, and by degrees, as that fwells, the fkin cleaves, and at length falls off, with its thorny top and all (which is a part of it) and leaves the feed Cafe to ripen, and by degrees, to fhatter out its feed at a place underneath this cap, B, which before the feed is ripe, appears like a flat barr'd button, without any hole in the middle; but as it ripens, the button grows bigger, and a hole appears in the middle of it, E, out of which, in all probability, the feed falls: For as it ripens by a provifion of Nature, that end of this Cafe turns downward after the fame manner as the ears of Wheat and Barley ufually do; and opening feveral of thefe dry red Cafes, F, I found them to be
quite

quite hollow, without any thing at all in them; whereas when I cut them afunder with a fharp Pen-knife when green, I found in the middle of this great Cafe, another fmaller round Cafe, between which two, the *interftices* were fill'd with multitudes of ftringie *fibres*, which feem'd to fufpend the leffer Cafe in the middle of the other, which (as farr as I was able to dif-cern) feem'd full of exceeding fmall white feeds, much like the feed-bagg in the knop of a Carnation, after the flowers have been two or three days, or a week, fallen off; but this I could not fo perfectly difcern, and therefore cannot pofitively affirm it.

After the feed was fallen away, I found both the Cafe, Stalk, and Plant, all grow red and wither, and from other parts of the root continually to fpring new branches or flips, which by degrees increafed, and grew as bigg as the former, feeded, ripen'd, fhatter'd, and wither'd.

I could not find that it obferv'd any particular feafons for thefe feveral kinds of growth, but rather found it to be fpringing, mature, ripe, feedy, and wither'd at all times of the year; But I found it moft to flourifh and increafe in warm and moift weather.

It gathers its nourifhments, for the moft part, out of fome *Lapidefcent*, or other fubftance corrupted or chang'd from its former texture, or fub-ftantial form; for I have found it to grow on the rotten parts of Stone, of Bricks, of Wood, of Bones, of Leather, &c.

It oft grows on the barks of feveral Trees, fpreading it felf, fometimes from the ground upwards, and fometimes from fome chink or cleft of the bark of the Tree, which has fome *putrify'd* fubftance in it; but this feems of a diftinct kind from that which I obferv'd to grow on *putrify'd* inanimate bodies, and rotten earth.

There are alfo great varieties of other kinds of Moffes, which grow on Trees, and feveral other Plants, of which I fhall here make no mention, nor of the Mofs growing on the fkull of a dead man, which much re-fembles that of Trees.

Whether this Plant does fometimes originally fpring or rife out of cor-ruption, without any diffeminated feed, I have not yet made trials enough to be very much, either pofitive or negative; for as it feems very hard to conceive how the feed fhould be generally difpers'd into all parts where there is a corruption begun, unlefs we may rationally fuppofe, that this feed being fo exceeding fmall, and confequently exceeding light, is there-by taken up, and carried to and fro in the Air into every place, and by the falling drops of rain is wafh'd down out of it, and fo difpers'd into all places, and there onely takes root and propagates, where it finds a con-venient foil or matrix for it to thrive in; fo if we will have it to proceed from corruption, it is not lefs difficult to conceive,

Firft, how the corruption of any Vegetable, much lefs of any Stone or Brick, fhould be the Parent of fo curioufly figur'd, and fo perfect a Plant as this is. But here indeed, I cannot but add, that it feems rather to be a product of the Rain in thofe bodies where it is ftay'd, then of the very bodies themfelves, fince I have found it growing on Marble, and Flint; but always the *Microfcope*, if not the naked eye, would difcover fome little hole of Dirt in which it was rooted.   Next,

Next, how the corruption of each of thofe exceedingly differing bodies fhould all confpire to the production of the fame Plant, that is, that Stones, Bricks, Wood, or vegetable fubftances, and Bones, Leather, Horns, or animate fubftances, unlefs we may with fome plaufiblenefs fay, that Air and Water are the coadjutors, or *menftruums*, in all kinds of *putrifactions*, and that thereby the bodies (though whilft they retain'd their fubftantial forms, were of exceeding differing natures, yet) fince they are diffolv'd and mixt into another, they may be very *Homogeneous*, they being almoft refolv'd again into Air, Water, and Earth; retaining, perhaps, one part of their vegetative faculty yet entire, which meeting with congruous affiftants, fuch as the heat of the Air, and the fluidity of the Water, and fuch like coadjutors and conveniences, acquires a certain vegetation for a time, wholly differing perhaps from that kind of vegetation it had before.

To explain my meaning a little better by a grofs Similitude:

Suppofe a curious piece of Clock-work, that had had feveral motions and contrivances in it, which, when in order. would all have mov'd in their defign'd methods and Periods. We will further fuppofe, by fome means, that this Clock comes to be broken, brufed, or otherwife difordered, fo that feveral parts of it being diflocated, are impeded, and fo ftand ftill, and not onely hinder its own progreffive motion, and produce not the effect which they were defign'd for, but becaufe the other parts alfo have a dependence upon them, put a ftop to their motion likewife; and fo the whole Inftrument becomes unferviceable,, and not fit for any ufe. This Inftrument afterwards, by fome fhaking and tumbling, and throwing up and down, comes to have feveral of its parts fhaken out, and feveral of its curious motions, and contrivances, and particles all fallen afunder; here a Pin falls out, and there a Pillar, and here a Wheel, and there a Hammer, and a Spring, and the like, and among the reft, away falls thofe parts alfo which were brufed and diforder'd, and had all this while impeded the motion of all the reft; hereupon feveral of thofe other motions that yet remain, whofe fprings were not quite run down, being now at liberty, begin each of them to move, thus or thus, but quite after another method then before, there being many regulating parts and the like, fallen away and loft. Upon this, the Owner, who chances to hear and obferve fome of thefe effects, being ignorant of the Watch-makers Art, wonders what is betid his Clock, and prefently imagines that fome Artift has been at work, and has fet his Clock in order, and made a new kind of Inftrument of it, but upon examining circumftances, he finds there was no fuch matter, but that the cafual flipping out of a Pin had made feveral parts of his Clock fall to pieces, and that thereby the obftacle that all this while hindred his Clock, together with other ufefull parts were fallen out, and fo his Clock was fet at liberty. And upon winding up thofe fprings again when run down, he finds his Clock to go, but quite after another manner then it was wont heretofore.

And thus may it be perhaps in the bufinefs of Mofs and Mould, and Mufhroms, and feveral other fpontaneous kinds of vegetations, which

may

may be caus'd by a vegetative principle, which was a coadjutor to the life and growth of the greater Vegetable, and was by the deſtroying of the life of it ſtopt and impeded in performing its office; but afterwards, upon a further corruption of ſeveral parts that had all the while impeded it, the heat of the Sun winding up, as it were, the ſpring, ſets it again into a vegetative motion, and this being ſingle, and not at all regulated as it was before(when a part of that greater *machine* the priſtine vegetable)is mov'd after quite a differing manner, and produces effects very differing from thoſe it did before.

But this I propound onely as a conjecture, not that I am more enclin'd to this *Hypotheſis* then the ſeminal, which upon good reaſon I gheſs to be Mechanical alſo, as I may elſewhere more fully ſhew : But becauſe I may, by this, hint a poſſible way how this appearance may be ſolv'd; ſuppoſing we ſhould be driven to confeſs from certain Experiments and Obſervations made, that ſuch or ſuch Vegetables were produc'd out of the corruption of another, without any concurrent ſeminal principle (as I have given ſome reaſon to ſuppoſe, in the deſcription of a *Microſcopical* Muſhrome) without derogating at all from the infinite wiſdom of the Creator. For this accidental production, as I may call it, does manifeſt as much, if not very much more, of the excellency of his contrivance as any thing in the more perfect vegetative bodies of the world, even as the accidental motion of the *Automaton* does make the owner ſee, that there was much more contrivance in it then at firſt he imagin'd. But of this I have added more in the deſcription of Mould, and the Vegetables on Roſe leaves, &c. thoſe being much more likely to have their original from ſuch a cauſe then this which I have here deſcribed, in the 13. *Scheme*, which indeed I cannot conceive otherwiſe of, then as of a moſt perfect Vegetable, wanting nothing of the perfections of the moſt conſpicuous and vaſteſt Vegetables of the world, and to be of a rank ſo high, as that it may very properly be reckon'd with the tall Cedar of *Lebanon*, as that Kingly Botaniſt has done.

We know there may be as much curioſity of contrivance, and excellency of form in a very ſmall Pocket-clock, that takes not up an Inch ſquare of room, as there may be in a Church-clock that fills a whole room; And I know not whether all the contrivances and *Mechaniſms* requiſite to a perfect Vegetable, may not be crowded into an exceedingly leſs room then this of Moſs, as I have heard of a ſtriking Watch ſo ſmall, that it ſerv'd for a Pendant in a Ladies ear; and I have already given you the deſcription of a Plant growing on Roſe leaves, that is abundantly ſmaller then Moſs; inſomuch, that neer 1000. of them would hardly make the bigneſs of one ſingle Plant of Moſs. And by comparing the bulk of Moſs, with the bulk of the biggeſt kind of Vegetable we meet with in Story (of which kind we find in ſome hotter climates, as *Guine*, and *Braſile*, the ſtock or body of ſome Trees to be twenty foot in Diameter, whereas the body or ſtem of Moſs, for the moſt part, is not above one ſixtieth part of an Inch) we ſhall find that the bulk of the one will exceed the bulk of the other, no leſs then 2985984 Millions,

or

or 2985984000000, and suppofing the production on a Rofe leaf to be a Plant, we fhall have of thofe *Indian* Plants to exceed a production of the fame Vegetable kingdom no lefs then 1000 times the former number; fo prodigioufly various are the works of the Creator, and fo All-fufficient is he to perform what to man would feem unpoffible, they being both alike eafie to him, even as one day, and a thoufand years are to him as one and the fame time.

I have taken notice of fuch an infinite variety of thofe fmaller kinds of vegetations, that fhould I have defcribed every one of them, they would almoft have fill'd a Volume, and prov'd bigg enough to have made a new Herbal, fuch multitudes are there to be found in moift hot weather, efpecially in the Summer time, on all kind of putrifying fubftances, which, whether they do more properly belong to the *Claffis* of *Mufhroms*, or *Moulds*, or *Mofses*, I fhall not now difpute, there being fome that feem more properly of one kind, others of another, their colours and magnitudes being as much differing as their Figures and fubftances.

Nay, I have obferv'd, that putting fair Water (whether Rain-water or Pump-water, or *May-dew*, or Snow-water, it was almoft all one) I have often obferv'd, I fay, that this Water would, with a little ftanding, tarnifh and cover all about the fides of the Glafs that lay under water, with a lovely green; but though I have often endeavour'd to difcover with my *Microfcope* whether this green were like Mofs, or long ftriped Sea-weed, or any other peculiar form, yet fo ill and imperfect are our *Microfcopes*, that I could not certainly difcriminate any.

Growing Trees alfo, and any kinds of Woods, Stones, Bones, &c. that have been long expos'd to the Air and Rain, will be all over cover'd with a greenifh fcurff, which will very much foul and green any kind of cloaths that are rubb'd againft it; viewing this, I could not certainly perceive in many parts of it any determinate form, though in many I could perceive a Bed as 'twere of young Mofs, but in other parts it look'd almoft like green bufhes, and very confus'd, but always of what ever irregular Figures the parts appear'd of, they were always green, and feem'd to be either fome Vegetable, or to have fome vegetating principle.

Obferv. XXII. *Of common* Sponges, *and feveral other* Spongie *fibrous bodies.*

A Sponge is commonly reckon'd among the *Zoophyts*, or Plant Animals; and the *texture* of it, which the *Microfcope* difcovers, feems to confirm it; for it is of a form whereof I never obferv'd any other Vegetable, and indeed, it feems impoffible that any fhould be of it, for it confifts of an infinite number of fmall fhort *fibres*, or nervous parts, much of the fame bignefs, curioufly jointed or contex'd together in the form of a Net, as is more plainly manifeft by the little Draught which I have
added

added, in the third *Figure* of the I X. *Scheme*, of a piece of it, which you may perceive reprefents a confus'd heap of the fibrous parts curiouſly jointed and implicated. The joints are, for the moſt part, where three *fibres* onely meet, for I have very feldom met with any that had four.

At thefe joints there is no one of the three that feems to be the ſtock whereon the other grow, but each of the *fibres* are, for the moſt part, of an equal bigneſs, and feem each of them to have an equal ſhare in the joint; the *fibres* are all of them much about the fame bigneſs, not ſmaller towards the top of the Sponge, and bigger neerer the bottom or root, as is ufuall in Plants, the length of each between the joints, is very irregular and different; the diſtance between fome two joints, being ten or twelve times more then between fome others.

Nor are the joints regular, and of an *equitriagonal Figure*, but for the moſt part, the three *fibres* ſo meet, that they compoſe three angles very differing all of them from one another.

The meſhes likewiſe, and holes of this reticulated body, are not leſs various and irregular : fome *bilateral*, others *trilateral*, and *quadrilateral* Figures; nay, I have obſerv d fome meſhes to have 5, 6, 7, 8, or 9. fides, and fome to have onely one, ſo exceeding various is the *Lufus Naturæ* in this body.

As to the outward appearance of this Vegetative body, they are ſo ufuall every where, that I need not defcribe them, confiſting of a foft and porous fubſtance, repreſenting a Lock, fometimes a fleece of Wooll; but it has befides thefe fmall *microfcopical* pores which lie between the *fibres*, a multitude of round pores or holes, which, from the top of it, pierce into the body, and fometimes go quite through to the bottom.

I have obferv'd many of thefe Sponges, to have included likewiſe in the midſt of their fibrous contextures, pretty large friable ſtones, which muſt either have been incloſ'd whil'ſt this Vegetable was in formation, or generated in thoſe places after it was perfectly ſhap'd. The later of which feems the more improbable, becauſe I did not find that any of thefe ſtony fubſtances were perforated with the *fibres* of the Sponge.

I have never feen nor been enform'd of the true manner of the growing of Sponges on the Rock; whether they are found to increafe from little to great, like Vegetables, that is, part after part, or like Animals, all parts equally growing together; or whether they be *matrices* or feed-baggs of any kind of Fiſhes, or fome kind of watry Infect ; or whether they are at any times more foft and tender, or of another nature and texture, which things, if I knew, I ſhould much defire to be informed of: but from a curfory view that I at firſt made with my *Microſcope*, and fome other trials, I fuppoſed it to be fome Animal fubſtance caſt out, and faſtned upon the Rocks in the form of a froth, or *congeries* of bubbles, like that which I have often obferv'd on Rofemary, and other Plants (wherein is included a little Infect) that all the little films which divide thefe bubbles one from another, did prefently, almoſt after the fubſtance began to grow a little harder, break, and leave onely the thread behind, which might be, as 'twere, the angle or thread between the bubbles, that the

great

great holes or pores obſervable in theſe Sponges were made by the eruption of the included *Heterogeneous* ſubſtance (whether air, or ſome other body, for many other fluid bodies will do the ſame thing) which breaking out of the leſſer, were collected into very large bubbles, and ſo might make their way out of the Sponge, and in their paſſage might leave a round cavity ; and if it were large, might carry up with it the adjacent bubbles, which may be perceiv'd at the outſide of the Sponge, if it be firſt throughly wetted, and ſuffer'd to plump it ſelf into its natural form, or be then wrung dry, and ſuffer'd to expand it ſelf again, which it will freely do whil'ſt moiſt : for when it has thus plump'd it ſelf into its natural ſhape and dimenſions, 'tis obvious enough that the mouths of the larger holes have a kind of lip or riſing round about them, but the other ſmaller pores have little or none. It may further be found, that each of theſe great pores has many other ſmall pores below, that are united unto it, and help to conſtitute it, almoſt like ſo many rivulets or ſmall ſtreams that contribute to the maintenance of a large River. Nor from this *Hypotheſis* would it have been difficult to explicate, how thoſe little branches of *Coral*, ſmal *Stones*, *Shells*, and the like, come to be included by theſe frothy bodies : But this indeed was but a conjecture ; and upon a more accurate enquiry into the form of it with the *Microſcope*, it ſeems not to be the true origine of them ; for whereas Sponges have onely three arms which join together at each knot, if they had been generated from bubbles they muſt have had four.

But that they are Animal Subſtances, the *Chymical* examination of them ſeems to manifeſt, they affording a volatil Salt and ſpirit, like *Harts-Horn*, as does alſo their great ſtrength and toughneſs, and their ſmell when burn'd in the Fire or a Candle, which has a kind of fleſhy ſent, not much unlike to hair. And having ſince examin'd ſeveral Authors concerning them, among others, I find this account given by *Bellonius*, in the XI. *Chap.* of his 2ᵈ Book, *De Aquatilibus. Spongiæ recentes,* ſays he, *à ſiccis longe diverſæ, ſcopulis aquæ marinæ ad duos vel tres cubitos, nonnunquam quatuor tantum digitos immerſis, ut fungi arboribus adhærent, ſordido quodam ſucco aut mucoſa potius ſanie reſærtæ, uſque adeò fætida, ut vel eminus nauſeam excitet, continetur autem iis cavernis, quas inanes in ſiccis & lotis Spongiis cernimus : Putris pulmonis modo nigræ conſpiciuntur, verum quæ in ſublimi aquæ naſcuntur multo magis opaca nigredine ſuffuſæ ſunt. Vivere quidem Spongias adhærendo* Ariſtoteles *cenſet : abſolute vero minime : ſenſumque aliquem habere, vel eo argumento (inquit) credantur, quod difficillime abſtrahantur, niſi clanculum agatur: Atqꝫ ad avulſoris acceſſum ita contrahantur, ut eas evellere difficile ſit, quod idem etiam faciunt quoties flatus tempeſtatéſque urgent. Puto autem illis ſuccum ſordidum quem ſupra diximus carnis loco à natura attributum fuiſſe : atque meatibus latioribus tanquam inteſtinis aut interaneis uti. Cæterum pars ea quæ Spongiæ cautibus adhærent eſt tanquam folii petiolus, à quo veluti collum quoddam gracile incipit : quod deinde in latitudinem diffuſum capitis globum facit. Recentibus nihil eſt fiſtuloſum, hæſitantque tanquam radicibus. Superne omnes propemodum meatus concreti latent : inferne verò quaterni aut quini patent, per quos*

V *eas*

eas fugere exiſtimamus. From which Deſcription, they ſeem to be a kind of Plant-Animal that adheres to a Rock, and theſe ſmall *fibres* or threads which we have deſcribed, ſeem to have been the Veſſels which ('tis very probable) were very much bigger whil'ſt the *Interſtitia* were fill'd (as he affirms) with a mucous, pulpy or fleſhy ſubſtance; but upon the drying were ſhrunk into the bigneſs they now appear.

The texture of it is ſuch, that I have not yet met with any other body in the world that has the like, but onely one of a larger ſort of Sponge (which is preſerv'd in the *Muſeum Harveanum* belonging to the moſt Illuſtrious and moſt learned Society of the *Phyſicians* of *London*) which is of a horney, or rather of a *petrify'd* ſubſtance. And of this indeed, the texture and make is exactly the ſame with common Sponges, but onely that both the holes and the *fibres*, or texture of it is exceedingly much bigger, for ſome of the holes were above an Inch and half over, and the *fibres* and *texture* of it was bigg enough to be diſtinguiſhed eaſily with ones eye, but conſpicuouſly with an ordinary ſingle *Microſcope*. And theſe indeed, ſeem'd to have been the habitation of ſome Animal; and examining *Ariſtotle*, I find a very conſonant account hereunto, namely, that he had known a certain little Animal, call'd *Pinnothera*, like a Spider, to be bred in thoſe caverns of a Sponge, from within which, by opening and cloſing thoſe holes, he inſnares and catches the little Fiſhes; and in another place he ſays, That 'tis very confidently reported, that there are certain Moths or Worms that reſide in the cavities of a Sponge, and are there nouriſhed: Notwithſtanding all which Hiſtories, I think it well worth the enquiring into the Hiſtory and nature of a Sponge, it ſeeming to promiſe ſome information of the Veſſels in Animal ſubſtances, which (by reaſon of the ſolidity of the interſerted fleſh that is not eaſily remov'd, without deſtroying alſo thoſe interſpers'd Veſſels) are hitherto undiſcover'd; whereas here in a Sponge, the *Parenchyma*, it ſeems, is but a kind of mucous gelly, which is very eaſily and cleerly waſh'd away.

The reaſon that makes me imagine, that there may probably be ſome ſuch texture in Animal ſubſtances, is, that examining the texture of the filaments of tann'd Leather, I find it to be much of the ſame nature and ſtrength of a Sponge; and with my *Microſcope*, I have obſerv'd many ſuch joints and knobs, as I have deſcribed in Sponges, the *fibres* alſo in the hollow of ſeveral ſorts of Bones, after the Marrow has been remov'd, I have found ſomewhat to reſemble this texture, though, I confeſs, I never yet found any texture exactly the ſame, nor any for curioſity comparable to it.

The filaments of it are much ſmaller then thoſe of Silk, and through the *Microſcope* appear very neer as tranſparent, nay, ſome parts of them I have obſerv'd much more.

Having examin'd alſo ſeveral kinds of Muſhroms, I finde their texture to be ſomewhat of this kind, that is, to conſiſt of an infinite company of ſmall filaments, every way contex'd and woven together, ſo as to make a kind of cloth, and more particularly, examining a piece of Touch-wood (which is a kind of *Jews-ear*, or Muſhrom, growing here in *England* alſo,

on

on feveral forts of Trees, fuch as Elders, Maples, Willows, &c. and is commonly call'd by the name of *Spunk*; but that we meet with to be fold in Shops, is brought from beyond Seas) I found it to be made of an exceeding delicate texture: For the fubftance of it feels, and looks to the naked eye, and may be ftretch'd any way, exactly like a very fine piece of *Chamois* Leather, or wafh'd Leather, but it is of fomewhat a browner hew, and nothing neer fo ftrong; but examining it with my *Microfcope*, I found it of fomewhat another make then any kind of Leather; for whereas both *Chamois*, and all other kinds of Leather I have yet view'd, confift of an infinite company of filaments, fomewhat like bufhes interwoven one within another, that is, of bigger parts or ftems, as it were, and fmaller branchings that grow out of them; or like a heap of Ropes ends, where each of the larger Ropes by degrees feem to fplit or untwift, into many fmaller Cords, and each of thofe Cords into fmaller Lines, and thofe Lines into Threads, &c. and thefe ftrangely intangled, or interwoven one within another : The texture of this Touch-wood feems more like that of a Lock or a Fleece of Wool, for it confifts of an infinite number of fmall filaments, all of them, as farr as I could perceive, of the fame bignefs like thofe of a Sponge, but that the *filaments* of this were not a twentieth part of the bignefs of thofe of a Sponge; and I could not fo plainly perceive their joints, or their manner of interweaving, though, as farr as I was able to difcern with that *Microfcope* I had, I fuppofe it to have fome kind of refemblance, but the joints are nothing neer fo thick, nor without much trouble vifible.

The filaments I could plainly enough perceive to be even, round, cylindrical, tranfparent bodies, and to crofs each other every way, that is, there were not more feem'd to lie *horizontally* then *perpendicularly* and thwartway, fo that it is fomewhat difficult to conceive how they fhould grow in that manner. By tearing off a fmall piece of it, and looking on the ragged edge, I could among feveral of thofe *fibres* perceive fmall joints, that is, one of thofe hairs fplit into two, each of the fame bignefs with the other out of which they feem'd to grow, but having not lately had an opportunity of examining their manner of growth, I cannot pofitively affirm any thing of them.

But to proceed, The fwelling of Sponges upon wetting, and the rifing of the Water in it above the furface of the Water that it touches, are both from the fame caufe, of which an account is already given in the fixth Obfervation.

The fubftance of them indeed, has fo many excellent properties, fcarce to be met with in any other body in the world, that I have often wondered that fo little ufe is made of it, and thofe onely vile and fordid; certainly, if it were well confider'd, it would afford much greater conveniencies.

That ufe which the Divers are faid to make of it, feems, if true, very ftrange, but having made trial of it my felf, by dipping a fmall piece of it in very good Sallet-oyl, and putting it in my mouth, and then keeping my mouth and nofe under water, I could not find any fuch thing; for I

was

was as foon out of breath,as if I had had no Sponge,nor could I fetch my breath without taking in water at my mouth; but I am very apt to think, that were there a contrivance whereby the expir'd air might be forc'd to pafs through a wet or oyly Sponge before it were again infpir'd, it might much cleanfe, and ftrain away from the Air divers fuliginous and other noifome fteams, and the dipping of it in certain liquors might, perhaps, fo renew that property in the Air which it lofes in the Lungs,by being breath'd, that one fquare foot of Air might laft a man for refpiration much longer, perhaps,then ten will now ferve him of common Air.

Obferv. XXIII. *Of the curious texture of* Sea-weeds.

FOr curiofity and beauty, I have not among all the Plants or Vegetables I have yet obferv'd, feen any one comparable to this Sea-weed I have here defcrib'd, of which I am able to fay very little more then what is reprefented by the fecond *Figure* of the ninth *Scheme*: Namely, that it is a Plant which grows upon the Rocks under the water, and increafes and fpreads it felf into a great tuft, which is not onely handfomely branch'd into feveral leaves, but the whole furface of the Plant is cover'd over with a moft curious kind of carv'd work, which confifts of a texture much refembling a Honey-comb; for the whole furface on both fides is cover'd over with a multitude of very fmall holes,being no bigger then fo many holes made with the point of a fmall Pinn, and rang'd in the neateft and moft delicate order imaginable, they being plac'd in the manner of a *Quincunx*, or very much like the rows of the eyes of a Fly, the rows or orders being very regular, which way foever they are obferv'd: what the texture was, as it appear'd through a pretty bigg Magnifying *Microfcope*, I have here adjoin'd in the firft *Figure* of the 14. *Scheme*. which round Area A B C D reprefents a part of the furface about one eighth part of an Inch in Diameter: Thofe little holes, which to the eye look'd round, like fo many little fpots, here appear'd very regularly fhap'd holes, reprefenting almoft the fhape of the fole of a round toed fhoe, the hinder part of which, is, as it were, trod on or cover'd by the toe of that next below it;thefe holes feem'd wall'd about with a very thin and tranfparent fubftance, looking of a pale ftraw-colour; from the edge of which, againft the middle of each hole, were fprouted out four fmall tranfparent ftraw-colour'd Thorns, which feem'd to protect and cover thofe cavities, from either fide two; neer the root of this Plant, were fprouted out feveral fmall branches of a kind of baftard *Coralline*, curioufly branch'd, though fmall.

And to confirm this, having lately the opportunity of viewing the large Plant (if I may fo call it) of a Sponge *petrify'd*, of which I made mention in the laft Obfervation, I found, that each of the Branches or Figures of it, did, by the range of its pores, exhibit juft fuch a texture, the

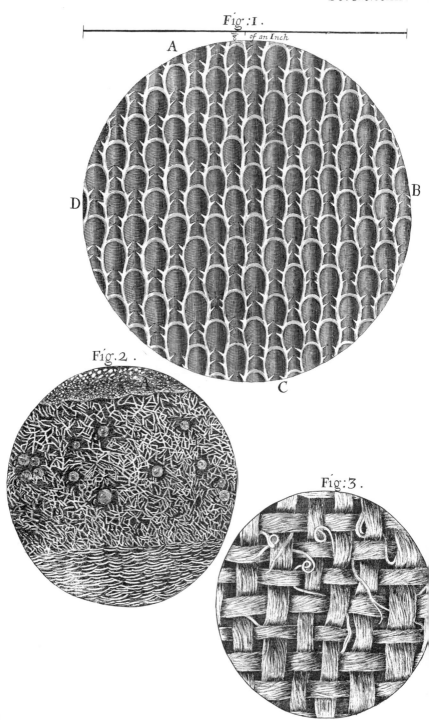

Schem: XIIII.

Fig: I.

⅛ of an Inch

A

D

B

C

Fig: 2.

Fig: 3.

the rows of pores croffing one another, much after the manner as the
rows of eyes do which are defcrib'd in the 26.*Scheme* : *Coralline* alfo, and
feveral forts of white *Coral*, I have with a *Microfcope* obferv'd very curioufly fhap'd. And I doubt not, but that he that fhall obferve thefe
feveral kinds of Plants that grow upon Rocks, which the Sea fometimes overflows, and thofe heaps of others which are vomited out of it upon the fhore, may find multitudes of little Plants, and other bodies,which
like this will afford very beautifull objects for the *Microfcope* ; and this
Specimen here is adjoin'd onely to excite their curiofities who have opportunity of obferving to examine and collect what they find worthy
their notice ; for the Sea, among terreftrial bodies, is alfo a *prolifick*
mother, and affords as many Inftances of *fpontaneous* generations as either the Air or Earth.

Obferv. XXIV. *Of the furfaces of* Rofemary, *and other leaves.*

THis which is delineated within the circle of the fecond *Figure* of the
14. *Scheme*, is a fmall part of the back or under fide of a leaf of
Rofemary, which I did not therefore make choice of, becaufe it had any
thing peculiar which was not obfervable with a *Microfcope* in feveral
other Plants, but becaufe it exhibits at one view,

Firft, a fmooth and fhining furface, namely, A B, which is a part of the
upper fide of the leaf, that by a kind of hem or doubling of the leaf appears on this fide. There are multitudes of leaves, whofe furfaces are
like this fmooth, and as it were quilted, which look like a curious quilted
bagg of green Silk, or like a Bladder, or fome fuch pliable tranfparent
fubftance, full ftuffed out with a green juice or liquor ; the furface of
Rue, or Herbgrafs, is polifh'd, and all over indented, or pitted, like the
Silk-worm's Egg,which I fhall anon defcribe ; the fmooth furfaces of other
Plants are otherwife quilted, Nature in this, as it were, expreffing her
Needle-work, or imbroidery.

Next a downy or bufhy furface, fuch as is all the under fide almoft,
appearing through the *Microfcope* much like a thicket of bufhes, and with
this kind of Down or Hair the leaves and ftalks of multitudes of Vegetables are covered ; and there feems to be as great a variety in the fhape,
bulk, and manner of the growing of thefe fecundary Plants, as I may call
them (they being, as it were, a Plant growing out of a Plant, or fomewhat like the hairs of Animals) as there is to be found amongft fmall
fhrubs that compofe bufhes ; but for the moft part, they confift of fmall
tranfparent parts, fome of which grow in the fhape of fmall Needles or
Bodkins,as on the Thiftle,Cowag-ecod and Nettle ; others in the form of
Cat's claws,as in Cliders, the beards of Barley, the edges of feveral forts
of Grafs and Reeds,&c. in other,as Coltsfoot,Rofe-campion, Aps, Poplar,
Willow, and almoft all other downy Plants, they grow in the form of
bufhes very much diverfify'd in each particular Plant. That which I have
before

before in the 19. Obfervation noted on Rofe-leaves, is of a quite differing kind, and feems indeed a real Vegetable, diftinct from the leaf.

Thirdly, among thefe fmall bufhes are obfervable an infinite company of fmall round Balls, exactly Globular, and very much refembling Pearls, namely, CCCC, of thefe there may be multitudes obferv'd in Sage, and feveral other Plants, which I fuppofe was the reafon why *Athanafius Kircher* fuppofed them to be all cover'd with Spiders Eggs, or young Spiders, which indeed is nothing elfe but fome kind of gummous exfudation, which is always much of the fame bignefs. At firft fight of thefe, I confefs, I imagin'd that they might have been fome kind of *matrices*, or nourifhing receptacles for fome fmall Infect, juft as I have found Oak-apples, and multitudes of fuch other large excrefcencies on the leaves and other parts of Trees and fhrubs to be for Flyes, and divers other Infects, but obferving them to be there all the year, and fcarce at all to change their magnitude, that conjecture feem'd not fo probable. But what ever be the ufe of it, it affords a very pleafant object through the *Microfcope*, and may, perhaps, upon further examination, prove very luciferous.

Obferv. XXV. *Of the ftinging points and juice of* Nettles, *and fome other venomous Plants.*

A Nettle is a Plant fo well known to every one, as to what the appearance of it is to the naked eye, that it needs no defcription; and there are very few that have not felt as well as feen it; and therefore it will be no news to tell that a gentle and flight touch of the fkin by a Nettle, does oftentime, not onely create very fenfible and acute pain, much like that of a burn or fcald, but often alfo very angry and hard fwellings and inflamations of the parts, fuch as will prefently rife, and continue fwoln divers hours. Thefe obfervations, I fay, are common enough; but how the pain is fo fuddenly created, and by what means continued, augmented for a time, and afterwards diminifh'd, and at length quite exftinguifh'd, has not, that I know, been explain'd by any.

And here we muft have recourfe to our *Microfcope*, and that will, if almoft any part of the Plant be looked on, fhew us the whole furface of it very thick fet with turn-Pikes, or fharp Needles, of the fhape of thofe reprefented in the 15. *Scheme* and firft *Figure* by A B, which are vifible alfo to the naked eye; each of which confifts of two parts very diftinct for fhape, and differing alfo in quality from one another. For the part A, is fhaped very much like a round Bodkin, from B tapering till it end in a very fharp point; it is of a fubftance very hard and ftiff, exceedingly tranfparent and cleer, and, as I by many trials certainly found, is hollow from top to bottom.

This I found by this Experiment, I had a very convenient *Micro-*

scope with a fingle Glafs which drew about half an Inch, this I had faftned into a little frame, almoft like a pair of Spectacles, which I placed before mine eyes, and fo holding the leaf of a Nettle at a convenient diftance from my eye, I did firft, with the thrufting of feveral of thefe briftles into my fkin, perceive that prefently after I had thruft them in I felt the burning pain begin; next I obferv'd in divers of them, that upon thrufting my finger againft their tops, the Bodkin (if I may fo call it) did not in the leaft bend, but I could perceive moving up and down within it a certain liquor, which upon thrufting the Bodkin againft its bafis, or bagg B, I could perceive to rife towards the top, and upon taking away my hand, I could fee it again fubfide, and fhrink into the bagg; this I did very often, and faw this *Phænomenon* as plain as I could ever fee a parcel of water afcend and defcend in a pipe of Glafs. But the bafis underneath thefe Bodkins on which they were faft, were made of a more pliable fubftance, and looked almoft like a little bagg of green Leather, or rather refembled the fhape and furface of a wilde Cucumber, or *cucumeris afinini,* and I could plainly perceive them to be certain little baggs, bladders, or receptacles full of water, or as I ghefs, the liquor of the Plant, which was poifonous, and thofe fmall Bodkins were but the Syringe-pipes, or Glyfter-pipes, which firft made way into the fkin, and then ferved to convey that poifonous juice, upon the preffing of thofe little baggs, into the interior and fenfible parts of the fkin, which being fo difcharg'd, does corrode, or, as it were, burn that part of the fkin it touches; and this pain will fometimes laft very long, according as the impreffion is made deeper or ftronger.

The other parts of the leaf or furface of the Nettle, have very little confiderable, but what is common to moft of thefe kinds of Plants, as the ruggednefs or indenting, and hairinefs, and other roughneffes of the furface or out-fide of the Plant, of which I may fay more in another place. As I fhall likewife of certain little pretty cleer Balls or Apples which I have obferved to ftick to the fides of thefe leaves, both on the upper and under fide, very much like the fmall Apples which I have often obferv'd to grow on the leaves of an Oak call'd *Oak-apples* which are nothing but the *Matrices* of an Infect, as I elfewhere fhew.

The chief thing therefore is, how this Plant comes, by fo flight a touch, to create fo great a pain; and the reafon of this feems to be nothing elfe, but the corrofive penetrant liquor contain'd in the fmall baggs or bladders, upon which grow out thofe fharp Syringe-pipes, as I before noted; and very confonant to this, is the reafon of the pain created by the fting of a Bee, Wafp, &c. as I elfewhere fhew: For by the Dart, which is likewife a pipe, is made a deep paffage into the fkin, and then by the anger of the Fly, is his gally poifonous liquor injected; which being admitted among the fenfible parts, and fo mix'd with the humours or *ftagnating* juices of that part, does create an Ebullition perhaps, or *effervefcens,* as is ufually obferv'd in the mingling of two differing *Chymical faline* liquors, by which means the parts become fwell'd, hard, and very painfull; for thereby the nervous and fenfible parts are not onely ftretch'd and ftrain'd

beyond

beyond their natural *tone*, but are alſo prick'd, perhaps, or corroded by the pungent and incongruous pores of the intruded liquor.

And this ſeems to be the reaſon, why *Aqua fortis*, and other *ſaline* liquors, if they come to touch the ſenſitive parts, as in a cut of the ſkin, or the like, do ſo violently and intollerably *excruciate* and torment the Patient. And 'tis not unlikely, but the Inventors of that Diabolical practice of poiſoning the points of Arrows and Ponyards, might receive their firſt hint from ſome ſuch Inſtance in natural contrivances, as this of the Nettle : for the ground why ſuch poiſon'd weapons kill ſo infallibly as they do, ſeems no other then this of our Nettle's ſtinging ; for the Ponyard or Dart makes a paſſage or entrance into the ſenſitive or vital parts of the body, whereby the contagious ſubſtance comes to be diſſolv'd by, and mix'd with the fluid parts or humours of the body, and by that means ſpreads it ſelf by degrees into the whole liquid part of the body, in the ſame manner, as a few grains of Salt, put into a great quantity of Water, will by degrees diffuſe it ſelf over the whole.

And this I take to be the reaſon of killing of Toads, Frogs, Effs, and ſeveral Fiſhes, by ſtrewing Salt on their backs (which Experiment was ſhewn to the *Royal Society* by a very ingenious Gentleman, and a worthy Member of it) for thoſe creatures having always a continual exſudation, as it were, of ſlimy and watry parts, ſweating out of the pores of their ſkin, the *ſaline* particles, by that means obtain a *vehicle*, which conveys them into the internal and vital parts of the body.

This ſeems alſo to be the reaſon why bathing in Mineral waters are ſuch ſoveraign remedies for multitudes of diſtempers, eſpecially chronical ; for the liquid & warm *vehicles* of the Mineral particles, which are known to be in very conſiderable quantities in thoſe healing baths, by the body's long ſtay in them, do by degrees ſteep and inſinuate themſelves into the pores and parts of the ſkin, and thereby thoſe Mineral particles have their ways and paſſages open'd to penetrate into the inner parts, and mingle themſelves with the *ſtagnant* juices of the ſeveral parts ; beſides, many of thoſe offenſive parts which were united with thoſe *ſtagnant* juices, and which were contrary to the natural conſtitution of the parts, and ſo become irkſome and painfull to the body, but could not be diſcharged, becauſe Nature had made no proviſion for ſuch accidental miſchiefs, are, by means of this ſoaking, and filling the pores of the ſkin with a liquor, afforded a paſſage through that liquor that fills the pores into the ambient fluid, and thereby the body comes to be diſcharged.

So that 'tis very evident, there may be a good as well as an evil application of this Principle. And the ingenious Invention of that Excellent perſon, Doctor *Wren*, of injecting liquors into the veins of an Animal, ſeems to be reducible to this head : I cannot ſtay, nor is this a fit place, to mention the ſeveral Experiments made of this kind by the moſt incomparable Mr. *Boyle*, the multitudes made by the lately mention'd *Phyſician* Doctor *Clark*, the Hiſtory whereof, as he has been pleas'd to communicate to the *Royal Society*, ſo he may perhaps be prevail'd with to make publique himſelf : But I ſhall rather hint, that certainly, if this Principle

were

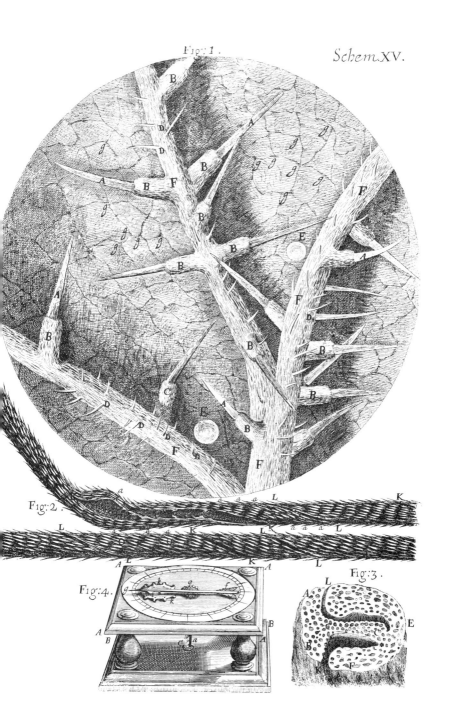

Fig: 1.

Schem.XV.

Fig: 2.

Fig: 4.

Fig: 3.

were well confider'd, there might, befides the further improving of Bathing and Syringing into the veins, be thought on feveral ways, whereby
feveral obftinate diftempers of a humane body, fuch as the Gout, Dropfie,
Stone, *&c.* might be mafter'd, and expell'd; and good men might make
as good a ufe of it, as evil men have made a perverfe and Diabolical.

And that the filling of the pores of the fkin with fome fluid *vehicle*, is
of no fmall efficacy towards the preparing a paffage for feveral kinds of
penetrant juices, and other diffoluble bodies, to infinuate themfelves
within the fkin, and into the fenfitive parts of the body, may be, I think,
prov'd by an Inftance given us by *Bellonius,* in the 26. *Chapter* of the
fecond Book of his *Obfervations,* which containing a very remarkable
Story I have here tranfcrib'd : *Cum Chamæleonis nigri radices* (fays he)
*apud Pagum quendam Livadochorio nuncupatum erui curaremus, plurimi
Græci & Turcæ fpectatum venerunt quid erueremus, eas vero fruftulatim fe-
cabamus, & filo trajiciebamus ut facilius exficcari poffent. Turcæ in eo ne-
gotio occupatos nos videntes, fimiliter eas radices tractare & fecare volue-
runt: at cum fummus effet æftus, & omnes fudore maderent, quicunque
eam radicem manibus tractaverant fudoremque abfterferant, aut faciem di-
gitis fcalpferant, tantam pruriginem iis locis quos attigerant poftea fenferunt,
ut aduri viderentur. Chamælconis enim nigri radix ea virtute pollet, ut cu-
ti applicata ipfam adeo inflammet, ut nec fquillæ, nec urticæ ullæ centefima
parte ita adurent : At prurigo non adeo celeriter fefe prodit. Poft unam aut alteram porro horam, finguli variis faciei locis cutem adeo inflammatam habere cæpimus ut tota fanguinea videretur, atque quo magis eam confricaba-
mus, tanto magis excitabatur prurigo. Fonti affidebamus fub platano, atque ini-
tio pro ludicro habebamus & ridebamus : at tandem illi plurimum indignati
funt, & nifi affeveraffemus nunquam expertos tali virtute eam plantam pollere,
haud dubie male nos multaffent. Attamen noftra excufatio fuit ab illis facilius accepta, cum eodem incommodo nos affectos confpicerent. Mirum fane quod
in tantillo radice tam ingentem efficaciam noftro malo experti fumus.*

By which obfervation of his, it feems manifeft, that their being all cover'd
with fweat who gather'd and cut this root of the black *Chameleon* Thiftle,
was the great reafon why they fuffer'd that inconvenience, for it feems the
like circumftance had not been before that noted, nor do I find any mention of fuch a property belonging to this Vegetable in any of the Herbals I have at prefent by me.

I could give very many Obfervations which I have made of this kind,
whereby I have found that the beft way to get a body to be infinuated
into the fubftance or infenfible pores of another, is firft, to find a fluid
vehicle that has fome congruity, both to the body to be infinuated, and to
the body into whofe pores you would have the other convey'd. And in
this Principle lies the great myftery of ftaining feveral forts of bodies, as
Marble, Woods, Bones, *&c.* and of Dying Silks, Cloaths, Wools, Feathers, *&c.* But thefe being digreffions, I fhall proceed to :

Obferv. XXVI. *Of Cowage, and the itching operation of fome bodies.*

THere is a certain Down of a Plant, brought from the *Eaft-Indies,* call'd
commonly, though very improperly, *Cow-itch,* the reafon of which

miftake

miftake is manifeft enough from the defcription of it, which Mr. *Parkinfon* fets down in his *Herbal,* Tribe XI. Chap. 2. *Phafiolus filiqua hirfuta* ; *The hairy Kidney-bean, called in* Zurratte *where it grows, Couhage : We have had* (fays he) *another of this kind brought us out of the* Eaft-Indies, *which being planted, was in fhew like the former, but came not to perfection, the unkindly feafon not fuffering it to fhew the flower ; but of the Cods that were brought, fome were fmaller, fhorter, and rounder then the Garden kind; others much longer, and many growing together, as it were in clufters, and cover'd all over with a brown fhort hairinefs, fo fine, that if any of it be rubb'd, or fall on the back of ones hand, or other tender parts of the skin, it will caufe a kind of itching, but not ftrong, nor long induring, but paffing quickly away, without either danger or harm ; the Beans were fmaller then ordinary, and of a black fhining colour.*

Having one of thefe Cods given me by a Sea-Captain, who had fre-quented thofe parts, I found it to be a fmall Cod, about three Inches long, much like a fhort Cod of *French Beans,* which had fix Beans in it, the whole furface of it was cover'd over with a very thick and fhining brown Down or Hair, which was very fine, and for its bignefs ftiff; taking fome of this Down, and rubbing it on the back of my hand, I found very little or no trouble, only I was fenfible that feveral of thefe little downy parts with rubbing did penetrate, and were funk, or ftuck pretty deep into my fkin. After I had thus rubb'd it for a pretty while, I felt very little or no pain, in fo much that I doubted, whether it were the true Couhage ; but whil'ft I was confidering, I found the Down begin to make my hand itch, and in fome places to fmart again, much like the ftinging of a Flea or Gnat, and this continued a pretty while, fo that by degrees I found my fkin to be fwell'd with little red puftules, and to look as if it had been itchie. But fuffering it without rubbing or fcratch-ing, the itching tickling pain quickly grew languid, and within an hour I felt nothing at all, and the little *protuberancies* were vanifh'd.

The caufe of which odd *Phænomenon*, I fuppofe to be much the fame with that of the ftinging of a Nettle, for by the *Microfcope*, I difcover'd this Down to confift of a multitude of fmall and flender conical bodies, much refembling Needles or Bodkins, fuch as are reprefented by A B. C D. E F. of the firft Figure of the XVI. *Scheme* ; that their ends A A A, were very fharp, and the fubftance of them ftiff and hard, much like the fubftance of feveral kinds of Thorns and crooks growing on Trees. And though they appear'd very cleer and tranfparent, yet I could not per-ceive whether they were hollow or not, but to me they appear'd like folid tranfparent bodies, without any cavity in them; whether, though they might not be a kind of Cane, fill'd with fome tranfpa-rent liquor which was hardned (becaufe the Cod which I had was very dry) I was not able to examine.

Now, being fuch ftiff, fharp bodies, it is eafie to conceive, how with rubbing they might eafily be thruft into the tender parts of the skin, and there, by reafon of their exceeding finenefs and drinefs, not create any confiderable trouble or pain, till by remaining in thofe places moiftned with the humours of the body, fome cauftick part fticking on them, or

refiding

refiding within them might be diffolv'd and mix'd with the ambient juices of that place, and thereby thofe *fibres* and tender parts adjoyning become affected, and as it were corroded by it; whence, while that action lafts, the pains created are pretty fharp and pungent, though fmall, which is the effential property of an itching one.

That the pain alfo caufed by the ftinging of a Flea, a Gnat, a Flie, a Wafp, and the like, proceeds much from the very fame caufe, I elfewhere in their proper places endeavour to manifeft. The ftinging alfo of fhred Horf-hair, which in meriment is often ftrew'd between the fheets of a Bed, feems to proceed from the fame caufe.

Obferv. XXVII. *Of the* Beard *of a wilde* Oat, *and the ufe that may be made of it for exhibiting always to the Eye the temperature of the Air, as to drinefß and moifture.*

THis Beard of a wild *Oat*, is a body of a very curious ftructure, though to the naked Eye it appears very flight, and inconfiderable, it being only a fmall black or brown Beard or Briftle, which grows out of the fide of the inner Husk that covers the Grain of a wild *Oat*; the whole length of it, when put in Water, fo that it may extend it felf to its full length, is not above an Inch and a half, and for the moft part fomewhat fhorter, but when the Grain is ripe, and very dry, which is ufualy in the Moneths of *July*, and *Auguft*, this Beard is bent fomewhat below the middle, namely, about ⅓ from the bottom of it, almoft to a right Angle, and the under part of it is wreath'd lik a With; the fubftance of it is very brittle when dry, and it will very eafily be broken from the husk on which it grows.

If you take one of thefe Grains, and wet the Beard in Water, you will prefently fee the fmall bended top to turn and move round, as if it were fenfible; and by degrees, if it be continued wet enough, the joint or knee will ftreighten it felf; and if it be fuffer'd to dry again, it will by degrees move round another way, and at length bend again into its former pofture.

If it be view'd with an ordinary fingle *Microfcope*, it will appear like a fmall wreath'd Sprig, with two clefts; and if wet as before, and then look'd on with this *Microfcope*, it will appear to unwreath it felf, and by degrees, to ftreighten its knee, and the two clefts will become ftreight, and almoft on oppofite fides of the fmall cylindrical body.

If it be continued to be look'd a little longer with a *Microfcope*, it will within a little while begin to wreath it felf again, and foon after return to its former pofture, bending it felf again neer the middle, into a kind of knee or angle.

Several of thofe bodies I examin'd with larger *Microfcopes*, and there found them much of the make of thofe two long wreath'd cylinders delineated in the fecond Figure of the 15. *Scheme*, which two cylinders reprefent present the wreathed part broken into two pieces, whereof the end A B is to be suppos'd to have join'd to the end C D, so that E A C F does represent the whole wreath'd part of the Beard, and E G a small piece of the upper part of the Beard which is beyond the knee, which as I had not room to insert, so was it not very considerable, either for its form, or any known property; but the under or wreathed part is notable for both: As to its form, it appear'd, if it were look'd on side-ways, almost like a Willow, or a small tapering rod of *Hazel*, the lower or bigger half of which onely, is twisted round several times, in some three, in others more, in others less, according to the bigness and maturity of the Grain on which it grew, and according to the driness and moisture of the ambient Air, as I shall shew more at large by and by.

The whole outward Superficies of this Cylindrical body is curiously adorned or fluted with little channels, and interjacent ridges, or little *protuberances* between them, which run the whole length of the Beard, and are streight where the Beard is not twisted, and wreath'd where it is, just after the same manner: each of those sides is beset pretty thick with small Bristles or Thorns, somewhat in form resembling that of *Porcupines* Quills, such as *a a a a a* in the Figure; all whose points are directed like so many Turn-pikes towards the small end or top of the Beard, which is the reason, why, if you endeavour to draw the Beard between your fingers the contrary way, you will find it to stick, and grate, as it were, against the skin.

The proportion of these small conical bodies *a a a a a* to that whereon they grow, the Figure will sufficiently shew, as also their manner of growing, their thickness, and neerness to each other, as, that towards the root or bottom of the Beard, they are more thin, and much shorter, insomuch that there is usually left between the top of the one, and the bottom of that next above it, more then the length of one of them, and that towards the top of the Beard they grow more thick and close (though there be fewer ridges) so that the root, and almost half the upper are hid by the tops of those next below them.

I could not perceive any *transverse* pores, unless the whole wreath'd part were separated and cleft, in those little channels, by the wreathing into so many little strings as there were ridges, which was very difficult to determine; but there were in the wreathed part two very conspicuous channels or clefts, which were continued from the bottom F to the elbow E H, or all along the part which was wreath'd, which seem'd to divide the wreath'd Cylinder into two parts, a bigger and a less; the bigger was that which was at the *convex* side of the knee, namely, on the side A, and was wreath'd by O O O O O; this, as it seem'd the broader, so did it also the longer, the other P P P P P, which was usually purs'd or wrinckled in the bending of the knee, as about E, seem'd both the shorter and narrower, so that at first I thought the wreathing and unwreathing of the Beard might have been caus'd by the shrinking or swelling of that part; but upon further examination, I found that the clefts, K K, L L, were stuft up with a kind of Spongie substance, which, for the most part, was

very

very confpicuous neer the knee, as in the cleft K K, when the Beard was dry ; upon the difcovery of which, I began to think, that it was upon the fwelling of this porous pith upon the accefs of moifture or water that the Beard, being made longer in the midft, was ftreightned, and by the fhrinking or fubfiding of the parts of that Spongie fubftance together, when the water or moifture was exhal'd or dried, the pith or middle parts growing fhorter, the whole became twifted.

But this I cannot be pofitive in, for upon cutting the wreath'd part in many places tranfverfly, I was not fo well fatisfy'd with the fhape and manner of the pores of the pith ; for looking on thefe tranfverfe Sections with a very good *Microfcope*, I found that the ends of thofe tranfverfe Sections appear'd much of the manner of the third Figure of the 15. *Scheme* A B C F E, and the middle or pith C C, feem'd very full of pores indeed, but all of them feem'd to run the long-ways.

This Figure plainly enough fhews in what manner thofe clefts, K and L divided the wreath'd Cylinder into two unequal parts, and alfo of what kind of fubftance the whole body confifts ; for by cutting the fame Beard in many places, with tranfverfe Sections, I found much the fame appearance with this exprefs'd ; fo that thofe pores feem to run, as in moft other fuch Cany bodies, the whole length of it.

The clefts of this body K K, and L L, feem'd (as is alfo exprefs'd in the Figure) to wind very oddly in the inner part of the wreath ; and in fome parts of them, they feem'd ftuffed, as it were, with that Spongie fubftance, which I juft now defcribed.

This fo oddly conftituted Vegetable fubftance, is firft (that I have met with) taken notice of by *Baptifta Porta*, in his *Natural Magick*, as a thing known to children and Juglers, and it has been call'd by fome of thofe laft named perfons, the better to cover their cheat, the Legg of an *Arabian Spider*, or the Legg of an inchanted *Egyptian Fly*, and has been ufed by them to make a fmall Index, Crofs, or the like, to move round upon the wetting of it with a drop of Water, and muttering certain words.

But the ufe that has been made of it, for the difcovery of the various conftitutions of the Air, as to drinefs and moiftnefs, is incomparably beyond any other ; for this it does to admiration : The manner of contriving it fo, as to perform this great effect, is onely thus :

Provide a good large Box of Ivory, about four Inches over, and of what depth you fhall judge convenient (according to your intention of making ufe of one, two, three, or more of thefe fmall Beards, ordered in the manner which I fhall by and by defcribe) let all the fides of this Box be turned of Bafket-work (which here in *London* is eafily enough procur'd) full of holes, in the manner almoft of a Lettice, the bigger, or more the holes are, the better, that fo the Air may have the more free paffage to the inclofed Beard, and may the more eafily pafs through the Inftrument ; it will be better yet, though not altogether fo handfom, if infteed of the Bafket-work on the fides of the Box, the bottom and top of the Box be join'd together onely with three or four fmall Pillars, after the manner reprefented sented in the 4. Figure of the 15. *Scheme*. Or, if you intend to make use of many of these small Beards join'd together, you may have a small long Case of Ivory, whose sides are turn'd of Basket-work, full of holes, which may be screw'd on to the underside of a broad Plate of Ivory, on the other side of which is to be made the divided Ring or Circle, to which divisions the pointing of the Hand or Index, which is moved by the conjoin'd Beard, may shew all the *Minute* variations of the Air.

There may be multitudes of other ways for contriving this small Instrument, so as to produce this effect, which any one may, according to his peculiar use, and the exigency of his present occasion, easily enough contrive and take, on which I shall not therefore insist. The whole manner of making any one of them is thus: Having your Box or frame A A B B, fitly adapted for the free passage of the Air through it, in the midst of the bottom B B B, you musthave a very small hole C, into which the lower end of the Beard is to be fi'xd, the upper end of which Beard *a b*, is to pass through a small hole of a Plate, or top A A, if you make use onely of a single one, and on the top of it *e*, is to be fix'd a small and very light *Index f g*, made of a very thin sliver of a Reed or Cane ; but if you make use of two or more Beards, they must be fix'd and bound together, either with a very fine piece of Silk, or with a very small touch of hard Wax, or Glew, which is better, and the *Index f g*, is to be fix'd on the top of the second, third, or fourth in the same manner as on the single one.

Now, because that in every of these contrivances, the *Index f g*, will with some temperatures of Air, move two, three, or more times round, which without some other contrivance then this, will be difficult to distinguish, therefore I thought of this Expedient : The *Index* or *Hand f g*, being rais'd a pretty way above the surface of the Plate A A, fix in at a little distance from the middle of it a small Pin *h*, so as almost to touch the surface of the Plate A A, and then in any convenient place of the surface of the Plate, fix a small Pin, on which put on a small piece of Paper, or thin Past-board, Vellom, or Parchment, made of a convenient cize, and shap'd in the manner of that in the Figure express'd by *i k*, so that having a convenient number of teeth every turn or return of the Pin *h*, may move this small indented Circle, a tooth forward or backwards, by which means the teeth of the Circle, being mark'd, it will be thereby very easie to know certainly, how much variation any change of weather will make upon the small wreath'd body. In the making of this Secundary Circle of Vellom, or the like, great care is to be had, that it be made exceeding light, and to move very easily, for otherwise a small variation will spoil the whole operation. The Box may be made of Brass, Silver, Iron, or any other substance, if care be taken to make it open enough, to let the Air have a sufficiently free access to the Beard. The *Index* also may be various ways contrived, so as to shew both the number of the revolutions it makes, and the *Minute* divisions of each revolution.

I have made several trials and Instruments for discovering the driness and moisture of the Air with this little wreath'd body, and find it to vary exceeding sensibly with the least change in the constitution of the Air, as

to

to drinefs and moifture, fo that with one breathing upon it, I have made it untwift a whole bout, and the *Index* or *Hand* has fhew'd or pointed to various divifions on the upper Face or Ring of the Inftrument, according as it was carried neerer and neerer to the fire, or as the heat of the Sun increafed upon it.

Other trials I have made with Gut-ftrings, but find them nothing neer fo fenfible, though they alfo may be fo contriv'd as to exhibit the changes of the Air, as to drinefs and moifture, both by their ftretching and fhrinking in length, and alfo by their wreathing and unwreathing themfelves; but thefe are nothing neer fo exact or fo tender, for their varying property will in a little time change very much. But there are feveral other Vegetable fubftances that are much more fenfible then even this Beard of a wilde *Oat*; fuch I have found the Beard of the feed of Mufk-grafs, or *Geranium mofchatum*, and thofe of other kinds of *Cranesbil* feeds, and the like. But always the fmaller the wreathing fubftance be, the more fenfible is it of the mutations of the Air, a conjecture at the reafon of which I fhall by and by add.

The lower end of this wreath'd Cylinder being ftuck upright in a little foft Wax, fo that the bended part or *Index* of it lay *horizontal*, I have obferv'd it always with moifture to unwreath it felf from the Eaft (For inftance) by the South to the Weft, and fo by the North to the Eaft again, moving with the Sun (as we commonly fay) and with heat and drouth to re-twift, and wreath it felf the contrary way, namely, from the Eaft, (for inftance) by the North to the Weft, and fo onwards.

The caufe of all which *Phænomena*, feems to be the differing texture of the parts of thefe bodies, each of them (efpecially the Beard of a wilde *Oat*, and of *Mosk-grafs* feed) feeming to have two kind of fubftances, one that is very porous, loofe, and fpongie, into which the watry fteams of the Air may be very eafily forced, which will be thereby fwell'd and extended in its dimenfions, juft as we may obferve all kind of Vegetable fubftance upon fteeping in water to fwell and grow bigger and longer. And a fecond that is more hard and clofe, into which the water can very little, or not at all penetrate, this therefore retaining always very neer the fame dimenfions, and the other ftretching and fhrinking, according as there is more or lefs moifture or water in its pores, by reafon of the make and fhape of the parts, the whole body muft neceffarily unwreath and wreath it felf.

And upon this Principle, it is very eafie to make feveral forts of contrivances that fhould thus wreath and unwreath themfelves, either by heat and cold, or by drinefs and moifture, or by any greater or lefs force, from whatever caufe it proceed, whether from gravity or weight, or from wind which is motion of the Air, or from fome fpringing body, or the like.

This, had I time, I fhould enlarge much more upon; for it feems to me to be the very firft footftep of *Senfation*, and Animate motion, the moft plain, fimple, and obvious contrivance that Nature has made ufe of to produce a motion, next to that of Rarefaction and Condenfation by heat
and

and cold. And were this Principle very well examin'd, I am very apt to think, it would afford us a very great help to find out the *Mechanifm* of the Mufcles, which indeed, as farr as I have hitherto been able to examine, feems to me not fo very perplex as one might imagine, efpecially upon the examination which I made of the Mufcles of *Crabs, Lobfters*, and feveral forts of large Shell-fifh, and comparing my Obfervations on them, with the circumftances I obferv'd in the mufcles of terreftrial Animals.

Now, as in this Inftance of the Beard of a wilde *Oat*, we fee there is nothing elfe requifite to make it wreath and unwreath it felf, and to ftreighten and bend its knee, then onely a little breath of moift or dry Air, or a fmall *atome* almoft of water or liquor, and a little heat to make it again evaporate; for, by holding this Beard, plac'd and fix'd as I before directed, neer a Fire, and dipping the tip of a fmall fhred of Paper in well rectify'd fpirit of Wine, and then touching the wreath'd *Cylindrical* part, you may perceive it to untwift it felf; and prefently again, upon the *avolation* of the fpirit, by the great heat, it will re-twift it felf, and thus will it move forward and backwards as oft as you repeat the touching it with the fpirit of Wine; fo may, perhaps, the fhrinking and relaxing of the mufcles be by the influx and evaporation of fome kind of liquor or juice. But of this Enquiry I fhall add more elfewhere.

Obferv. XXVIII. *Of the Seeds of* Venus *looking-glaß, or* Corn Violet.

FRom the Leaves, and Downs, and Beards of Plants, we come at laft to the Seeds; and here indeed feems to be the Cabinet of Nature, wherein are laid up its Jewels. The providence of Nature about Vegetables, is in no part manifefted more, then in the various contrivances about the feed, nor indeed is there in any part of the Vegetable fo curious carvings, and beautifull adornments, as about the feed; this in the larger forts of feeds is moft evident to the eye; nor is it lefs manifeft through the *Microfcope*, in thofe feeds whofe fhape and ftructure, by reafon of their fmalnefs, the eye is hardly able to diftinguifh.

Of thefe there are multitudes, many of which I have obferv'd through a *Microfcope*, and find, that they do, for the moft part, every one afford exceeding pleafant and beautifull objects. For befides thofe that have various kinds of carv'd furfaces, there are other that have fmooth and perfectly polifh'd furfaces, others a downy hairy furface; fome are cover'd onely with a fkin, others with a kind of fhell, others with both, as is obfervable alfo in greater feeds.

Of thefe feeds I have onely defcribed four forts which may ferve as a *fpecimen* of what the inquifitive obfervers are likely to find among the reft. The firft of thefe feeds which are defcribed in the 17. *fcheme*, are thofe of Corn-Violets, the feed is very fmall, black, and fhining, and, to the naked eye, looks almoft like a very fmall Flea; But through the
Microfcope

Microscope, it appears a large body, cover'd with a tough thick and bright reflecting skin very irregularly shrunk and pitted, infomuch that it is almost an impoffibility to find two of them wrinkled alike, so great a variety may there be even in this little feed.

This, though it appear'd one of the most promising feeds for beauty to the naked eye, yet through the *Microscope* it appear'd but a rude mishapen feed, which I therefore drew, that I might thereby manifest how unable we are by the naked eye to judge of beauteous or lefs curious *microscopical* Objects; cutting some of them in funder, I obferv'd them to be fill'd with a greenish yellow pulp, and to have a very thick huſk, in proportion to the pulp.

Obferv. XXIX. *Of the Seeds of* Tyme.

THefe pretty fruits here reprefented, in the 18. *Scheme*, are nothing elfe, but nine feveral feeds of Tyme; they are all of them in differing pofture, both as to the eye and the light; nor are they all of them exactly of the fame fhape, there being a great variety both in the bulk and figure of each feed; but they all agreed in this, that being look'd on with a *Microscope*, they each of them exactly refembled a Lemmon or Orange dry'd; and this both in fhape and colour. Some of them are a little rounder, of the fhape of an Orange, as A and B, they have each of them a very confpicuous part by which they were join'd to their little ftalk, and one of them had a little piece of ftalk remaining on; the oppofite fide of the feed, you may perceive very plainly by the Figure, is very copped and prominent, as is very ufual in Lemmons, which prominencies are exprefs'd in D, E and F.

They feem'd each of them a little creas'd or wrinckled, but E was very confpicuoufly furrow'd, as if the inward make of this feed had been fomewhat like that of a Lemmon alfo, but upon dividing feveral feeds with a very fharp Pen-knife, and examining them afterward, I found their make to be in nothing but bulk differing from that of Peas, that is, to have a pretty thick coat, and all the reft an indifferent white pulp, which feem'd very clofe; fo that it feems Nature does not very much alter her method in the manner of inclofing and preferving the vital Principle in the feed, in thefe very fmall grains, from that of Beans, Peas, &c.

The Grain affords a very pretty Object for the *Microscope*, namely, a Difh of Lemmons plac'd in a very little room; fhould a Lemmon or Nut be proportionably magnify'd to what this feed of Tyme is, it would make it appear as bigg as a large Hay-reek and it would be no great wonder to fee *Homers Iliads*, and *Homer* and all, cramm'd into fuch a Nut-fhell. We may perceive even in thefe fmall Grains, as well as in greater, how curious and carefull Nature is in preferving the feminal principle of Vegetable bodies, in what delicate, ftrong and moft convenient Cabinets fhe

Y   lays

lays them and clofes them in a pulp for their fafer protection from outward dangers, and for the fupply of convenient alimental juice, when
the heat of the Sun begins to animate and move thefe little *automatons*
or Engines; as iffhe would,from the ornaments wherewith fhe has deckt
thefe Cabinets, hint to us, that in them fhe has laid up her Jewels
and Mafter-pieces. And this, if we are but diligent in obferving, we
fhall find her method throughout. There is no curiofity in the Elemental
kingdom, if I may fo call the bodies of Air, Water, Earth, that are comparable in form to thofe of Minerals; Air and Water having no form at
all, unlefs a potentiality to be form'd into Globules; and the clods and
parcels of Earth are all irregular, whereas in Minerals fhe does begin to
Geometrize, and practife, as 'twere, the firft principles of *Mechanicks*,
fhaping them of plain regular figures, as triangles, fquares, &c. and *te-
traedrons*, cubes, &c. But none of their forms are comparable to the
more compounded ones of Vegetables; For here fhe goes a ftep further,
forming them both of more complicated fhapes, and adding alfo multitudes of curious Mechanick contrivances in their ftructure;for whereas in
Vegetables there was no determinate number of the leaves or branches,
nor no exactly certain figure of leaves, or flowers,or feeds, in Animals all
thofe things are exactly defin'd and determin'd; and where-ever there
is either an excefs or defect of thofe determinate parts or limbs,there has
been fome impediment that has fpoil'd the principle which was moft regular: Here we fhall find, not onely moft curioufly compounded fhapes,
but moft ftupendious Mechanifms and contrivances, here the ornaments
are in the higheft perfection, nothing in all the Vegetable kingdom that
is comparable to the deckings of a Peacock; nay,to the curiofity of any
feather, as I elfewhere fhew; nor to that of the fmalleft and moft defpicable Fly. But I muft not ftay on thefe fpeculations, though perhaps it
were very well worth while for one that had leifure,to fee what Information may be learn'd of the nature,or ufe,or virtues of bodies,by their feveral forms and various excellencies and properties. Who knows but *Adam*
might from fome fuch contemplation, give names to all creatures? If at
leaft his names had any fignificancy in them of the creature's nature on
which he impos'd it; as many (upon what grounds I know not) have
fuppos'd: And who knows,but the Creator may,in thofe characters,have
written and engraven many of his moft myfterious defigns and counfels,
and given man a capacity, which, affifted with diligence and induftry,
may be able to read and underftand them. But not to multiply my di-
greffion more then I can the time, I will proceed to the next, which is,

---

Obferv. XXX. *Of the Seeds of* Poppy.

THe fmall feeds of Poppy, which are defcribed in the 19. *Scheme*, both
for their fmalnefs, multiplicity and prettinefs, as alfo for their admirable foporifick quality, deferve to be taken notice of among the
other

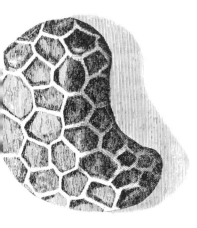

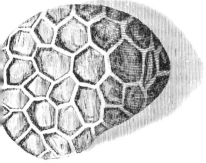

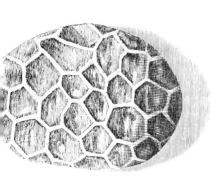

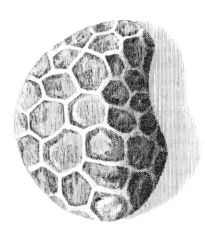

other *microscopical* seeds of Vegetables: For first, though they grow in a Case or Hive oftentimes bigger then one of these Pictures of the *microscopical* appearance, yet are they for the most part so very little, that they exceed not the bulk of a small Nitt, being not above $\frac{1}{32}$ part of an Inch in Diameter, whereas the Diameter of the Hive of them oftentimes exceeds two Inches, so that it is capable of containing neer two hundred thousand, and so in all likelihood does contain a vast quantity, though perhaps not that number. Next, for their prettiness, they may be compar'd to any *microscopical* seed I have yet seen; for they are of a dark brownish red colour, curiously Honey-comb'd all over with a very pretty variety of Net-work, or a small kind of imbosment of very orderly rais'd ridges, the surface of them looking not unlike the inside of a Beev's stomack. But that which makes it most considerable of all, is, the medicinal virtues of it, which are such as are not afforded us by any Mineral preparation; and that is for the procuring of sleep, a thing as necessary to the well-being of a creature as his meat, and that which refreshes both the voluntary and rational faculties, which, whil'st this affection has seis'd the body, are for the most part unmov'd, and at rest. And, methinks, Nature does seem to hint some very notable virtue or excellency in this Plant from the curiosity it has bestow'd upon it. First, in its flower, it is of the highest scarlet-Dye, which is indeed the prime and chiefest colour, and has been in all Ages of the world most highly esteem'd: Next, it has as much curiosity shew'd also in the husk or case of the seed, as any one Plant I have yet met withall; and thirdly, the very seeds themselves, the *Microscope* discovers to be very curiously shap'd bodies; and lastly, Nature has taken such abundant care for the propagation of it, that one single seed grown into a Plant, is capable of bringing some hundred thousands of seeds.

It were very worthy some able man's enquiry whether the intention of Nature, as to the secundary end of Animal and Vegetable substances might not be found out by some such characters and notable impressions as these, or from divers other circumstances, as the figure, colour, place, time of flourishing, springing and fading, duration, taste, smell, &c. For if such there are (as an able *Physician* upon good grounds has given me cause to believe) we might then, insteed of studying Herbals (where so little is deliver'd of the virtues of a Plant, and less of truth) have recourse to the Book of Nature it self, and there find the most natural, usefull, and most effectual and specifick Medicines, of which we have amongst Vegetables, two very noble Instances to incourage such a hope, the one of the *Jesuite powder* for the cure of *intermitting Feavers*, and the other of the juice of *Poppy* for the curing the defect of sleeping.

Obſerv. XXXI. *Of* Purſlane-ſeed.

THe Seeds of *Purſlane* ſeem of very notable ſhapes, appearing through the *Microſcope* ſhap'd ſomewhat like a *nautilus* or *Porcelane* ſhell, as may be ſeen in the XX. *Scheme*, it being a ſmall body, coyl'd round in the manner of a Spiral; at the greater end whereof, which repreſents the mouth or orifice of the Shell, there is left a little white tranſparent ſubſtance, like a ſkin, repreſented by B B B B, which ſeems to have been the place whereunto the ſtem was join'd. The whole ſurface of this *Coclea* or Shell, is cover'd over with abundance of little *prominencies* or buttons very orderly rang'd into Spiral rows, the ſhape of each of which ſeem'd much to reſemble a Wart upon a mans hand. The order, variety, and curioſity in the ſhape of this little ſeed, makes it a very pleaſant object for the *Microſcope*, one of them being cut aſunder with a very ſharp Pen-knife, diſcover'd this carved Caſket to be of a browniſh red, and ſome-what tranſparent ſubſtance, and manifeſted the inſide to be fill'd with a whitiſh green ſubſtance or pulp, the Bed wherein the ſeminal principle lies *invelop'd*.

There are multitudes of other ſeeds which in ſhape repreſent or imitate the forms of divers other ſorts of Shells: as the ſeed of *Scurvy-graſ*, very much reſembles the make of a *Concha Venerea*, a kind of Purce-lane Shell; others repreſent ſeveral ſorts of larger fruits, ſweat Marje-rome and Pot-marjerome repreſent Olives. Carret ſeeds are like a cleft of a Coco-Nut Huſk; others are like Artificial things, as Succory ſeeds are like a Quiver full of Arrows, the ſeeds of *Amaranthus* are of an ex-ceeding lovely ſhape, ſomewhat like an Eye: The ſkin of the black and ſhrivled ſeeds of Onyons and Leeks, are all over knobbed like a Seals ſkin. Sorrel has a pretty black ſhining three-ſquare ſeed, which is picked at both ends with three ridges, that are bent the whole length of it. It were al-moſt endleſ to reckon up the ſeveral ſhapes, they are ſo many and ſo va-rious; Leaving them therefore to the curious obſerver, I ſhall proceed to the Obſervations on the parts of Animals.

Obſerv. XXXII. *Of the Figure of ſeveral ſorts of* Hair, *and of the texture of the* ſkin.

Viewing ſome of the Hairs of my Head with a very good *Microſcope*, I took notice of theſe particulars:

1. That they were, for the moſt part, *Cylindrical*, ſome of them were ſomewhat *Priſmatical*, but generally they were veryneer round, ſuch as are repreſented in the ſecond Figure of the 5. *Scheme*, by the *Cylinders* E E E. nor could I find any that had ſharp angules.

2. That

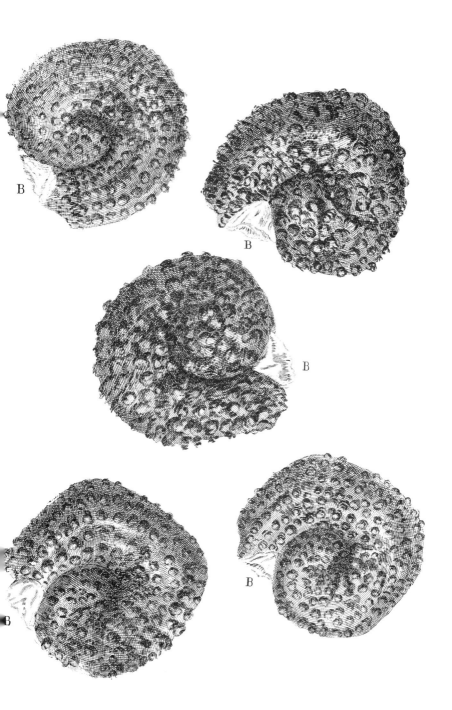

2. That that part which was next the top, was bigger then that which was neerer the root.

3. That they were all along from end to end tranfparent, though not very cleer, the end next the root appearing like a black tranfparent piece of Horn, the end next the top more brown, fomewhat like tranfparent Horn.

4. That the root of the Hairs were pretty fmooth, tapering inwards, almoft like a Parfneb; nor could I find that it had any filaments, or any other veffels, fuch as the *fibres* of Plants.

5. That the top when fplit (which is common in long Hair) appear'd like the end of a ftick, beaten till it be all flitter'd, there being not onely two fplinters, but fometimes half a fcore and more.

6. That they were all, as farr as I was able to find, folid *Cylindrical* bodies, not pervious, like a Cane or Bulrufh; nor could I find that they had any Pith, or diftinction of Rind, or the like, fuch as I had obferv'd in Horfe-hairs, the Briftles of a Cat, the *Indian* Deer's Hair, &c.

Obfervations on feveral other forts of Hair.

For the Brifles of a Hogg, I found them to be firft a hard tranfparent horny fubftance, without the leaft appearance of pores or holes in it; and this I try'd with the greateft care I was able, cutting many of them with a very fharp Razor, fo that they appear'd, even in the Glafs, to have a pretty fmooth furface, but fomewhat waved by the fawing to and fro of the Razor, as is vifible in the end of the *Prifmatical* body A of the fame Figure; and then making trials with caufing the light to be caft on them all the various ways I could think of, that was likely to make the pores appear, if there had been any, I was not able to difcover any.

Next, the Figure of the Brifles was very various, neither perfectly round, nor fharp edg'd, but *Prifmatical*, with divers fides, and round angles, as appears in the Figure A. The bending of them in any part where they before appear'd cleer, would all flaw them, and make them look white.

The Muftacheos of a Cat (part of one of which is reprefented by the fhort *Cylinder* B of the fame Figure) feem'd to have, all of them that I obferv'd, a large pith in the middle, like the pith of an Elder, whofe texture was fo clofe, that I was not able to difcover the leaft fign of pores; and thofe parts which feem to be pores, as they appear'd in one pofition to the light, in another I could find a manifeft reflectiom to be caft from them.

This I inftance in, to hint that it is not fafe to conclude any thing to be pofitively this or that, though it appear never fo plain and likely when look'd on with a *Microfcope* in one pofture, before the fame be examin'd by placing it in feveral other pofitions.

And this I take to be the reafon why many have believed and afferted the Hairs of a man's head to be hollow, and like fo many fmall pipes perforated from end to end.

Now, though I grant that by an *Analogie* one may fuppofe them fo,

and

and from the *Polonian* difeafe one may believe them fuch, yet I think we have not the leaft encouragement to either from the *Microfcope*, much lefs pofitively to affert them fuch. And perhaps the very effence of the *Plica Polonica* may be the hairs growing hollow, and of an unnatural conftitution.

And as for the *Analogie*, though I am apt enough to think that the hairs of feveral Animals may be perforated fomewhat like a Cane, or at leaft have a kind of pith in them, firft, becaufe they feem as 'twere a kind of Vegetable growing on an Animal, which growing, they fay, remains a long while after the Animal is dead, and therefore fhould like other Vegetables have a pith ; and fecondly, becaufe Horns and Feathers, and Porcupine's Quils, and Cats Brifles, and the long hairs of Horfes, which come very neer the nature of a mans hair, feem all of them to have a kind of pith, and fome of them to be porous, yet I think it not (in thefe cafes, where we have fuch helps for the fenfe as the *Microfcope* affords) fafe concluding or building on more then we fenfibly know, fince we may, with examining, find that Nature does in the make of the fame kind of fubftance, often vary her method in framing of it : Inftances enough to confirm this we may find in the Horns of feveral creatures : as what a vaft difference is there between the Horns of an Oxe, and thofe of fome forts of Staggs as to their fhape ? and even in the hairs of feveral creatures, we find a vaft difference ; as the hair of a man's head feems, as I faid before, long, *Cylindrical* and fometime a little *Prifmatical*, folid or impervious, and very fmall ; the hair of an *Indian* Deer (a part of the middle of which is defcribed in the third Figure of the fifth *Scheme*, marked with F) is bigger in compafs through all the middle of it, then the Brifle of an Hogg, but the end of it is fmaller then the hair of any kind of Animal (as may be feen by the Figure G) the whole belly of it, which is about two or three Inches long, looks to the eye like a thread of courfe Canvafs, that has been newly unwreath'd, it being all wav'd or bended to and fro, much after that manner, but through the *Microfcope*, it appears all perforated from fide to fide, and Spongie, like a fmall kind of fpongy Coral, which is often found upon the *Englifh* fhores ; but though I cut it tranfverfly, I could not perceive that it had any pores that ran the long-way of the hair : the long hairs of Horfes C C and D, feem *Cylindrical* and fomewhat pithy ; the Brifles of a Cat B, are conical and pithy : the Quils of Porcupines and Hedghoggs, being cut tranfverfly, have a whitifh pith, in the manner of a Starr, or Spur-rowel : Piggs-hair (A) is fomewhat *triagonal*, and feems to have neither pith nor pore : And other kinds of hair have quite a differing ftructure and form. And therefore I think it no way agreeable to a true natural Hiftorian, to pretend to be fo fharp-fighted, as to fee what a pre-conceiv'd *Hypothefis* tells them fhould be there, where another man, though perhaps as feeing, but not foreftall'd, can difcover no fuch matter.

But to proceed ; I obferv'd feveral kind of hairs that had been Dyed, and found them to be a kind of horny *Cylinder*, being of much about the tranfparency of a pretty cleer piece of Oxe horn ; thefe appear'd quite

through-

throughout 'ting'd with the colours they exhibited. And tis likely, that those hairs being boyl'd or fteep'd in those very hot ting'd liquors in the Dye-fat, And the fubftance of the hair being much like that of an Oxes Horn, the penetrant liquor does fo far mollifie and foften the fubftance, that it finks into the very center of it, and fo the ting'd parts come to be mix'd and united with the very body of the hair, and do not (as fome have thought) only ftick on upon the outward furface. And this, the boiling of Horn will make more probable; for we fhall find by that action, that the water will infinuate it felf to a pretty depth within the furface of it, efpecially if this penetrancy of the water be much helped by the Salts that are ufually mix'd with the Dying liquors. Now, whereas Silk may be dyed or ting'd into all kind of colours without boiling or dipping into hot liquors, I ghefs the reafon to be two-fold : Firft, becaufe the filaments, or fmall cylinders of Silk, are abundantly fmaller and finer, and fo have a much lefs depth to be penetrated then moft kind of hairs ; and next, becaufe the fubftance or matter of Silk, is much more like a Glew then the fubftance of Hair is. And that I have reafon to fuppofe : Firft, becaufe when it is fpun or drawn out of the Worm, it is a perfect glutinous fubftance, and very eafily fticks and cleaves to any adjacent body, as I have feveral times obferved, both in Silk-worms and Spiders. Next, becaufe that I find that water does eafily diffolve and mollifie the fubftance again, which is evident from their manner of ordering thofe bottoms or pods of the Silk-worm before they are able to unwind them. It is no great wonder therefore, if thofe Dyes or ting'd liquors do very quickly mollifie and tinge the furfaces of fo fmall and fo glutinous a body. And we need not wonder that the colours appear fo lovely in the one, and fo dull in the other, if we view but the ting'd cylinders of both kinds with a good *Microfcope*; for whereas the fubftance of Hair, at beft, is but a dirty dufkifh white fomewhat tranfparent, the filaments of Silk have a moft lovely tranfparency and cleernefs, the difference between thofe two being not much lefs then that between a piece of Horn, and a piece of Cryftal; the one yielding a bright and vivid reflection from the concave fide of the cylinder, that is, from the concave furface of the Air that incompaffes the back-part of the cylinder; the other yielding a dull and perturb'd reflection from the feveral *Heterogeneous* parts that compofe it. And this difference will be manifeft enough to the eye, if you get a couple of fmall Cylinders, the fmaller of Cryftal Glafs, the other of Horn, and then varnifhing them over very thinly with fome tranfparent colour, which will reprefent to the naked eye much the fame kind of object which is reprefented to it from the filaments of Silk and Hair by the help of the *Microfcope*. Now, fince the threads of Silk and Serge are made up of a great number of thefe filaments, we may henceforth ceafe to wonder at the difference. From much the fame reafon proceeds the vivid and lovely colours of Feathers, wherein they very farr exceed the natural as well as Artificial colours of hair, of which I fhall fay more in its proper place.

The Teguments indeed of creatures are all of them adapted to the peculiar ufe and convenience of that Animal which they inwrap; and very
much

much alfo for the ornament and beauty of it, as will be moft evident to
any one that fhall attentively confider the various kinds of cloathings
wherewith moft creatures are by Nature invefted and cover'd. Thus I
have obferved, that the hair or furr of thofe Northern white Bears that
inhabite the colder Regions, is exeeeding thick and warm : the like have
I obferv'd of the hair of a *Greenland* Deer, which being brought alive to
London, I had the opportunity of viewing ; its hair was fo exceeding thick,
long and foft, that I could hardly with my hand, grafp or take hold of
his fkin, and it feem'd fo exceeding warm, as I had never met with any
before. And as for the ornamentative ufe of them, it is moft evident in a
multitude of creatures, not onely for colour, as the Leopards, Cats, Rhein
Deer, &c. but for the fhape, as in Horfes manes, Cats beards, and feveral
other of the greater fort of terreftrial Animals, but is much more confpicuous, in the Veftments of Fifhes, Birds, Infects, of which I fhall by and
by give fome Inftances.

As for the fkin, the *Microfcope* difcovers as great a difference between
the texture of thofe feveral kinds of Animals, as it does between their
hairs ; but all that I have yet taken notice of, when tann'd or drefs'd, are
of a Spongie nature, and feem to be conftituted of an infinite company
of fmall long *fibres* or hairs, which look not unlike a heap of Tow or
Okum ; every of which *fibres* feem to have been fome part of a Mufcle,
and probably, whil'ft the Animal was alive, might have its diftinct function, and ferve for the contraction and relaxation of the fkin, and for the
ftretching and fhrinking of it this or that way.

And indeed, without fuch a kind of texture as this, which is very like
that of *Spunk*, it would feem very ftrange, how any body fo ftrong as the
fkin of an Animal ufually is, and fo clofe as it feems, whil'ft the Animal is
living, fhould be able to fuffer fo great an extenfion any ways, without at
all hurting or dilacerating any part of it. But, fince we are inform'd by the
Microfcope, that it confifts of a great many fmall filaments, which are implicated, or intangled one within another, almoft no otherwife then the
hairs in a lock of Wool, or the flakes in a heap of Tow, though not altogether fo loofe ; but the filaments are here and there twifted, as twere, or
interwoven, and here and there they join and unite with one another, fo as
indeed the whole fkin feems to be but one piece, we need not much wonder: And though thefe *fibres* appear not through a *Microfcope*, exactly jointed and contex'd, as in Sponge ; yet, as I formerly hinted, I am apt to think,
that could we find fome way of difcovering the texture of it, whil'ft it in-
vefts the living Animal, or had fome very eafie way of feparating the pulp
or intercurrent juices, fuch as in all probability fill thofe *Interftitia*, without dilacerating, brufing, or otherwife fpoiling the texture of it (as it
feems to be very much by the ways of tanning and dreffing now us'd) we
might difcover a much more curious texture then I have hitherto been
able to find ; perhaps, fomewhat like that of Sponges.

That of *Chamoife* Leather is indeed very much like that of *Spunk*, fave
onely that the *filaments* feem nothing neer fo even and round, nor altogether fo fmall, nor has it fo curious joints as *Spunk* has, fome of which I
have

have lately difcover'd like thofe of a Sponge, and perhaps all thefe three bodies may be of the fame kind of fubftance, though two of them indeed are commonly accounted Vegetable (which, whether they be fo or no, I fhall not now difpute) But this feems common to all three, that they undergo a tanning or drefling, whereby the interfpers'd juices are wafted and wafh'd away before the texture of them can be difcover'd.

What their way is of drefling, or curing Sponges, I confefs, I cannot learn; but the way of drefling *Spunk*, is, by boiling it a good while in a ftrong *Lixivium*, and then beating it very well; and the manner of drefling Leather is fufficiently known.

It were indeed extremely defirable, if fuch a way could be found whereby the *Parenchyma* or flefh of the Mufcles, and feveral other parts of the bod, y might be wafh'd, or wafted clean away, without vitiating the form of the *fibrous* parts or veflells of it, for hereby the texture of thofe parts, by the help of a good *Microfcope*, might be moft accurately found.

But to digrefs no further, we may, from this difcovery of the *Microfcope*, plainly enough underftand how the fkin, though it looks fo clofe as it does, comes to give a pafage to fo vaft a quantity of *excrementitious* fubftances, as the diligent *Sanctorius* has excellently obferved it to do, in his *medicina ftatica*; for it feems very probable, from the texture after drefling, that there are an infinit of pores that every way pierce it, and that thofe pores are onely fill'd with fome kind of juice, or fome very pulpy fc ft fubftance, and thereby the fteams may almoft as eafily find a paflage through fuch a fluid *vehicle* as the vaporous bubbles which are generated at the bottom of a Kettle of hot water do find a paflage through that fluid *medium* into the ambient Air.

Nor is the fkin of animals only thus pervious, but even thofe of vegetables alfo feem to be the fame; for otherwife I cannot conceive why, if two fprigs of Rofemary (for Inftance) be taken as exactly alike in all particulars as can be, and the one be fet with the bottom in a Glafs of water, and the other be fet juft without the Glafs, but in the Air onely, though you ftop the lower end of that in the Air very carefully with Wax, yet fhall it prefently almoft wither, whereas the other that feems to have a fupply from the fubjacent water by its fmall pipes, or *microfcopical* pores, preferves its greennefs for many days, and fometimes weeks.

Now, this to me, feems not likely to proceed from any other caufe then the *avolation* of the juice through the fkin; for by the Wax, all thofe other pores of the ftem are very firmly and clofely ftop'd up. And from the more or lefs poroufnefs of the fkins or rinds of Vegetables may, perhaps, be fomewhat of the reafon given, why they keep longer green, or fooner wither; for we may obferve by the bladdering and craking of the leaves of Bays, Holly, Laurel, &c. that their fkins are very clofe, and do not fuffer fo free a paflage through them of the included juices.

But of this, and of the Experiment of the Rofemary, I fhall elfewhere more fully confider, it feeming to me an extreme luciferous Experiment, fuch as feems indeed very plainly to prove the *Schematifm* or ftructure

Z   of

of Vegetables altogether *mechanical*, and as neceſſary, that (water and warmth being apply'd to the bottom of the ſprig of a Plant) ſome of it ſhould be carried upwards into the ſtem, and thence diſtributed into the leaves, as that the water of the *Thames* covering the bottom of the Mills at the Bridge foot of *London*, and by the ebbing and flowing of it, paſ-ſing ſtrongly by them, ſhould have ſome part of it convey'd to the Ceſterns above, and thence into ſeveral houſes and Ceſterns up and down the City.

Obſerv. XXXIII. *Of the* Scales *of a* Soal, *and other Fiſhes.*

HAving hinted ſomewhat of the ſkin and covering of terreſtrial Ani-mals, I ſhall next add an Obſervation I made on the ſkin and Scales of a *Soal*, a ſmall Fiſh, commonly enough known; and here in Fiſhes, as well as other Animals, Nature follows its uſual method, framing all parts ſo, as that they are both uſefull and ornamental in all its compoſures, mingling *utile* and *dulce* together; and both theſe deſigns it ſeems to follow, though our unaſſiſted ſenſes are not able to peceive them : This is not onely manifeſt in the covering of this Fiſh only, but in multitudes of others, which it would be too long to enumerate, witneſs particularly that ſmall Sand Shell, which I mention'd in the XI. Obſervation, and infinite other ſmall Shells and Scales, divers of which I have view'd. This ſkin I view'd, was flead from a pretty large *Soal*, and then expanded and dry'd, the inſide of it, when dry, to the naked eye, look'd very like a piece of Canvaſs, but the *Microſcope* diſcover'd that texture to be nothing elſe, but the inner ends of thoſe curious Scolop'd Scales I, I, I, in the ſecond *Figure* of the XXI. *Scheme*, namely, the part of GGGG (of the larger repre-ſentation of a ſingle Scale, in the firſt *Figure* of the ſame *Scheme*) which on the back ſide, through an ordinary ſingle Magnifying Glaſs, look'd not unlike the Tyles on an houſe.

The outſide of it, to the naked eye, exhibited nothing more of orna-ment, ſave the uſual order of ranging the Scales into a *triagonal* form, onely the edges ſeem'd a little to ſhine, the finger being rubb'd from the tail-wards towards the head, the Scales ſeem'd to ſtay and raze it; But through an ordinary Magnifying glaſs, it exhibited a moſt curiouſly carved and adorned ſurface, ſuch as is viſible in the ſecond *Figure*, each of thoſe (formerly almoſt imperceptible) Scales appearing much of the ſhape I, I, I, that is, they were round, and protuberant, and ſomewhat ſhap'd like a Scolop, the whole Scale being creas d with curiouſly wav'd and indented ridges, with proportionable furrows between; each of which was terminated with a very ſharp tranſparent bony ſubſtance, which, like ſo many ſmall Turnpikes, ſeem'd to arm the edges.

The back part KKK was the ſkin into which each of theſe Scales were very deeply fix'd, in the curious regular order, viſible in the ſecond
Figure.

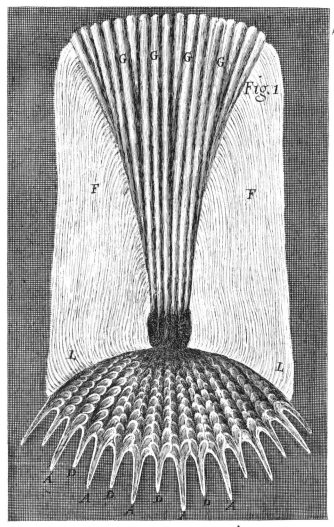

*Fig.*1

*Fig.*2

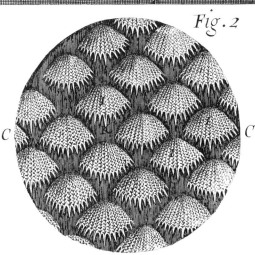

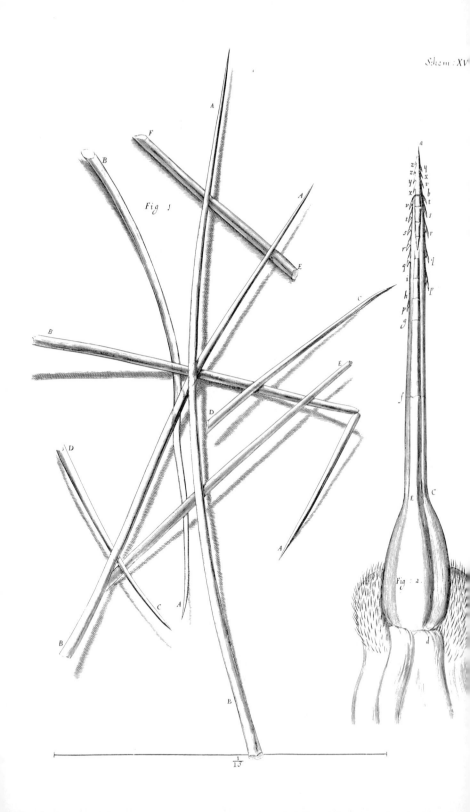

Fig: 1

Fig: 2

$\frac{1}{10}$

Figure. The length and ſhape of the part of the Scale which was buried by the ſkin,is evidenced by the firſt *Figure*; which is the repreſentation of one of them pluck'd out and view'd through a good *Microſcope,* namely, the part L F G G F L, wherein is alſo more plainly to be ſeen,the manner of carving of the ſeolopt part of every particular Scale,how each ridge or barr E E E is alternately hollowed or engraven, and how every gutter between them is terminated with very tranſparent and hard pointed ſpikes, and how every other of theſe, as A A A A, are much longer then the interjacent ones, D D D.

The texture or form alſo of the hidden part appears, namely, the middle part, G G G, ſeems to conſiſt of a great number of ſmall quills or pipes, by which, perhaps, the whole may be nouriſhed ; and the ſide parts F F conſiſt of a more fibrous texture, though indeed the whole Scale ſeem'd to be of a very tough griſly ſubſtance,like the larger Scales of other Fiſhes.

The Scales of the ſkin of a Dog-fiſh (which is us'd by ſuch as work in Wood,for the ſmoothing of their work,and conſiſts plainly enough to the naked eye,of a great number of ſmall horny points)through the *Microſcope* appear'd each of them curiouſly ridg'd, and very neatly carved ; and indeed, you can hardly look on the ſcales of any Fiſh, but you may diſcover abundance of curioſity and beautifying; and not only in theſe Fiſhes,but in the ſhells and cruſts or armour of moſt ſorts of *Marine* Animals ſo inveſted.

Obſerv. X X X I V. *Of the Sting of a* Bee.

THe Sting of a *Bee,* delineated in the ſecond Figure of the XVI.*Scheme,* ſeems to be a weapon of offence, and is as great an Inſtance, that Nature did realy intend revenge as any, and that firſt, becauſe there ſeems to be no other uſe of it. Secondly, by reaſon of its admirable ſhape, ſeeming to be purpoſely ſhap'd for that very end. Thirdly,from the virulency of the liquor it ejects, and the ſad effects and ſymptoms that follow it.

But whatever be the uſe of it,certain it is,that the ſtructure of it is very admirable ; what it appears to the naked eye, I need not deſcribe, the thing being known almoſt to every one,but it appears through the *Microſcope,* to conſiſt of two parts, the one a ſheath, without a chape or top, ſhap'd almoſt like the Holſter of a Piſtol, beginning at *d,* and ending at *b,* this ſheath I could moſt plainly perceive to be hollow, and to contain in it, both a Sword or Dart, and the poiſonous liqnor that cauſes the pain. The ſheath or caſe ſeem d to have ſeveral joints or ſettings together, marked by *f g h i k l m n o,* it was arm'd moreover neer the top, with ſeveral crooks or forks (*p q r ſ t*) on one ſide, and (*p q r ſ t u*) on the other, each of which ſeem'd like ſo many Thorns growing on a briar, or rather like ſo many Cat's Claws; for the crooks themſelves ſeem'd to be little ſharp tranſparent points or claws, growing out of little *protuberancies* on

the

the fide of the fheath, which, by obferving the Figure diligently, is eafie enough to be perceiv'd ; and from feveral particulars, I fuppofe the Animal has a power of difplaying them, and fhutting them in again as it pleafes, as a Cat does its claws, or as an Adder or Viper can its teeth or fangs.

The other part of the Sting was the Sword, as I may fo call it, which is fheath'd, as it were, in it, the top of which *a b* appears quite through at the fmaller end, juft as if the chape of the fheath of a Sword were loft, and the end of it appear'd beyond the Scabbard ; the end of this Dart (*a*) was very fharp, and it was arm'd likewife with the like Tenterhooks or claws with thofe of the fheath, fuch as (*v x y, x y z z*) thefe crooks, I am very apt to think, can be clos'd up alfo, or laid flat to the fides of the Sword when it is drawn into the Scabbard, as I have feveral times obferv'd it to be, and can be fpred again or extended when ever the Animal pleafes.

The confideration of which very pretty ftructure, has hinted to me, that certainly the ufe of thefe claws feems to be very confiderable, as to the main end of this Inftrument, for the drawing in, and holding the fting in the flefh ; for the point being very fharp, the top of the Sting or Dagger (*a b*) is very eafily thruft into an Animal's body, which being once entred, the Bee, by endeavouring to pull it into the fheath, draws (by reafon of the crooks (*v x y*) and (*x y z z*) which lay hold of the fkin on either fide) the top of the fheath (*t ſ r v*) into the fkin after it, and the crooks *t, s,* and *r, v,* being entred, when the Bee endeavours to thruft out the top of the fting out of the fheath again, they lay hold of the fkin on either fide, and fo not onely keep the fheath from fliding back, but helps the top inwards, and thus, by an alternate and fucceffive retracting and emitting of the Sting in and out of the fheath, the little enraged creature by degrees makes his revengfull weapon pierce the tougheft and thickeft Hides of his enemies, in fo much that fome few of thefe ftout and refolute foldiers with thefe little engines, do often put to flight a huge mafty Bear, one of their deadly enemies, and thereby fhew the world how much more confiderable in Warr a few fkilfull Engineers and refolute foldiers politickly order'd, that know how to manage fuch engines, are, then a vaft unweildy rude force, that confides in, and acts onely by, its ftrength. But (to proceed) that he thus gets in his Sting into the fkin, I conjecture, becaufe, when I have obferv'd this creature living, I have found it to move the Sting thus, to and fro, and thereby alfo, perhaps, does, as 'twere, pump or force out the poifonous liquor, and make it hang at the end of the fheath about *b* in a drop. The crooks, I fuppofe alfo to be the caufe why thefe angry creatures, haftily removing themfelves from their revenge, do often leave thefe weapons behind them, fheath'd, as 'twere, in the flefh, and, by that means, caufe the painfull fymptoms to be greater, and more lafting, which are very probably caus'd, partly by the piercing and tearing of the fkin by the Sting, but chiefly by the corrofive and poifonous liquor that is by this Syringe-pipe convey'd among the fenfitive parts thereof and thereby more eafily gnaws

and

and corrodes thofe tender *fibres :* As I have fhewed in the defcription
of a Nettle and of Cowhage.

Obferv. X X X V. *Of the contexture and fhape of the particles of*
Feathers.

EXamining feveral forts of *Feathers,* I took notice of thefe particulars
in all forts of wing-Feathers, efpecially in thofe which ferv'd for the
beating of the air in the action of flying.

That the outward furface of the Quill and Stem was of a very hard, ftiff,
and horny fubftance, which is obvious enough, and that the part above
the Quill was fill'd with a very white and light pith, and, with the *Microfcope,* I found this pith to be nothing elfe, but a kind of natural *congeries*
of fmall bubbles, the films of which feem to be of the fame fubftance with
that of the Quill, that is, of a ftiff tranfparent horny fubftance.

Which particular feems to me, very worthy a more ferious confideration;
For here we may obferve Nature, as 'twere, put to its fhifts, to make a fubftance, which fhall be both light enough, and very ftiff and ftrong, without
varying from its own eftablifh d principles, which we may obferve to be
fuch, that very ftrong bodies are for the moft part very heavie alfo, a
ftrength of the parts ufually requiring a denfity, and a denfity a gravity;
and therefore fhould Nature have made a body fo broad and fo ftrong as
a Feather, almoft, any other way then what it has taken, the gravity of it
muft neceffarily have many times exceeded this; for this pith feems to be
like fo many ftops or crofs pieces in a long optical tube, which do very
much contribute to the ftrength of the whole, the pores of which were
fuch, as that they feem'd not to have any communication with one another, as I have elfewhere hinted.

But the Mechanifm of Nature is ufually fo excellent, that one and the
fame fubftance is adapted to ferve for many ends. For the chief ufe of
this, indeed, feems to be for the fupply of nourifhment to the downy or
feathery part of the ftem; for 'tis obvious enough in all forts of Feathers,
that 'tis plac'd juft under the roots of the branches that grow out of either fide of the quill or ftalk, and is exactly fhap'd according to the rank-
ing of thofe branches, coming no lower into the quill, then juft the beginning of the downy branches, and growing onely on the under fide of
of the quill where thofe branches do fo. Now, in a ripe Feather (as one
may call it) it feems difficult to conceive how the *Succus nutritius* fhould
be convey'd to this pith; for it cannot, I think, be well imagin'd to pafs
through the fubftance of the quill, fince, having examin'd it with the
greateft diligence I was able, I could not find the leaft appearance of
pores; but he that fhall well examine an unripe or pinn'd Feather, will
plainly enough perceive the Veffel for the conveyance of it to be the thin
filmy pith (as tis call'd) which paffes through the middle of the quill.

As for the make and contexture of the Down it felf, it is indeed very
rare

rare and admirable, and such as I can hardly believe, that the like is to be discover'd in any other body in the world; for there is hardly a large Feather in the wing of a Bird, but contains neer a million of distinct parts, and every one of them shap d in a most regular & admirable form, adapted to a particular Design : For examining a middle ciz'd Goose-quill, I easily enough found with my naked eye, that the main stem of it contain'd about 300. longer and more Downy branchings upon one side, and as many on the other of more stiff but somewhat shorter branchings. Many of these long and downy branchings, examining with an ordinary *Microscope*, I found divers of them to contain neer 1200. small leaves (as I may call them, such as E F of the first Figure of the 23. *scheme*) and as many stalks; on the other side, such as I K of the same Figure, each of the leaves or branchings, E F, seem'd to be divided into about sixteen or eighteen small joints, as may be seen plainly enough in the Figure, out of most of which there seem to grow small long *fibres*, such as are expres'd in the Figure, each of them very proportionably shap'd according to its position, or plac'd on the stalk E F; those on the under side of it, namely, 1, 2, 3, 4, 5, 6, 7, 8, 9, &c. being much longer then those directly opposite to them on the upper; and divers of them, such as 2,3,4,5,6,7,8,9, &c. were terminated with small crooks, much resembling those small crooks, which are visible enough to the naked eye, in the seed-buttons of Bur-docks. The stalks likewise, I K on the other side, seem'd divided into neer as many small knotted joints, but without any appearance of strings or crooks, each of them about the middle K, seem'd divided into two parts by a kind of fork, one side of which, namely, K L, was extended neer the length of K I, the other, M, was very short.

The transverse Sections of the stems of these branchings, manifested the shape or figure of it to be much like I N O E, which consisted of a horny skin or covering, and a white seemingly frothy pith, much like the make of the main stem of a Feather.

The use of this strange kind of form, is indeed more admirable then all the rest, and such as deserves to be much more seriously examin'd and consider'd, then I have hitherto found time or ability to do; for certainly, it may very much instruct us in the nature of the Air, especially as to some properties of it.

The stems of the Downy branches I N O E, being rang'd in the order visible enough to the naked eye, at the distance of I F, or somewhat more, the *collateral* stalks and leaves (if I may so call those bodies I newly described) are so rang'd, that the leaves or hairy stalks of the one side lie at top, or are incumbent on the stalks of the other, and cross each other, much after the manner expres'd in the second Figure of the 23. *scheme*, by which means every of those little hooked *fibres* of the leaved stalk get between the naked stalks, and the stalks being full of knots, and a pretty way dif-join'd so as that the *fibres* can easily get between them, the two parts are so closely and admirably woven together, that it is able to impede, for the greatest part, the transcursion of the Air; and though they are so exceeding small, as that the thickness of one of

these

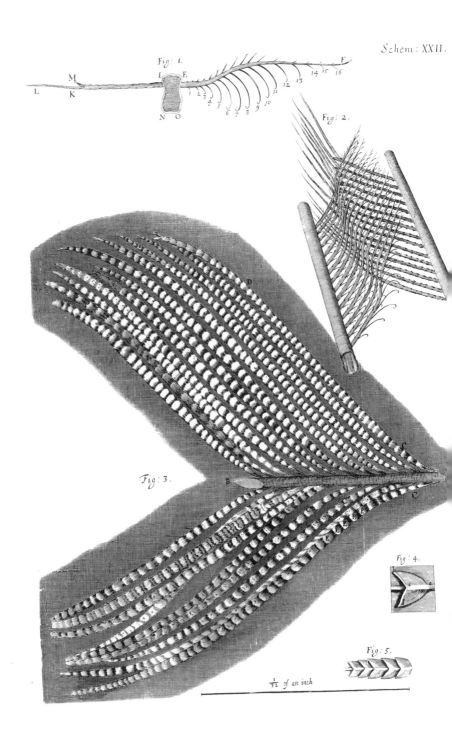

Fig: 1.

M

L   K

I   E

N O

1 2 3 4 5 6 7 8 9 10 11 12 13 14 15 16

E

Fig: 2.

D

Fig: 3.

B

C

Fig: 4.

a

E

Fig: 5.

$\frac{1}{32}$ of an inch

thefe ftalks amounts not to a 500. part of an Inch, yet do they compofe fo ftrong a texture, as, notwithftanding the exceeding quick and violent beating of them againft the Air, by the ftrength of the Birds wing, they firmly hold together. And it argues an admirable providence of Nature in the contrivance and fabrick of them; for their texture is fuch, that though by any external injury the parts of them are violently dif-joyn'd, fo as that the leaves and ftalks touch not one another, and confequently feveral of thefe rents would impede the Bird's flying; yet, for the moft part, of themfelves they readily re-join and re-contex themfelves, and are eafily by the Birds ftroking the Feather, or drawing it through its Bill, all of them fettled and woven into their former and natural pofture; for there are fuch an infinite company of thofe fmall *fibres* in the under fide of the leaves, and moft of them have fuch little crooks at their ends, that they readily catch and hold the ftalks they touch.

From which ftrange contexture, it feems rational to fuppofe that there is a certain kind of mefh or hole fo fmall, that the Air will not very eafily pafs through it, as I hinted alfo in the fixth Obfervation about fmall Glafs Canes, for otherwife it feems probable, that Nature would have drawn over fome kind of thin film which fhould have covered all thofe almoft fquare mefhes or holes, there feeming through the *Microfcope* to be more then half of the furface of the Feather which is open and vifibly pervious; which conjecture will yet feem more probable from the texture of the brufhie wings of the *Tinea argentea*, or white Feather wing'd moth, which I fhall anone defcribe. But Nature, that knows beft its own laws, and the feveral properties of bodies, knows alfo beft how to adapt and fit them to her defigned ends, and whofo would know thofe properties, muft endeavour to trace Nature in its working, and to fee what courfe fhe obferves. And this I fuppofe will be no inconfiderable advantage which the *Schematifms* and Structures of Animate bodies will afford the diligent enquirer, namely, moft fure and excellent inftructions, both as to the practical part of *Mechanicks* and to the *Theory* and knowledge of the nature of the bodies and motions.

Obferv. XXXVI. *Of* Peacoks, Ducks, *and other* Feathers *of* changeable colours.

THe parts of the Feathers of this glorious Bird appear, through the *Microfcope*, no lefs gaudy then do the whole Feathers; for, as to the naked eye 'tis evident that the ftem or quill of each Feather in the tail fends out multitudes of *Lateral* branches, fuch as A B in the third Figure of the 23. *Scheme* reprefents a fmall part of about $\frac{1}{32}$ part of an Inch long, and each of the *lateral* branches emit multitudes of little fprigs, threads or hairs on either fide of them, fuch as C D, C D, C D, fo each of thofe threads in the *Microfcope* appears a large long body, confifting of a multitude

tude of bright reflecting parts, whose Figure 'tis no easie matter to determine, as he that examines it shall find; for every new position of it to the light makes it perfectly seem of another form and shape, and nothing what it appear'd a little before; nay, it appear'd very differing oftentimes from so seemingly inconsiderable a circumstance, that the interposing of ones hand between the light and it, makes a very great change, and the opening or shutting a Casement and the like, very much diversifies the appearance. And though, by examining the form of it very many ways, which would be tedious here to enumerate, I suppose I have discover'd the true Figure of it, yet oftentimes, upon looking on it in another posture, I have almost thought my former observations deficient, though indeed, upon further examination, I have found even those also to confirm them.

These threads therefore I find to be a *congeries* of small *Laminæ* or plates, as *e e e e e*, &c. each of them shap'd much like this of *a b c d*, in the fourth *Figure*, the part *a c* being a ridge, prominency, or stem, and *b* and *d* the corners of two small thin Plates that grow unto the small stalk in the middle, so that they make a kind of little feather; each of these Plates lie one close to another, almost like a company of sloping ridge or gutter Tyles; they grow on each side of the stalk opposite to one another, by two and two, from top to bottom, in the manner express'd in the fifth Figure, the tops of the lower covering the roots of the next above them; the under side of each of these laminated bodies, is of a very dark and opacous substance, and suffers very few Rays to be trajected, but reflects them all toward that side from whence they come, much like the foil of a Looking-glass; but their upper sides seem to me to consist of a multitude of thin plated bodies, which are exceeding thin, and lie very close together, and thereby, like mother of Pearl shells, do not onely reflect a very brisk light, but tinge that light in a most curious manner; and by means of various positions, in respect of the light, they reflect back now one colour, and then another, and those most vividly.

Now, that these colours are onely *fantastical* ones, that is, such as arise immediately from the refractions of the light, I found by this, that water wetting these colour'd parts, destroy'd their colours, which seem'd to proceed from the alteration of the reflection and refraction. Now, though I was not able to see those hairs at all transparent by a common light, yet by looking on them against the Sun, I found them to be ting'd with a darkish red colour, nothing a-kin to the curious and lovely greens and blues they exhibited.

What the reason of colour seems to be in such thin plated bodies, I have elsewhere shewn. But how water cast upon those threads destroys their colours, I suppose to be perform'd thus; The water falling upon these plated bodies from its having a greater congruity to Feathers then the Air, insinuates it self between those Plates, and so extrudes the strong reflecting Air, whence both these parts grow more transparent, as the *Microscope* informs, and colourless also, at best retaining a very faint and dull

dull colour. But this wet being wasted away by the continual evapora-
tions and steams that pass through them from the Peacock, whil'st that
Bird is yet alive, the colours again appear in their former luster, the *interstitia* of these Plates being fill'd with the strongly reflecting Air.

The beauteous and vivid colours of the Feathers of this Bird, being
found to proceed from the curious and exceeding smalness and fineness
of the reflecting parts, we have here the reason given us of all those gau-
deries in the apparel of other Birds also, and how they come to exceed
the colours of all other kinds of Animals, besides Insects; for since (as we
here, and elsewhere also shew) the vividness of a colour, depends upon
the fineness and transparency of the reflecting and refracting parts; and
since our *Microscope* discovers to us, that the component parts of feathers
are such, and that the hairs of Animals are otherwise; and since we find
also by the Experiment of that Noble and most Excellent Person I formerly named; that the difference between Silk and Flax, as to its colour, is
nothing else (for Flax reduc'd to a very great fineness of parts, both
white and colour'd, appears as white and as vivid as any Silk, but loses
that brightness and its Silken aspect as soon as it is twisted into thread, by
reason that the component parts, though very small and fine, are yet pliable flakes, and not cylinders, and thence, by twisting, become united into one opacous body, whereas the threads of Silk and Feathers retain
their lustre, by preserving their cylindrical form intire without mixing; so that each reflected and refracted beam that composes the gloss
of Silk, preserves its own property of modulating the light intire); And
since we find the same confirm'd by many other Experiments elsewhere
mentioned, I think we may safely conclude this for an Axiome, that
wheresoever we meet with transparent bodies, spun out into very fine
parts, either cleer, or any ways ting'd, the colours resulting from such a
composition must necessarily be very glorious, vivid, and cleer, like those
of Silk and Feathers. This may perhaps hint some usefull way of making
other bodies, besides Silk, be susceptible of bright tinctures, but of this
onely by the by.

The changeable colour'd Feathers also of Ducks, and several other
Birds, I have found by examination with my *Microscope*, to proceed from
much the same causes and textures.

Observ. XXXVII. *Of the Feet of Flies, and several other In-sects.*

THe foot of a Fly (delineated in the first *Figure* of the 23. *Scheme*,
which represents three joints, the two Tallons, and the two Pattens
in a flat posture; and in the second *Figure* of the same *Scheme*, which represents onely one joint, the Tallons and Pattens in another posture) is
of a most admirable and curious contrivance, for by this the Flies are inabled to walk against the sides of Glass, perpendicularly upwards, and to

contain themselves in that posture as long as they please; nay, to walk and suspend themselves against the under surface of many bodies, as the ceiling of a room, or the like, and this with as great a seeming facility and firmness, as if they were a kind of *Antipodes*, and had a *tendency* upwards, as we are sure they have the contrary, which they also evidently discover, in that they cannot make themselves so light, as to stick or suspend themselves on the under surface of a Glass well polish'd and cleans'd; their suspension therefore is wholly to be ascrib'd to some Mechanical contrivance in their feet; which, what it is, we shall in brief explain, by shewing, that its Mechanism consists principally in two parts, that is, first its two Claws, or Tallons, and secondly, two Palms, Pattens, or Soles.

The two Tallons are very large, in proportion to the foot, and handsomly shap'd in the manner describ'd in the *Figures*, by A B, and A C, the bigger part of them from A to d d, is all hairy, or brisled, but toward the top, at C and B smooth, the tops or points, which seem very sharp turning downwards and inwards, are each of them mov'd on a joint at A, by which the Fly is able to open or shut them at pleasure, so that the points B and C being entered in any pores, and the Fly endeavouring to shut them, the Claws not onely draw one against another, and so fasten each other, but they draw the whole foot, G G A D D forward, so that on a soft footing, the tenters or points G G G G, (whereof a Fly has about ten in each foot, to wit, two in every joint) run into the pores, if they find any, or at least make their way; and this is sensible to the naked eye, in the feet of a *Chafer*, which, if he be suffer'd to creep over the hand, or any other part of the skin of ones body, does make his steps as sensible to the touch as the sight.

But this contrivance, as it often fails the *Chafer*, when he walks on hard and close bodies, so would it also our Fly, though he be a much lesser, and nimbler creature, and therefore Nature has furnish'd his foot with another *additament* much more curious and admirable, and that is, with a couple of Palms, Pattens or Soles D D, the structure of which is this:

From the bottom or under part of the last joint of his foot, K, arise two small thin plated horny substances, each consisting of two flat pieces, D D, which seem to be flexible, like the covers of a Book, about F F, by which means, the plains of the two sides E E, do not always lie in the same plain, but may be sometimes shut closer, and so each of them may take a little hold themselves on a body; but that is not all, for the under sides of these Soles are all beset with small brisles, or tenters, like the Wire teeth of a Card used for working Wool, the points of all which tend forwards, hence the two Tallons drawing the feet forwards, as I before hinted, and these being applied to the surface of the body with all the points looking the contrary way, that is, forwards and outwards, if there be any irregularity or yielding in the surface of the body, the Fly suspends it self very firmly and easily, without the access or need of any such Sponges fill'd with an imaginary *gluten*, as many have, for want of good Glasses, perhaps, or a troublesome and diligent examination, suppos'd.

Now, that the Fly is able to walk on Glass, proceeds partly from some
ruggedness

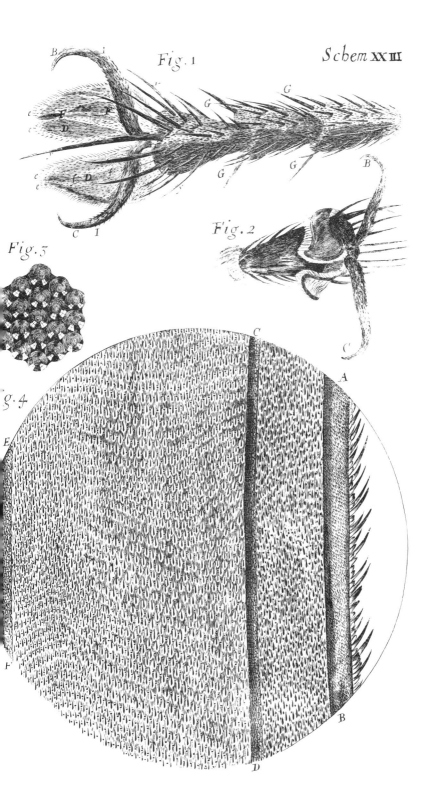

Schem **XXIII**

Fig. 1

Fig. 2

Fig. 3

g. 4

ruggedneſs of the ſurface: and chiefly from a kind of tarniſh, or dirty ſmoaky ſubſtance, which adheres to the ſurface of that very hard body; and though the pointed parts cannot penetrate the ſubſtance of Glaſs, yet may they find pores enough in the tarniſh, or at leaſt make them.

This Structure I ſomewhat the more diligently ſurvey'd, becauſe I could not well comprehend, how, if there were ſuch a glutinous matter in thoſe ſuppoſed Sponges, as moſt (that have obſerv'd that Object in a *Microſcope*) have hitherto believ'd, how, I ſay, the Fly could ſo readily un-glew and looſen its feet: and, becauſe I have not found any other crea-ture to have a contrivance any ways like it; and chiefly, that we might not be caſt upon unintelligible explications of the *Phænomena* of Nature, at leaſt others then the true ones, where our ſenſes were able to furniſh us with an intelligible, rationall and true one.

Somewhat a like contrivance to this of Flies ſhall we find in moſt other Animals, ſuch as all kinds of Flies and caſe-wing'd creatures; nay, in a Flea, an Animal abundantly ſmaller then this Fly. Other creatures, as Mites, the Land-Crab, &c. have onely one ſmall very ſharp Tallon at the end of each of their legs, which all drawing towards the center or middle of their body, inable theſe exceeding light bodies to ſuſpend and faſten themſelves to almoſt any ſurface.

Which how they are able to do, will not ſeem ſtrange, if we conſider, firſt, how little body there is in one of theſe creatures compar'd to their ſuperficies, or outſide, their thickneſs, perhaps, oftentimes, not amounting to the hundredth part of an Inch: Next, the ſtrength and agility of theſe creatures compar'd to their bulk, being, proportionable to their bulk, perhaps, an hundred times ſtronger then an Horſe or Man. And thirdly, if we conſider that Nature does always appropriate the inſtruments, ſo as they are the moſt fit and convenient to perform their offices, and the moſt ſimple and plain that poſſibly can be; this we may ſee further veri-fy'd alſo in the foot of a Louſe which is very much differing from thoſe I have been deſcribing, but more convenient and neceſſary for the place of its habitation, each of his leggs being footed with a couple of ſmall claws which he can open or ſhut at pleaſure, ſhap'd almoſt like the claws of a Lobſter or Crab, but with appropriated contrivances for his peculiar uſe, which being to move its body to and fro upon the hairs of the crea-ture it inhabits, Nature has furniſh'd one of its claws with joints, almoſt like the joints of a man's fingers, ſo as thereby it is able to encompaſs or graſp a hair as firmly as a man can a ſtick or rope.

Nor, is there a leſs admirable and wonderfull *Mechaniſm* in the foot of a Spider, whereby he is able to ſpin, weave, and climb, or run on his curious tranſparent clew, of which I ſhall ſay more in the deſcription of that Animal.

And to conclude, we ſhall in all things find, that Nature does not onely work Mechanically, but by ſuch excellent and moſt compendi-ous, as well as ſtupendious contrivances, that it were impoſſible for all the reaſon in the world to find out any contrivance to do the ſame thing that ſhould have more convenient properties. And can any be ſo ſottiſh, as to think all thofe things the productions of chance? Certainly, either their Ratiocination muft be extremely depraved, or they did never attentively confider and contemplate the Works of the Al-mighty.

Obferv. XXXVIII. *Of the Structure and motion of the Wings of* Flies.

THe Wings of all kinds of Infects, are, for the moft part, very beautifull Objects, and afford no lefs pleafing an Object to the mind to fpeculate upon, then to the eye to behold. This of the blue Fly, among the reft, wants not its peculiar ornaments and contrivances; it grows out of the *Thorax,* or middle part of the body of a Fly, and is feated a little beyond the center of gravity in the body towards the head, but that *Excentricly* is curioufly balanc'd; firft, by the expanded *Area* of the wings which lies all more backwards then the root, by the motion of them, whereby the center of their vibration is much more backwards towards the tail of the Fly then the root of the wing is. What the vibrative motion of the wings is, and after what manner they are moved, I have endeavoured by many trials to find out: And for the firft manner of their motion, I endeavoured to obferve feveral of thofe kind of fmall fpinning Flies, which will naturally fufpend themfelves, as it were, pois'd and fteady in one place of the air, without rifing or falling, or moving forwards or backwards; for by looking down on thofe, I could by a kind of faint fhadow, perceive the utmoft extremes of the vibrative motion of their wings, which fhadow, whil'ft they fo endeavoured to fufpend themfelves, was not very long, but when they endeavour'd to flie forwards, it was fomewhat longer; next, I tried it, by fixing the leggs of a Fly upon the top of the ftalk of a feather, with Glew, Wax, *&c.* and then making it endeavour to flie away; for being thereby able to view it in any pofture, I collected that the motion of the wing was after this manner. The extreme limits of the vibrations were ufually fomewhat about the length of the body diftant from one another, oftentimes fhorter, and fometimes alfo longer; that the formoft limit was ufually a little above the back, and the hinder fomwhat beneath the belly; between which two limits, if one may ghefs by the found, the wing feem'd to be mov'd forwards and backwards with an equal velocity: And if one may (from the fhadow or faint reprefentation the wings afforded, and from the confideration of the nature of the thing) ghefs at the pofture or manner of the wings moving betweeen them, it feem'd to be this: The wing being fuppos'd placed in the upmoft limit, feems to be put fo that the plain of it lies almoft *horizontal,* but onely the forepart does dip a little, or is fomewhat more depreft; in this pofition is the wing vibrated or mov'd to the lower limit, being almoft arrived at the lower limit, the hinder part of the wing moving fomewhat fafter then the

former,

former, the *Area* of the wing begins to dip behind, and in that posture seems it to be mov'd to the upper limit back again, and thence back again in the first posture, the former part of the *Area* dipping again, as it is moved downwards by means of the quicker motion of the main stem which terminates or edges the forepart of the wing. And these vibrations or motions to and fro between the two limits seem so swift, that 'tis very probable (from the sound it affords, if it be compar'd with the vibration of a musical string, tun'd unison to it) it makes many hundreds, if not some thousands of vibrations in a second minute of time. And, if we may be allow'd to ghess by the sound, the wing of a Bee is yet more swift, for the tone is much more acute, and that, in all likelihood, proceeds from the exceeding swift beating of the air by the small wing. And it seems the more likely too, because the wing of a Bee is less in proportion to its body, then the other wing to the body of a Fly; so that for ought I know, it may be one of the quickest vibrating *spontaneous* motions of any in the world; and though perhaps there may be many Flies in other places that afford a yet more shrill noise with their wings, yet 'tis most probable that the quickest vibrating *spontaneous* motion is to be found in the wing of some creature. Now, if we consider the exceeding quickness of these Animal spirits that must cause these motions, we cannot chuse but admire the exceeding vividness of the governing faculty or *Anima* of the Insect, which is able to dispose and regulate so the the motive faculties, as to cause every peculiar organ, not onely to move or act so quick, but to do it also so regularly.

Whil'st I was examining and considering the curious *Mechanism* of the wings, I observ'd that under the wings of most kind of Flies, Bees, &c. there were plac'd certain *pendulums* or extended drops (as I may so call them from their resembling motion and figure) for they much resembled a long hanging drop of some transparent viscous liquor; and I observed them constantly to move just before the wings of the Fly began to move, so that at the first sight I could not but ghess, that there was some excellent use, as to the regulation of the motion of the wing, and did phancy, that it might be something like the handle of a Cock, which by vibrating to and fro, might, as 'twere, open and shut the Cock, and thereby give a passage to the determinate influences into the Muscles; afterwards, upon some other trials, I suppos'd that they might be for some use in respiration, which for many reasons I suppose those Animals to use, and, me thought, it was not very improbable, but that they might have convenient passages under the wings for the emitting, at least, of the air, if not admitting, as in the gills of Fishes is most evident; or, perhaps, this *Pendulum* might be somewhat like the staff to a Pump, whereby these creatures might exercise their *Analogus* lungs, and not only draw in, but force out, the air they live by: but these were but conjectures, and upon further examination seem'd less probable.

The fabrick of the wing, as it appears through a moderately magnifying *Microscope*, seems to be a body consisting of two parts, as is visible in the 4.*Figure* of the 23.*Scheme*; and by the 2.*Figure* of the 26.*Scheme*; the one is
a quilly

a quilly or finny fubftance, confifting of feveral long, flender and varioufly bended quills or wires, fomething refembling the veins of leaves; thefe are, as 'twere, the finns or quills which ftiffen the whole *Area*, and keep the other part diftended, which is a very thin tranfparent fkin or membrane varioufly folded, and platted, but not very regularly, and is befides exceeding thickly beftuck with innumerable fmall brifles, which are onely perceptible by the bigger magnifying *Microfcope*, and not with that neither, but with a very convenient augmentation of fkylight projected on the Object with a burning Glafs, as I have elfewhere fhew'd, or by looking through it againft the light.

In fteed of thefe fmall hairs, in feveral other Flies, there are infinite of fmall Feathers, which cover both the under and upper fides of this thin film as in almoft all the forts of Butterflies and Moths: and thofe fmall parts are not onely fhap'd very much like the feathers of Birds, but like thofe variegated with all the variety of curious bright and vivid colours imaginable; and thofe feathers are likewife fo admirably and delicately rang'd, as to compofe very fine flourifhings and ornamental paintings, like *Turkie* and *Perfian* Carpets, but of far more furpaffing beauty, as is evident enough to the naked eye, in the painted wings of Butterflies, but much more through an ordinary *Microfcope*.

Intermingled likewife with thefe hairs, may be perceived multitudes of little pits, or black fpots, in the exended membrane, which feem to be the root of the hairs that grow on the other fide; thefe two bodies feem difpers'd over the whole furface of the wing.

The hairs are beft perceiv'd, by looking through it againft the light, or, by laying the wing upon a very white piece of Paper, in a convenient light, for thereby every little hair moft manifeftly appears; a *Specimen* of which you may obferve drawn in the fourth *Figure* of the 23. *Scheme*, A B, C D, E F whereof reprefent fome parts of the bones or quills of the wing, each of which you may perceive to be cover'd over with a multitude of fcales, or brifles, the former A B, is the biggeft ftem of all the wing, and may be properly enough call'd the cut-air, it being that which terminates and ftiffens the formoft edge of the wing; the fore-edge of this is arm'd with a multitude of little brifles, or Tenter-hooks, in fome ftanding regular and in order, in others not; all the points of which are directed from the body towards the tip of the wing, nor is this edge onely thus fring'd, but even all the whole edge of the wing is cover'd with a fmall fringe, confifting of fhort and more flender brifles.

This Subject, had I time, would afford excellent matter for the contemplation of the nature of wings and of flying; but, becaufe I may, perhaps, get a more convenient time to profecute that fpeculation, and recollect feveral Obfervations that I have made of that particular. I fhall at prefent proceed to

Obferv. XXXIX. *Of the Eyes and Head of a* Grey drone-Fly, *and of feveral other creatures.*

I took a large grey *Drone-Fly*, that had a large head, but a fmall and flender body in proportion to it, and cutting off its head, I fix'd it with the forepart or face upwards upon my Object Plate (this I made choice of rather then the head of a great blue Fly, becaufe my enquiry being now about the eyes, I found this Fly to have, firft the biggeft clufters of eyes in proportion to his head, of any fmall kind of Fly that I have yet feen, it being fomewhat inclining towards the make of the large *Dragon-Flies.* Next, becaufe there is a greater variety in the knobs or balls of each clufter, then is of any fmall Fly) Then examining it according to my ufual manner, by varying the degrees of light, and altering its pofition to each kinde of light, I drew that reprefentation of it which is delineated in the 24. *Scheme,* and found thefe things to be as plain and evident, as notable and pleafant.

Firft, that the greateft part of the face, nay, of the head, was nothing elfe but two large and *protuberant* bunches, or *prominent* parts, A B C D E A, the furface of each of which was all cover'd over, or fhap'd into a multitude of fmall *Hemifpheres,* plac'd in a *triagonal* order, that being the clofeft and moft compacted, and in that order, rang'd over the whole furface of the eye in very lovely rows between each of which, as is neceffary, were left long and regular trenches, the bottoms of every of which, were perfectly intire and not at all perforated or drill'd through, which I moft certainly was affured of, by the regularly reflected Image of certain Objects which I mov'd to and fro between the head and the light. And by examining the *Cornea* or outward fkin, after I had ftript it off from the feveral fub-ftances that lay within it, and by looking both upon the infide and againft the light.

Next, that of thofe multitudes of *Hemifpheres,* there were obfervable two degrees of bignefs, the half of them that were lowermoft, and look'd toward the ground or their own leggs, namely, C D E, C D E being a pretty deal fmaller then the other, namely, A B C E, A B C E, that look'd upward, and fide-ways, or foreright, and backward, which variety I have not found in any other fmall Fly.

Thirdly, that every one of thefe *Hemifpheres,* as they feem'd to be pret-ty neer the true fhape of a *Hemifphere,* fo was the furface exceeding fmooth and regular, reflecting as exact, regular, and perfect an Image of any Object from the furface of them, as a fmall Ball of Quick-filver of that bignefs would do, but nothing neer fo vivid, the reflection from thefe being very languid, much like the reflection from the outfide of Water, Glafs, Cryftal, *&c.* In fo much that in each of thefe *Hemifpheres,* I have been able to difcover a Land-fcape of thofe things which lay before my window,

window, one thing of which was a large Tree, whose trunk and top I could plainly difcover, as I could alfo the parts of my window, and my hand and fingers, if I held it between the Window and the Object; a fmall draught of nineteen of which, as they appear'd in the bigger Magnifying-glafs to reflect the Image of the two windows of my Chamber, are delineated in the third *Figure* of the 23. *Scheme*.

Fourthly, that thefe rows were fo difpos'd, that there was no quarter vifible from his head that there was not fome of thefe *Hemifpheres* directed againft; fo that a Fly may be truly faid to have *an eye every way*, and to be really *circumfpect*. And it was further obfervable, that that way where the trunk of his body did hinder his profpect backward, thefe *protuberances* were elevated, as it were, above the plain of his fhoulders and back, fo that he was able to fee backwards alfo over his back.

Fifthly, in living Flies, I have obferv'd, that when any fmall mote or duft, which flies up and down the air, chances to light upon any part of thefe knobs, as it is fure to ftick firmly to it and not fall, though through the *Microfcope* it appears like a large ftone or ftick (which one would admire, efpecially fince it is no ways probable that there is any wet or glutinous matter upon thefe *Hemifpheres*, but I hope I fhall render the reafon in another place) fo the Fly prefently makes ufe of his two fore-feet in ftead of eye-lids, with which, as with two Brooms or Brufhes, they being all beftuck with Brifles, he often fweeps or brufhes off what ever hinders the profpect of any of his *Hemifpheres*, and then, to free his leggs from that dirt, he rubs them one againft another, the pointed Brifles or Tenters of which looking both one way, the rubbing of them to and fro one againft another, does cleanfe them in the fame manner as I have obferv'd thofe that Card Wool, to cleanfe their Cards, by placing their Cards, fo as the teeth of both look the fame way, and then rubbing them one againft another. In the very fame manner do they brufh and cleanfe their bodies and wings, as I fhall by and by fhew; other creatures have other contrivances for the cleanfing and cleering their eyes.

Sixthly, that the number of the *Pearls* or *Hemifpheres* in the clufters of this Fly, was neer 14000. which I judged by numbering certain rows of them feveral ways, and cafting up the whole content, accounting each clufter to contain about feven thoufand Pearls, three thoufand of which were of a cize, and confequently the rows not fo thick, and the foure thoufand I accounted to be the number of the fmaller Pearls next the feet and *probofcis*. Other Animals I obferv'd to have yet a greater number, as the *Dragon-Fly* or *Adderbolt*: And others to have a much lefs company, as an *Ant*, &c. and feveral other fmall Flies and Infects.

Seventhly, that the order of thefe eies or *Hemifpheres* was altogether curious and admirable, they being plac'd in all kind of Flies, and *aerial* animals, in a moft curious and regular ordination of triangular rows, in which order they are rang'd the neereft together that poffibly they can, and confequently leave the leaft pits or trenches between them. But in *Shrimps*, *Crawfifhes*, *Lobfters*, and fuch kinds of *Cruftaceous* water Animals, I have

yet

yet obferv'd them rang'd in a quadrangular order, the rows cutting each other at right angles,which as it admits of a lefs number of Pearls in equal furfaces ; fo have thofe creatures a recompence made them, by having their eyes a little movable in their heads, which the other altogether want. So infinitely wife and provident do we find all the Difpenfations in Nature, that certainly *Epicurus*, and his followers, muft very little have confider'd them, who afcrib'd thofe things to the production of chance, that wil.to a more attentive confiderer,appear the products of the higheft Wifdom and Providence.

Upon the Anatomy or Diffection of the Head, I obferv'd thefe particulars :

Firft, that this outward fkin, like the *Cornea* of the eyes of the greater Animals, was both flexible and tranfparent, and feem'd, through the *Microfcope*, perfectly to refemble the very fubftance of the *Cornea* of a man's eye ; for having cut out the clufter, and remov'd the dark and *mucous* ftuff that is fubjacent to it, I could fee it tranfparent like a thin piece of fkin, having as many cavities in the infide of it, and rang'd in the fame order as it had *protuberances* on the outfide,and this propriety,I found the fame in all the Animals that had it, whether Flies or Shell-Fifh.

Secondly, I found that all Animals that I have obferv'd with thofe kind of eyes, have within this *Cornea*, a certain cleer liquor or juice, though in a very little quantity, and,

I obferv'd thirdly, that within that cleer liquor, they had a kind of dark *mucous* lining, which was all fpread round within the cavity of the clufter, and feem'd very neer adjoining to it, the colour of which, in fome Flies,was grey ; in others, black; in others red ; in others,of a mix'd colour ; in others,fpotted ; and that the whole clufters, when look'd on whil'ft the Animal was living, or but newly kill'd, appear'd of the fame colour that this coat (as I may fo call it) appear'd of, when that outward fkin, or *Cornea*,was remov'd.

Fourthly, that the reft of the capacity of the clufters was in fome, as in Dragon Flies, *&c.* hollow, or empty ; in others fill'd with fome kind of fubftance ; in blue Flies,with a reddifh mufculous fubftance, with*fibres* tending from the center or bottom outwards ; and divers other,with various and differing kinds of fubftances.

That this curious contrivance is the organ of fight to all thofe various *Cruftaceous* Animals, which are furnifh'd with it, I think we need not doubt, if we confider but the feveral congruities it has with the eyes of greater creatures.

As firft, that it is furnifh'd with a *Cornea*,with a *tranfparent humour*,and with a *uvea* or *retina*, that the Figure of each of the fmall *Hemifpheres* are very *Spherical*, exactly polifh'd, and moft vivid, lively and plump,when the Animal is living,as in greater Animals,and in like manner dull,flaccid, and irregular, or fhrunk,when the Animal is dead.

Next, that thofe creatures that are furnifh'd with it, have no other organs that have any refemblance to the known eyes of other creatures.

Thirdly,

Thirdly, that thofe which they call the eyes of Crabs, Lobfters, Shrimps, and the like, and are really fo, are *Hemifpher'd*, almoft in the fame manner as thefe of Flies are. And that they really are fo, I have very often try'd, by cutting off thefe little movable knobs, and putting the creature again into the water, that it would fwim to and fro, and move up and down as well as before, but would often hit it felf againft the rocks or ftones; and though I put my hand juft before its head, it would not at all ftart or fly back till I touch'd it, whereas whil'ft thofe were remaining, it would ftart back, and avoid my hand or a ftick at a good diftance before it touch'd it. And if in *cruftaceous* Sea-animals, then it feems very probable alfo, that thefe knobs are the eyes in *cruftaceous* Infects, which are alfo of the fame kind, onely in a higher and more active Element; this the conformity or congruity of many other parts common to either of them, will ftrongly argue, their *cruftaceous* armour, their number of leggs, which are fix, befide the two great claws, which anfwer to the wings in Infects; and in all kind of Spiders, as alfo in many other Infects that want wings, we fhall find the compleat number of them, and not onely the number, but the very fhape, figure, joints, and claws of Lobfters and Crabs, as is evident in Scorpions and Spiders, as is vifible in the fecond *Figure* of the 31. *Scheme*, and in the little Mite-worm, which I call a Land-crab, defcrib'd in the fecond Figure of the 33. *Scheme*, but in their manner of generation being oviparous, *&c.* And it were very worthy obfervation, whether there be not fome kinds of transformation and metamorphofis in the feveral ftates of *cruftaceous* water-animals, as there is in feveral forts of Infects; for if fuch could be met with, the progrefs of the variations would be much more confpicuous in thofe larger Animals, then they can be in any kind of Infects our colder Climate affords.

Thefe being their eyes, it affords us a very pretty Speculation to contemplate their manner of vifion, which, as it is very differing from that of *biocular* Animals, fo is it not lefs admirable.

That each of thefe Pearls or *Hemifpheres* is a perfect eye, I think we need not doubt, if we confider onely the outfide or figure of any one of them, for they being each of them cover'd with a tranfparent protuberant *Cornea*, and containing a liquor within them, refembling the watry or glaffie humours of the eye, muft neceffarily refract all the parallel Rays that fall on them out of the air, into a point not farr diftant within them, where (in all probability) the *Retina* of the eye is placed, and that opacous, dark, and mucous inward coat that (I formerly fhew'd) I found to fubtend the concave part of the clufter is very likely to be that *tunicle* or coat, it appearing through the *Microfcope* to be plac'd a little more than a Diameter of thofe Pearls below or within the *tunica cornea*. And if fo, then is there in all probability, a little Picture or Image of the objects without, painted or made at the bottom of the *Retina* againft every one of thofe Pearls, fo that there are as many impreffions on the *Retina* or opacous fkin, as there are Pearls or *Hemifpheres* on the clufter. But becaufe it is impoffible for any protuberant furface whatfoever, whether *fphærial* or other, fo to refract the Rays that come from farr remote

lateral

lateral points of any Object as to collect them again, and unite them each in a diſtinct point, and that onely thoſe Rays which come from ſome point that lies in the *Axis* of the Figure produc'd, are ſo accurately re-fracted to one and the ſame point again, and that the *lateral* Rays, the fur-ther they are remov'd, the more imperfect is their refracted confluence; It follows therefore, that onely the Picture of thoſe parts of the external objects that lie in, or neer, the *Axis* of each *Hemiſphere*, are diſcernably painted or made on the *Retina* of each *Hemiſphere*, and that therefore each of them can diſtinctlyſenſate or ſee onely thoſe parts which are very neer perpendicularly oppos'd to it, or lie in or neer its optick *Axis*. Now, though there may be by each of theſe eye-pearls, a repreſentation to the Animal of a whole *Hemiſphere* in the ſame manner as in a man's eye there is a picture or ſenſation in the *Retina* of all the objects lying almoſt in an *Hemiſphere*; yet, as in a man's eye alſo, there are but ſome very few points which liyng in, or neer, the optick *Axis* are diſtinctly diſ-cern'd: So there may be multitudes of Pictures made of an Object in the ſeveral Pearls, and yet but one, or ſome very few that are diſtinct; The repreſentation of any object that is made in any other Pearl, but that which is directly, or very neer directly, oppos'd, being altogether confuſ'd and unable to produce a diſtinct viſion.

So that we ſee, that though it has pleas'd the All-wiſe Creator, to in-due this creature with ſuch multitudes of eyes, yet has he not indued it with the faculty of ſeeing more then another creature; for whereas this cannot move his head, at leaſt can move itvery little, without moving his whole body, *biocular* creatures can in an inſtant (or *the twinkling of an eye*, which, being very quick, is vulgarly uſed in the ſame ſignification) move their eyes ſo as to direct the optick *Axis* to any point; nor is it probable, that they are able to ſee attentively at one time more then one Phyſical point; for though there be a diſtinct Image made in every eye, yet 'tis very likely, that the obſerving faculty is only imploy'd about ſome one object for which they have moſt concern.

Now, as we accurately diſtinguiſh the ſite or poſition of an Object by the motion of the Muſcles of the eye requiſite to put the optick Line in a direct poſition, and confuſedly by the poſition of the imperfect Picture of the object at the bottom of the eye; ſo are theſe *cruſtaceous* creatures able to judge confuſedly of the poſition of objects by the Picture or im-preſſion made at the bottom of the oppoſite Pearl, and diſtinctly by the removal of the attentive or obſerving faculty, from one Pearl to another, but what this faculty is, as it requires another place, ſo a much deeper ſpe-culation. Now, becauſeit were impoſſible, even with this multitude of eye-balls, to ſee any object diſtinct(for as I hinted before, onely thoſe parts that lay in, or veryneer, the optick Lines could be ſo)the Infinitely wiſe Creator has not left the creature without a power of moving the head a little in *Aerial cruſtaceous* animals, and the very eyes alſo in *cruſtaceous* Sea-animals; ſo that by theſe means they are inabled to direct ſome optick line or other againſt any object, and by that means they have the viſive faculty as com-pleat as any Animal that can move its eyes.

Diſtances

Diſtances of Objeƈts alſo, 'tis very likely they diſtinguiſh, partly by the conſonant impreſſions made in ſome two convenient Pearls, one in each cluſter; for, according as thoſe congruous impreſſions affeƈt, two Pearls neerer approach'd to each other, the neerer is the Objeƈt, and the farther they are diſtant, the more diſtant is the Objeƈt: partly alſo by the alteration of each Pearl, requiſite to make the Senſation or Piƈture perfeƈt; for 'tis impoſſible that the Piƈtures of two Objeƈts, variouſly diſtant, can be perfeƈtly painted, or made on the ſame *Retina* or bottom of the eye not altered, as will be very evident to any one that ſhall attentively conſider the nature of refraƈtion. Now, whether this alteration may be in the Figure of the *Cornea*, in the motion of acceſs or receſs of the *Retina* towards the *Cornea*, or in the alteration of a cruſtaline humour, if ſuch there be, I pretend not to determine; though I think we need not doubt, but that there may be as much curioſity of contrivance and ſtruƈture in every one of theſe Pearls, as in the eye of a Whale or Elephant, and the almighty's *Fiat* could as eaſily cauſe the exiſtence of the one as the other; and as one day and a thouſand years are the ſame with him, ſo may one eye and ten thouſand.

This we may be ſure of, that the filaments or ſenſative parts of the *Retina* muſt be moſt exceedingly curious and minute, ſince the whole Piƈture it ſelf is ſuch; what muſt needs the component parts be of that *Retina* which diſtinguiſhes the part of an objeƈt's Piƈture that muſt be many millions of millions leſs then that in a man's eye? And how exceeding curious and ſubtile muſt the component parts of the *medium* that conveys light be, when we find the inſtrument made for its reception or refraƈtion to be ſo exceedingly ſmall? we may, I think, from this ſpeculation be ſufficiently diſcouraged from hoping to diſcover by any optick or other inſtrument the determinate bulk of the parts of the *medium* that conveys the pulſe of light, ſince we find that there is not leſs accurateneſs ſhewn in the Figure and poliſh of thoſe exceedingly minute lenticular ſurfaces, then in thoſe more large and conſpicuous ſurfaces of our own eyes. And yet can I not doubt, but that there is a determinate bulk of thoſe parts, ſince I find them unable to enter between the parts of Mercury, which being in motion, muſt neceſſarily have pores, as I ſhall elſewhere ſhew, and here paſs by, as being a digreſſion.

As concerning the horns F F, the feelers or ſmellers, G G, the *Probaſcis* H H, and I, the hairs and briſles, K K, I ſhall indeavour to deſcribe in the 42. *Obſervation.*

Obſerv. XL. *Of the Teeth of a* Snail.

I Have little more to add of the Teeth of a Snail, beſides the Piƈture of it, which is repreſented in the firſt *Figure* of the 25. *Scheme*, ſave that his bended body, A B C D E F, which ſeem'd faſhioned very much like a row of ſmall teeth, orderly plac'd in the Gums, and looks as if it
were

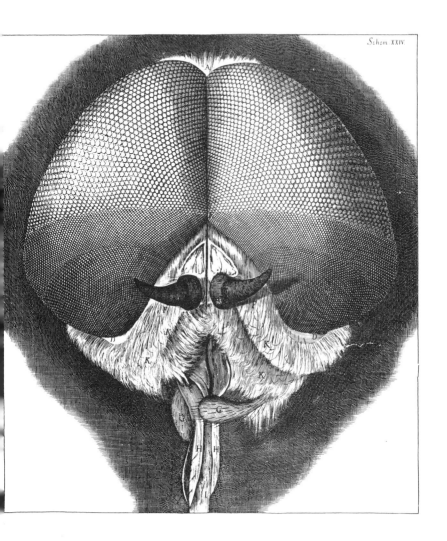

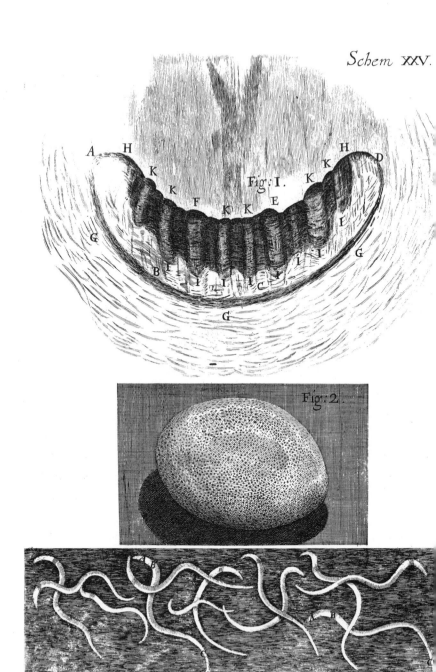

Fig: I.

Fig: 2.

Fig: 3.

were divided into feveral fmaller and greater black teeth, was nothing but one fmall bended hard bone,which was plac'd in the upper jaw of the mouth of a Houfe-Snail, with which I obferv'd this very Snail to feed on the leaves of a Rofe-tree, and to bite out pretty large and half round bits, not unlike the Figure of a (C) nor very much differing from it in bignefs, the upper part A B C D of this bone, I found to be much whiter, and to grow out of the upper chap of the Snail G G G, and not to be any thing neer fo much creas'd as the lower and blacker part of it H I I H K K H which was exactly fhap'd like teeth, the bone growing thinner, or taper-ing to an edge towards K K K. It feem'd to have nine teeth, or prominent parts I K, I K, I F &c. which were join'd together by the thinner inter-pos'd parts of the bone. The Animal to which thefe teeth belong, is a very *anomalous* creature, and feems of a kind quite diftinct from any other terreftrial Animal or Infect, the Anatomy whereof exceedingly dif-fering from what has been hitherto given of it I fhould have inferted, but that it will be more proper in another place. I have never met with any kind of Animal whofe teeth are all join'd in one, fave onely that I lately obferv'd, that all the teeth of a Rhinocerot, which grow on either fide of its mouth, are join'd into one large bone, the weight of one of which I found to be neer eleven pound *Haverdupois.* So that it feems one of the biggeft fort of terreftrial Animals, as well as one of the fmalleft, has his teeth thus fhap'd.

Obferv. XLI. *Of the Eggs of* Silk-worms, *and other Infects.*

THe Eggs of Silk-worms (one of which I have defcrib'd in the fecond *Figure* of 25. *Scheme*) afford a pretty Object for a *Microfcope* that magnifies very much, efpecially if it be bright weather, and the light of a window be caft or collected on it by a deep *Convex-glafs,* or Water-ball. For then the whole furface of the Shell may be perceiv'd all cover'd over with exceeding fmall pits or cavities with interpofed edges, almoft in the manner of the furface of a Poppy-feed, but that thefe holes are not an hun-dredth part fcarce of their bignefs; the Shell, when the young ones were hatch'd (which I found an eafie thing to do, if the Eggs were kept in a warm place) appear'd no thicker in proportion to its bulk, then that of an Hen's or Goos's Egg is to its bulk, and all the Shell appear'd very white (which feem'd to proceed from its tranfparency) whence all thofe pit-tings did almoft vanifh, fo that they could not, without much difficulty, be difcern'd, the infide of the Shell feem'd to be lin'd alfo with a kind of thin film, not unlike (keeping the proportion to its Shell) that with which the fhell of an Hen-egg is lin'd ; and the fhell it felf feem'd like common Egg-fhells, very brittle, and crack'd. In divers other of thefe Eggs I could plainly enough, through the fhell, perceive the fmall Infect lie coyled round the edges of the fhell. The fhape of the Egg it felf, the Figure pretty well reprefents (though by default of the Graver it does

not

not appear fo rounded, and lying above the Paper, as it were, as it ought to do) that is, it was for the moft part pretty oval end-ways, fomewhat like an Egg, but the other way it was a little flatted on two oppofite fides. Divers of thefe Eggs, as is common to moft others, I found to be barren, or addle, for they never afforded any young ones. And thofe I ufually found much whiter then the other that were prolifick. The Eggs of other kinds of Oviparous Infects I have found to be perfectly round every way, like fo many Globules, of this fort I have obferv'd fome forts of Spiders Eggs; and chancing the laft Summer to inclofe a very large and curioufly painted Butterfly in a Box, intending to examine its gaudery with my *Microfcope*, I found within a day or two after I inclos'd her, almoft all the inner furface of the Box cover'd over with an infinite of exactly round Eggs, which were ftuck very faft to the fides of it, and in fo exactly regular and clofe an order, that made me call to mind my *Hypothefis*, which I had formerly thought on for the making out of all the regular Figures of Salt, which I have elfewhere hinted ; for here I found all of them rang'd into a moft exact *triagonal* order, much after the manner as the *Hemifpheres* are place on the eye of a Fly ; all which Eggs I found after a little time to be hatch'd, and out of them to come a multitude of fmall Worms, very much refembling young Silk-worms, leaving all their thin hollow fhells behind them, fticking on the Box in their *triagonal* pofture ; thefe I found with the *Microfcope* to have much fuch a fubftance as the Silk-worms Eggs, but could not perceive them pitted. And indeed, there is as great a variety in the fhape of the Eggs of Oviparous Infects as among thofe of Birds.

Of thefe Eggs, a large and lufty Fly will at one time lay neer four or five hundred, fo that the increafe of thefe kind of Infects muft needs be very prodigious, were they not prey'd on by multitudes of Birds, and deftroy'd by Frofts and Rains ; and hence 'tis thofe hotter Climates between the *Tropicks* are infefted with fuch multitudes of Locufts, and fuch other Vermine.

Obferv. XLII. *Of a blue* Fly.

THis kind of Fly, whereof a *Microfcopical* Picture is delineated in the firft *Figure* of the 26. *Scheme*, is a very beautifull creature, and has many things about it very notable ; divers of which I have already partly defcrib'd, namely, the feet, wings, eyes, and head, in the preceding Obfervations.

And though the head before defcrib'd be that of a grey *Drone-Fly*, yet for the main it is very agreeable to this. The things wherein they differ moft, will be eafily enough found by the following particulars :

Firft, the clufters of eyes of this Fly, are very much fmaller then thofe of the *Dron-Fly*, in proportion to the head.

And

Schem:XXVI.

Fig:1.

Fig:2.

And next, all the eyes of each clufter feem'd much of the fame bignefs one with another, not differing as the other, but rang'd in the fame *tri-agonal* order.

Thirdly, between thefe two clufters, there was a fcaly prominent *front* B, which was arm'd and adorn'd with large tapering fharp black brifles, which growing out in rows on either fide, were fo bent toward each other neer the top, as to make a kind of arched arbour of Brifles, which almoft cover'd the former *front.*

Fourthly, at the end of this Arch, about the middle of the face, on a prominent part C, grew two fmall oblong bodies, D D, which through a *Microfcope* look'd not unlike the Pendants in Lillies, thefe feem'd to be jointed on to two fmall parts at C, each of which feem'd again jointed into the front.

Fifthly, out of the upper part and outfides of thefe horns (as I may call them, from the Figure they are of, in the 24. *Scheme*, where they are marked with F F) there grows a fingle feather, or brufhy Brifle, E E, fomewhat of the fame kind with the tufts of a Gnat, which I have before defcribed.

What the ufe of thefe kind of horned and tufted bodies fhould be, I cannot well imagine, unlefs they ferve for fmelling or hearing, though how they are adapted for either, it feems very difficult to defcribe they are in almoft every feveral kind of Flies of fo various a fhape; though certain-ly they are fome very effential part of the head, and have fome very notable office affign'd them by Nature, fince in all Infects they are to be found in one or other form.

Sixthly, at the under part of the face F F, were feveral of the former fort of bended Brifles; and below all, the mouth, out of the middle of which, grew the *probofcis* G H I, which, by means of feveral joints, where-of it feem'd to confift, the Fly was able to move to and fro, and thruft it in and out as it pleas'd; the end of this hollow body (which was all over cover'd with fmall fhort hairs or brifles) was, as 'twere, bent at H, and the outer or formoft fide of the bended part H I, flit, as it were, into two chaps, H I, H I, all the outfide of which where cover'd with hairs, and pretty large brifles; thefe he could, like two chaps, very readily open and fhut, and when he feem'd to fuck any thing from the furface of a body, he would fpread abroad thofe chaps, and apply the hollow part of them very clofe to it.

From either fide of the *Probofcis*, within the mouth, grew two other fmall horns, or fingers, K K, which were hairy, but fmall in this Figure; but of another fhape, and bigger in proportion, in the 24. *Scheme*, where they are marked with G G, which two indeed feem'd a kind of fmellers, but whether fo or not, I cannot pofitively determine.

The *Thorax* or middle part of this Fly, was cas'd, both above and be-neath, with a very firm cruft of armour, the upper part more round, and covered over with long *conical* brifles, all whofe ends pointed backwards; out of the hinder and under part of this grew out in a clufter fix leggs, three of which are apparent in the Figure, the other three were hid by the
body

body plac'd in that posture. The leggs were all much of the same make, being all of them cover'd with a strong hairy scale or shel, just like the legs of a Crabb or Lobster, and the contrivance of the joints seem'd much the same; each legg seem'd made up of eight parts, 1, 2, 3, 4, 5, 6, 7, 8, to the eighth or last of which, grew the soles and claws, described before in the 38. *Observation.*

Out of the upper part of this trunck grew the two wings, which I mention'd in the 38. *Observation,* consisting of a film, extended on certain small stiff wires or bones: these in a blue Fly, were much longer then the body, but in other kind of Flies they are of very differing proportions to the body. These films, in many Flies, were so thin, that, like several other plated bodies (mention'd in the ninth *Observation*) they afforded all varieties of fantastical or transient colours (the reason of which I have here endeavoured to explain) they seem'd to receive their nourishment from the stalks or wires, which seem'd to be hollow, and neer the upper part of the wing L L several of them seem'd jointed, the shape of which will sufficiently appear by the black lines in the second Figure of the 26. *Scheme,* which is a delineation of one of those wings expanded directly to the eyes.

All the hinder part of its body is cover'd with a most curious blue shining armour, looking exactly like a polish'd piece of steel brought to that blue colour by annealing, all which armour is very thick bestuck with abundance of tapering brisles, such as grow on its back, as is visible enough by the Figure.

Nor was the inside of this creature less beautifull then its outside, for cutting off a part of the belly, and then viewing it, to see if I could discover any Vessels, such as are to be found in a greater Animals, and even in Snails exceeding manifestly, I found, much beyond my expectation, that there were abundance of branchings of Milk-white vessels, no less curious then the branchings of veins and arteries in bigger terrestrial Animals, in one of which, I found two notable branches, joining their two main stocks, as it were, into one common *ductus*; now, to what veins or arteries these Vessells were *analogus,* whether to the *vena porta,* or the *meseraick vessels,* or the like, or indeed, whether they were veins and arteries, or *vasa lactea,* properly so called, I am not hitherto able to determine, having not yet made sufficient enquiry; but in all particulars, there seems not to be any thing less of curious contrivance in these Insects, then in those larger terrestrial Animals, for I had never seen any more curious branchings of Vessells, then those I observ'd in two or three of these Flies thus opened.

It is a creature active and nimble, so as there are very few creatures like it, whether bigger or smaller, in so much, that it will scape and avoid a small body, though coming on it exceeding swiftly, and if it sees any thing approaching it, which it fears, it presently squats down, as it were, that it may be the more ready for its rise.

Nor is it less hardy in the Winter, then active in the Summer, induring all the Frosts, and surviving till the next Summer, notwithstanding the

bitter

bitter cold of our Climate ; nay, this creature will indure to be frozen, and yet not be deſtroy'd,for I have taken one of them out of the Snow whereon it has been frozen almoſt white, with the Ice about it, and yet by thawing it gently by the warmth of a fire, it has quickly reviv'd and flown about.

This kind of Fly ſeems by the ſteams or taſte of fermenting and putrifying meat (which it often kiſſes,as'twere, with its *proboſcis* as it trips over it) to be ſtimulated or excited to eject its Eggs or Seed on it, perhaps, from the ſame reaſon as Dogs,Cats,and many other brute creatures are excited to their particular luſts, by the ſmell of their females, when by Nature prepared for generation ; the males ſeeming by thoſe kind of ſmells, or other incitations, to be as much neceſſitated thereto, as *Aqua Regis* ſtrongly impregnated with a ſolution of Gold,is forced to precipitate it by the affuſion of ſpirit of *Vrine,* or a ſolution of *Salt* of *Tartar.*

One of theſe put in ſpirit of *Wine,* was very quickly ſeemingly kill'd, and both its eys and mouth began to look very red, but upon the taking of it out, and ſuffering it to lie three or four hours, and heating it with the Sun beams caſt through a Burning-glaſs, it again reviv'd, ſeeming, as it were, to have been all the intermediate time, but dead drunk, and after certain hours to grow freſh again and ſober.

Obſerv. XLIII. *Of the* Water-Inſect *or* Gnat.

THis little creature, deſcribed in the firſt *Figure* of the 27. *Scheme,* was a ſmall ſcaled or cruſted Animal, which I have often obſerv'd to be generated in Rain-water ; I have alſo obſerv'd it both in Pond and River-water. It is ſuppos'd by ſome, to deduce its firſt original from the putrifaction of Rain-water, in which,if it have ſtood any time open to the air, you ſhall ſeldom miſs,all the Summer long, of ſtore of them friſking too and fro.

'Tis a creature, wholly differing in ſhape from any I ever obſerv'd ; nor is its motion leſs ſtrange : It has a very large head, in proportion to its body, all covered with a ſhell, like other *teſtaceous* Animals, but it differs in this, that it has, up and down ſeveral parts of it, ſeveral tufts of hairs, or briſles, plac'd in the order expreſs'd in the Figure ; It has two horns,whichſeem'd almoſt like the horns of an Oxe,inverted,and, as neer as I could gheſs,were hollow,with tufts of briſles,likewiſe at the top ; theſe horns they could move eaſily this or that way,and might, perchance, be their noſtrils. It has a pretty large mouth, which ſeem'd contriv'd much like thoſe of Crabs and Lobſters,by which, I have often obſerv'd them to feed on water, or ſome imperceptible nutritive ſubſtance in it.

I could perceive, through the tranſparent ſhell,while the Animal ſurviv'd, ſeveral motions in the head, thorax, and belly, very diſtinctly,

of

of differing kinds which I may, perhaps, elſewhere endeavour more accurately to examine, and to ſhew of how great benefit the uſe of a *Microſcope* may be for the diſcovery of Nature's courſe in the operations perform'd in Animal bodies, by which we have the opportunity of obſerving
her through theſe delicate and pellucid teguments of the bodies of Inſects
acting according to her uſual courſe and way, undiſturbed, whereas,
when we endeavour to pry into her ſecrets by breaking open the doors
upon her, and diſſecting and mangling creatures whil'ſt there is life yet
within them, we find her indeed at work, but put into ſuch diſorder by
the violence offer'd, as it may eaſily be imagin'd how differing a thing we
ſhould find, if we could, as we can with a *Microſcope*, in theſe ſmaller creatures, quietly peep in at the windows, without frighting her out of her
uſual byas.

The form of the whole creature, as it appear'd in the *Microſcope*, may,
without troubling you with more deſcriptions, be plainly enough perceiv'd by the *Scheme*, the hinder part or belly conſiſting of eight ſeveral
jointed parts, namely, A B C D E F G H, of the firſt *Figure*, from the
midſt of each of which, on either ſide, iſſued out three or four ſmall briſles
or hairs, I, I, I, I, I, the tail was divided into two parts of very differing
make; one of them, namely, K, having many tufts of hair or briſles, which
ſeem'd to ſerve both for the finns and tail, for the Oars and Ruder of this
little creature, wherewith it was able, by friſking and bending its body
nimbly to and fro, to move himſelf any whither, and to ſkull and ſteer himſelf as he pleas'd; the other part, L, ſeem'd to be, as 'twere, the ninth diviſion of his belly, and had many ſingle briſles on either ſide. From the end V,
of which, through the whole belly, there was a kind of Gut of a darker
colour, M M M, wherein, by certain *Periſtaltick* motions there was a kind
of black ſubſtance mov'd upwards and downwards through it from the
orbicular part of it, N, (which ſeem'd the *Ventricle*, or ſtomach) to the tail
V, and ſo back again, which *periſtaltick* motion I have obſerv'd alſo in a
Louſe, a Gnat, and ſeveral other kinds of tranſparent body'd Flies. The
Thorax or cheſt of this creature O O O O, was thick and ſhort, and pretty tranſparent, for through it I could ſee the white heart (which is the
colour alſo of the bloud in theſe, and moſt other Inſects) to beat, and
ſeveral other kind of motions. It was beſtuck and adorn'd up and down
with ſeveral tufts of briſles, ſuch as are pointed out by P, P, P, P, the
head Q was likewiſe beſtuck with ſeveral of thoſe tufts, S S S; it was
broad and ſhort, had two black eyes, T T, which I could not perceive at
all pearl'd, as they afterwards appear'd, and two ſmall horns, R R, ſuch
as I formerly deſcrib'd.

Both its motion and reſt is very ſtrange, and pleaſant, and differing
from thoſe of moſt other creatures I have obſerv'd; for, where it ceaſes
from moving its body, the tail of it ſeeming much lighter then the reſt
of its body, and a little lighter then the water it ſwims in, preſently buoys
it up to the top of the water, where it hangs ſuſpended with the head always downward; and like our *Antipodes*, if they do by a friſk get below that ſuperficies, they preſently aſcend again unto it, if they ceaſe
moving,

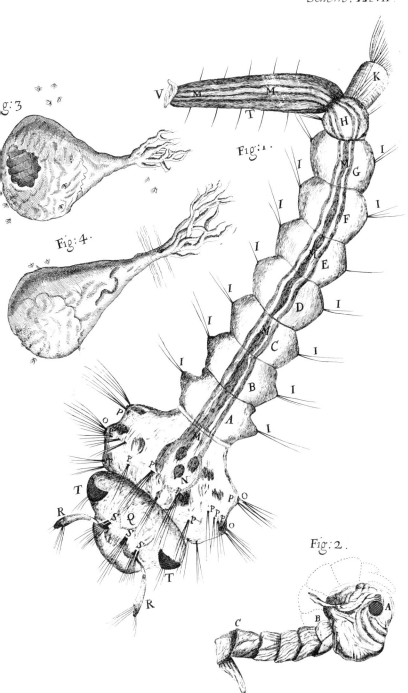

Fig: 3.

Fig: 4.

Fig: 1.

Fig: 2.

moving, until they tread, as it were, under that superficies with their tails ; the hanging of thefe in this pofture, put me in mind of a certain creature I have feen in *London*, that was brought out of *America*, which would very firmly fufpend it felf by the tail, with the head downwards,and was faid to fleep in that pofture, with her young ones in her falfe belly, which is a Purfe, provided by Nature for the produ&ion, nutrition, and prefervation of her young ones, which is defcribed by *Pifo* in the 24. Chapter of the fifth Book of his Natural Hiftory of *Brafil*.

The motion of it was with the tail forwards,drawing its felf backwards, by the frifking to and fro of that tuft which grew out of one of the ftumps of its tail. It had another motion,which was more futable to that of other creatures, and that is, with the head forward ; for by the moving of his chaps (if I may fo call the parts of his mouth) it was able to move it felf downwards very gently towards the bottom, and did,as twere,eat up its way through the water.

But that which was moft obfervable in this creature, was, its Metamorphofis or change;for having kept feveral of thefe Animals in a Glafs of Rain-water,in which they were produc'd, I found,after about a fortnight or three weeks keeping, that feveral of them flew away in Gnats,leaving their hufks behind them in the water floating under the furface, the place where thefe Animals were wont to refide, whil'ft they were inhabitants of the water : this made me more diligently to watch them, to fee if I could find them at the time of their transformation ; and not long after, I obferv'd feveral of them to be changed into an unufual fhape, wholly differing from that they were of before, their head and body being grown much bigger and deeper, but not broader, and their belly, or hinder part fmaller, and coyl'd, about this great body much of the fafhion reprefented by the prick'd line in the fecond *Figure* of the 27. *Scheme*, the head and horns now fwam uppermoft, and the whole bulk of the body feem'd to be grown much lighter ; for when by my frighting of it, it would by frifking out of its tail (in the manner exprefs'd in the Figure by B C) fink it felf below the furface towards the bottom ; the body would more fwiftly re-afcend, then when it was in its former fhape.

I ftill marked its progrefs from time to time,and found its body ftill to grow bigger and bigger, Nature, as it were, fitting and accoutring it for the lighter Element, of which it was now going to be an inhabitant ; for,by obferving one of thefe with my *Microfcope*, I found the eyes of it to be altogether differing from what they feem'd before, appearing now all over pearl'd or knobb'd, like the eyes of Gnats. as is vifible in the fecong *Figure* by A. At length, I faw part of this creature to fwim above, and part beneath the furface of the water, below which though it would quickly plunge it felf if I by any means frighted it,and prefently re-afcend into its former pofture; after a little longer expe&ation, I found that the head and body of a Gnat,began to appear and ftand cleer above the furface, and by degrees it drew out its leggs, firft the two formoft,then the other,at length its whole body perfe& and entire appear'd out of the hufk (which it left in the water) ftanding on its leggs upon the top of the water, and by degrees it began to move, and after flew about the Glafs a perfect Gnat.

I have been the more particular, and large in the relation of the tranf-formation of divers of thefe little Animals which I obferv'd, becaufe I have not found that any Authour has obferv'd the like; and becaufe the thing it felf is fo ftrange and heterogeneous from the ufual progrefs of other Animals, that I judge it may not onely be pleafant, but very ufefull and neceffary towards the compleating of Natural Hiftory.

There is indeed in *Pifo*, a very odd Hiftory, which this relation may make the more probable; and that is in the 2. Chapter of the 4. Book of his Natural Hiftory of *Brafil*, where he fays, *Porro præter tot documenta fer-tilitatis circa vegetabilia & fenfitiva marina telluris æmula, accidit & illud, quod paucis à Paranambucenfi milliaribus pifcatoris uncum citra intentionem contingat infigi vadis petrofis, & loco pifcis fpongia, coralla, aliafque arbufculas marinas capi. Inter hæc inufitatæ formæ prodit fpongiofa arbufcula, fefquipedis longitudinis, brevioribus radicibus, lapideis nitens vadis, & rupibus infixa, erigiturque in corpus fpongiofum molle oblongum rotundum turbinatum: intus miris cancellis & alveis fabricatum, extus autem tenaci glutine inftar Apum propolis undique veftitum, oftio fatis patulo & profundo in fummitate relicto, ficut ex altera iconum probe depicta videre licet* (fee the third and fourth Figures of the 27. Scheme.) *Ita ut Apiarium marinum vere dixeris; primo enim intuitu è Mare ad Terram delatum, vermiculis fcatebat cæruleis parvis, qui mox à calore folis in Mufcas, vel Apes potius, eafq; exiguas & nigras tranf-formebantur, circumvolantefque evanefcebant, ita ut de eorum mellificatione nihil certi confpici datum fuerit, cum tamen cærofa materia propolis Apum-que cellæ manifefte apparerent, atque ipfa mellis qualifcunque fubftantia procul-dubio urinatoribus patebit, ubi curiofius inquifiverint hæc apiaria, eaque in natali folo & falo diverfis temporibus penitius luftrarint.*

Which Hiftory contains things fufficiently ftrange to be confider'd, as whether the hufk were a Plant, growing at the bottom of the Sea before, of it felf, out of whofe putrifaction might be generated thefe ftrange kind of Magots; or whether the feed of certain Bees, finking to the bottom, might there naturally form it felf that vegetable hive, and take root; or, whether it might not be placed there by fome diving Fly; or, whether it might not be fome peculiar propriety of that Plant, whereby it might ripen or form its vegetable juice into an Animal fubftance; or, whether it may not be of the nature of a Sponge, or rather a Sponge of the nature of this, according to fome of thofe relations and conjectures I formerly made of that body, is a matter very difficult to be determined. But indeed, in this defcription, the Excellent *Pifo* has not been fufficiently particular in the fetting down the whole procefs, as it were to be wifh'd: There are indeed very odd progreffes in the production of feveral kinds of Infects, which are not lefs inftructive then pleafant, feveral of which, the diligent *Goedartius* has carefully obferv'd and recorded, but among all his Obfervations, he has none like this, though that of the *Hemerobius* be fomewhat of this kind, which is added as an Appendix by *Johannes Mey.*

I have

I have, for my own particular, besides several of those mention'd by him, observ'd divers other circumstances, perhaps, not much taken notice of, though very common, which do indeed afford us a very *coercive* argument to admire the goodness and providence of the infinitely wise Creator in his most excellent contrivances and dispensations. I have observ'd, at several times of the Summer, that many of the leaves of divers Plants have been spotted, or, as it were scabbed, and looking on the undersides of those of them that have been but a litte irregular, I have perceiv'd them to be sprinkled with divers sorts of little Eggs, which letting alone, I have found by degrees to grow bigger, and become little Worms with leggs, but still to keep their former places, and those places of the leaves, of their own accords, to be grown very protuberant upwards, and very hollow, and arched underneath, whereby those young creatures are, as it were, shelter'd and housed from external injury ; divers leaves I have observ'd to grow and swell so farr, as at length perfectly to inclose the Animal, which, by other observations I have made, I ghess to contain it, and become, as it were a womb to it, so long, till it be fit and prepar'd to be translated into another state, at what time, like (what they say of) Vipers, they gnaw their way through the womb that bred them ; divers of these kinds I have met with upon Goosberry leaves, Rose-tree leaves, Willow leaves, and many other kinds.

There are often to be found upon Rose-trees and Brier bushes, little red tufts, which are certain knobs or excrescencies, growing out from the Rind, or barks of those kinds of Plants, they are cover'd with strange kinds of threads or red hairs, which feel very soft, and look not unpleasantly. In most of these, if it has no hole in it, you shall find certain little Worms, which I suppose to be the causes of their production ; for when that Worm has eat its way through, they, having performed what they were design'd by Nature to do, by degrees die and wither away.

Now, the manner of their production, I suppose to be thus ; that the Alwise Creator has as well implanted in every creature a faculty of knowing what place is convenient for the hatching, nutrition, and preservation of their Eggs and of-springs, whereby they are stimulated and directed to convenient places, which becom, as 'twere the wombs that perform those offices : As he has also suited and adapted a property to those places wherby they grow and inclose those seeds, and having inclosed them, provide a convenient nourishment for them, but as soon as they have done the office of a womb, they die and wither.

The progress of inclosure I have often observ'd in leaves, which in those places where those seeds have been cast, have by degrees swell'd and inclos'd them, so perfectly round, as not to leave any perceptible passage out.

From this same cause, I suppose that Galls, Oak-apples, and several other productions of that kind, upon the branches and leaves of Trees, have their original ; for if you open any of them, when almost ripe, you shall find a little Worm in them. Thus, if you open never so many dry Galls, you shall find either a hole whereby the Worm has eat its passage out,

out, or if you find no paſſage, you may,by breaking or cutting the Gall, find in the middle of it a ſmall cavity, and in it a ſmall body, which does plainly enough yet retain a ſhape, to manifeſt it once to have been a Worm, though it dy'd by a too early ſeparation from the Oak on which it grew,its navel-ſtring,as 'twere,being broken off from the leaf or branch by which the Globular body that invelop'd it, received its nouriſhment from the Oak.

And indeed,if we conſider the great care of the Creator in the diſpen-ſations of his providences for the propagation and increaſe of the race,not onely of all kind of Animals, but even of Vegetables, we cannot chuſe but admire and adore him for his Excellencies, but we ſhall leave off to admire the creature, or to wonder at the ſtrange kind of acting in ſeveral Animals, which ſeem to favour ſo much of reaſon; it ſeeming to me moſt manifeſt,that thoſe are but actings according to their ſtructures, and ſuch operations as ſuch bodies, ſo compos'd, muſt neceſſarily, when there are ſuch and ſuch circumſtances concurring, perform : thus,whenwe find Flies ſwarming,about any piece of fleſh that does begin a little to ferment;But-terflies about Colworts,and ſeveral other leaves,which will ſerve to hatch and nouriſh their young; Gnats, and ſeveral other Flies about the Wa-ters, and mariſhy places,or any other creatures, ſeeking and placing their Seeds in convenient repoſitories, we may, if we attentively conſider and examine it, find that there are circumſtances ſufficient,upon the ſuppoſals of the excellent contrivance of their machine,to excite and force them to act after ſuch or ſuch a manner; thoſe ſteams that riſe from theſe ſeveral places may, perhaps, ſet ſeveral parts of theſe little Animals at work,even as in the contrivance of killing a Fox or Wolf with a Gun, the moving of a ſtring, is the death of the Animal; for the Beaſt, by moving the fleſh that is laid to entrap him, pulls the ſtring which moves the trigger, and that lets go the Cock which on the ſteel ſtrikes certain ſparks of fire which kindle the powder in the pann, and that preſently flies into the barrel, where the powder catching fire rarifies and drives out the bullet which kills the Animal; in all which actions, there is nothing of intention or ratiocination to be aſcrib'd either to the Animal or Engine, but all to the ingeniouſneſs of the contriver.

But to return to the more immediate conſideration of our Gnat: We have in it an Inſtance, not uſual or common, of a very ſtange *amphi-bious* creature, that being a creature that inhabits the Air, does yet pro-duce a creature, that for ſome time lives in the water as a Fiſh, though afterward (which is as ſtrange) it becomes an inhabitant of the Air, like its Sire,in the form of a Fly. And this, me thinks, does prompt me to pro-poſe certain conjectures, as Queries, having not yet had ſufficient oppor-tunity and leiſure to anſwer them my ſelf from my own Experiments or Obſervations.

And the firſt is, Whether all thoſe things that we ſuppoſe to be bred from corruption and putrifaction, may not be rationally ſuppos'd to have their origination as natural as theſe Gnats, who, 'tis very probable, were firſt dropt into this Water, in the form of Eggs. Thoſe Seeds or

Eggs muſt certainly be very ſmall, which ſo ſmall a creature as a Gnat yields, and therefore we need not wonder that we find not the Eggs themſelves, ſome of the younger of them, which I have obſerv'd, having not exceeded a tenth part of the bulk they have afterwards come to; and next, I have obſerved ſome of thoſe little ones which muſt have been gene-rated after the Water was incloſed in the Bottle, and therefore moſt pro-bably from Eggs, whereas thoſe creatures have been ſuppos'd to be bred of the corruption of the Water, there being not formerly known any probable way how they ſhould be generated.

A ſecond is, whether theſe Eggs are immediately dropt into the Water by the Gnats themſelves, or, mediately, are brought down by the falling rain; for it ſeems not very improbable, but that thoſe ſmall ſeeds of Gnats may (being, perhaps, of ſo light a nature, and having ſo great a propor-tion of ſurface to ſo ſmall a bulk of body) be ejected into the Air, and ſo, perhaps, carried for a good while too and fro in it, till by the drops of Rain it be waſh'd out of it.

A third is, whether multitudes of thoſe other little creatures that are found to inhabit the Water for ſome time, do not, at certain times, take wing and fly into the Air, others dive and hide themſelves in the Earth, and ſo contribute to the increaſe both of the one and the other Element.

Poſtſcript.

A good while ſince the writing of this Deſcription, I was preſented by Doctor *Peter Ball*, an ingenious Member of the *Royal Society*, with a little Paper of Nuts, which he told me was ſent him from a Brother of his out of the Countrey, from *Mamhead* in *Devonſhire*, ſome of them were looſe, having been, as I ſuppoſe, broken off, others were ſtill growing faſt on upon the ſides of a ſtick, which ſeem'd by the bark, pliableneſs of it, and by certain ſtrings that grew out of it, to be ſome piece of the root of a Tree; they were all of them dry'd, and a little ſhrivell'd, others more round, of a brown colour; their ſhape was much like a Figg, but very much ſmaller, ſome being about the bigneſs of a Bay-berry others, and the biggeſt, of a Hazel-Nut. Some of theſe that had no hole in them, I care-fully opened with my Knife, and found in them a good large round white Maggot, almoſt as bigg as a ſmall Pea, which ſeem'd ſhap'd like other Maggots, but ſhorter. I could not find them to move, though I gheſs'd them to be alive, becauſe upon pricking them with a Pinn, there would iſ-ſue out a great deal of white *mucous* matter, which ſeem'd to be from a vo-luntary contraction of their ſkin; their huſk or matrix conſiſted of three Coats, like the barks of Trees, the outermoſt being more rough and ſpon-gie, and the thickeſt, the middlemoſt more cloſe, hard, white, and thin, the innermoſt very thin, ſeeming almoſt like the ſkin within an Egg's ſhell. The two outermoſt had root in the branch or ſtick, but the innermoſt had no ſtem or proceſs, but was onely a ſkin that cover'd the cavity of the Nut. All the Nuts that had no holes eaten in them, I found to con-tain theſe Maggots, but all that had holes, I found empty, the Maggots,

it

it feems, having eaten their way through, taken wings and flown away, as this following account (which I receiv'd in writing from the fame perfon, as it was fent him by his Brother) manifefts. *In a moorifh black Peaty mould, with fome fmall veins of whitifh yellow Sands, upon occafion of digging a hole two or three foot deep, at the head of a Pond or Pool, to fet a Tree in, at that depth, were found, about the end of* October 1663. *in thofe very veins of Sand, thofe Buttons or Nuts, fticking to a little loofe ftick, that is, not belonging to any live Tree, and fome of them alfo free by themfelves.*

Four or five of which being then open'd, fome were found to contvin live Infects come to perfection, moft like to flying Ants, *if not the fame; in others, Infects, yet imperfect, having but the head and wings form'd, the reft remaining a foft white pulpy fubftance.*

Now, as this furnifhes us with one odd Hiftory more, very agreeable to what I before hinted, fo I doubt not, but were men diligent obfervers, they might meet with multitudes of the fame kind, both in the Earth and in the Water, and in the Air, on Trees, Plants, and other Vegetables, all places and things being, as it were, *animarum plena.* And I have often, with wonder and pleafure, in the Spring and Summer-time, look'd clofe to, and diligently on, common Garden mould, and in a very fmall parcel of it, found fuch multitudes and diverfities of little *reptiles,* fome in hufks, others onely creepers, many wing'd, and ready for the Air; divers hufks or habitations left behind empty. Now, if the Earth of our cold Climate be fo fertile of animate bodies, what may we think of the fat Earth of hotter Climates? Certainly, the Sun may there, by its activity, caufe as great a parcel of Earth to fly on wings in the Air, as it does of Water in fteams and vapours. And what fwarms muft we fuppofe to be fent out of thofe plentifull inundations of water which are poured down by the fluces of Rain in fuch vaft quantities? So that we need not much wonder at thofe innumerable clouds of Locufts with which *Africa,* and other hot countries are fo peftred, fince in thofe places are found all the convenient caufes of their production, namely, genitors, or Parents, concurrent receptacles or matrixes, and a fufficient degree of natural heat and moifture.

I was going to annex a little draught of the Figure of thofe Nuts fent out of *Devonfhire,* but chancing to examine Mr. *Parkinfon's* Herbal for fomething elfe, and particularly about Galls and Oak-apples, I found among no lefs then 24. feveral kinds of excrefcencies of the Oak, which I doubt not, but upon examination, will be all found to be the *matrixes* of fo many feveral kinds of Infects; I having obferv'd many of them my felf to be fo, among 24. feveral kinds, I fay, I found one defcribed and Figur'd directly like that which I had by me, the *Scheme* is there to be feen, the defcription, becaufe but fhort, I have here adjoin'd *Theatri Botanici trib.* 16. *Chap.* 2. *There groweth at the roots of old Oaks in the Spring-time, and fometimes alfo in the very heat of Summer, a peculiar kind of Mufhrom or Excrefcence, call'd Uva* Quercina, *fwelling ont of the Earth, many growing one clofe unto another, of the fafhion of a Grape, and therefore took the name, the* Oak-Grape, *and is of a Purplifh colour on the outfide, and*

*Schem:*XXVIII.

and white within like Milk, and in the end of Summer becometh hard
and woody. Whether this be the very fame kind, I cannot affirm, but
both the Picture and Defcription come very neer to that I have,
but that he feems not to take notice of the hollownefs or Worm, for
which 'tis moft obfervable. And therefore 'tis very likely, if men
did but take notice, they might find very many differing Species of thefe
Nuts, *Ovaries,* or *Matrixes,* and all of them to have much the fame
defignation and office. And I have very lately found feveral kinds of Excrefcencies on Trees and Shrubs, which having endured the Winter, upon opening them, I found moft of them to contain little Worms, but
dead, thofe things that contain'd them being wither'd and dry.

Obferv. XLIV. *Of the tufted or Brufh-horn'd Gnat.*

THis little creature was one of thofe multitudes that fill our *Englifh*
air all the time that warm weather lafts, and is exactly of the fhape
of that I obferv'd to be generated and hatch'd out of thofe little Infects
that wriggle up and down in Rain-water. But, though many were of this
form, yet I obferv'd others to be of quite other kinds; nor were all
of this or the other kind generated out of Water Infects; for whereas I
obferv'd that thofe that proceeded from thofe Infects were at their full
growth, I have alfo found multitudes of the fame fhape, but much fmaller
and tenderer feeming to be very young ones, creep up and down upon
the leaves of Trees, and flying up and down in fmall clufters, in places
very remote from water; and this Spring, I obferv'd one day, when the
Wind was very calm, and the afternoon very fair, and pretty warm,
though it had for a long time been very cold weather, and the wind continued ftill in the Eaft, feveral fmall fwarms of them playing to and fro
in little clouds in the Sun, each of which were not a tenth part of the
bignefs of one of thefe I here have delineated, though very much of the
fame fhape, which makes me ghefs, that each of thefe fwarms might be
the of-fpring of one onely Gnat, which had been hoorded up in fome fafe
repofitory all this Winter by fome provident Parent, and were now, by
the warmth of the Spring-air, hatch'd into little Flies.

And indeed, fo various, and feemingly irregular are the generations or
productions of Infects, that he that fhall carefully and diligently obferve
the feveral methods of Nature therein, will have infinitely caufe further
to admire the wifdom and providence of the Creator; for not onely the
fame kind of creature may be produc'd from feveral kinds of ways, but
the very fame creature may produce feveral kinds: For, as divers Watches
may be made out of feveral materials, which may yet have all the fame
appearance, and move after the fame manner, that is, fhew the hour equally
true, the one as the other, and out of the fame kind of matter, like
Watches, may be wrought differing ways; and, as one and the fame Watch
D a   may,

may, by being diverfly agitated, or mov'd, by this or that agent, or after this or that manner, produce a quite contrary effect : So may it be with thefe moft curious Engines of Infect's bodies; the All-wife God of Nature, may have fo ordered and difpofed the little *Automatons*, that when nourifhed, acted, or enlivened by this caufe, they produce one kind of effect, or animate fhape, when by another they act quite another way, and another Animal is produc'd. So may he fo order feveral materials, as to make them, by feveral kinds of methods, produce fimilar *Automatons.*

But to come to the Defcription of this Infect, as it appears through a *Microfcope*, of which a reprefentation is made in the 28.*Scheme.* Its head A, is exceeding fmall, in proportion to its body, confifting of two clufters of pearl'd eyes B B, on each fide of its head, whofe pearls or eye-balls are curioufly rang'd like thofe of other Flies ; between thefe, in the forehead of it, there are plac'd upon two fmall black balls, C C, two long jointed horns, tapering towards the top, much refembling the long horns of Lobfters, each of whofe ftems or quills, D D, were brifled or brufhed with multitudes of fmall ftiff hairs, iffuing out every way from the feveral joints, like the ftrings or fproutings of the herb *Horfe-tail,* which is oft obferv'd to grow among Corn, and for the whole fhape, it does very much refemble thofe *brufhy Vegetables ;* befides thefe, there are two other jointed and brifled horns, or feelers, E E, in the forepart of the head, and a *probofcis,* F, underneath, which in fome Gnats are very long, ftreight hollow pipes, by which thefe creatures are able to drill and penetrate the fkin, and thence, through thofe pipes fuck fo much bloud as to ftuff their bellies fo full till they be ready to burft.

This fmall head, with its appurtenances, is faftned on by a fhort neck, G, to the middle of the *thorax,* which is large, and feems cafed with a ftrong black fhel, H I K, out of the under part of which, iffue fix long and flender legs, L L L L L L, fhap'd juft like the legs of Flies, but fpun or drawn out longer and flenderer, which could not be exprefs'd in the Figure, becaufe of their great length ; and from the upper part, two oblong, but flender tranfparent wings, M M, fhaped fomewhat like thofe of a Fly, underneath each of which, as I have obferv'd alfo in divers forts of Flies, and other kinds of Gnats, was placed a fmall body, N, much refembling a drop of fome tranfparent glutinous fubftance, hardned or cool'd, as it was almoft ready to fall, for it has a round knob at the end, which by degrees grows flenderer into a fmall ftem, and neer the infertion under the wing, this ftem again grows bigger ; thefe little *Pendulums,* as I may fo call them, the litle creature vibrates to and fro very quick when it moves its wings, and I have fometimes obferv'd it to move them alfo, whil'ft the wing lay ftill, but always their motion feem'd to further the motion of the wing ready to follow ; of what ufe they are, as to the moving of the wing, or otherwife, I have not now time to examine.

Its belly was large, as it is ufually in all Infects, and extended into nine lengths or partitions, each of which was cover'd with round armed rings or fhells ; fix of which, O P Q R S T were tranfparent, and divers kinds of *Periftaltick* motions might be very eafily perceiv'd, whil'ft the Animal

was

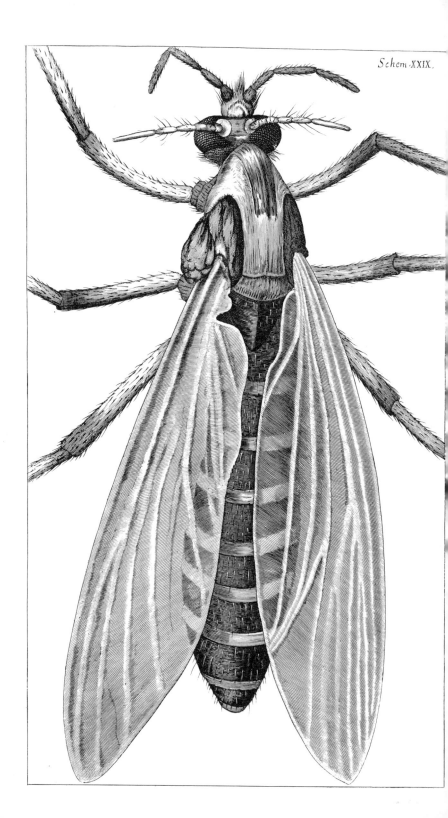

Was alive, but especially a small cleer white part V, seemed to beat like the heart of a larger Animal. The last three divisios, W X Y, were cover'd with black and opacous shells. To conclude, take this creature altogether, and for beauty and curious contrivances, it may be compared with the largest Animal upon the Earth. Nor doth the Alwise Creator seem to have shewn less care and providence in the fabrick of it, then in those which seem most considerable.

Observ. X L V. *Of the great Belly'd* Gnat *or female* Gnat.

THe second Gnat, delineated in the twenty ninth *Scheme*, is of a very differing shape from the former;but yet of this sort also,I found several of the Gnats, that were generated out of the Water Insect : the wings of this, were much larger then those of the other, and the belly much bigger, shorter and of an other shape ; and, from several particulars, I gheft it to be the Female Gnat, and the former to be the Male.

The *thorax* of this was much like that of the other,having a very strong and ridged back-piece, which went also on either side of its leggs; about the wings there were several joynted pieces of Armor, which seem'd curiously and conveniently contriv'd, for the promoting and strengthning the motion of the wings:its head was much differing from the other,being much bigger and neater shap'd, and the horns that grew out between his eyes on two little balls, were of a very differing shape from the tufts of the other Gnat, these having but a few knots or joynts, and each of those but a few, and those short and strong, brissles. The formost horns or feelers, were like those of the former Gnat.

One of these Gnats I have suffer'd to pierce the skin of my hand, with its *proboscis,* and thence to draw out as much blood as to fill its belly as full as it could hold, making it appear very red and transparent ; and this without any further pain, then whil'ft it was sinking in its *proboscis,* as it is also in the stinging of Fleas : a good argument, that these creatures do not wound the skin, and suck the blood out of enmity and revenge, but for meer necessity, and to satisfy their hunger. By what means this creature is able to suck, we shall shew in another place.

Observ. X L V I. *Of the white featherwing'd* Moth *or* Tinea Argentea.

THis white long wing'd Moth, which is delineated in the 30. *Scheme,* afforded a lovely object both to the naked Eye,and through a *Microscope* : to the Eye it appear'd a small Milk white Fly with four white

Wings

Wings, the two formoſt ſomewhat longer then the two hindermoſt, and the two ſhorter about half an Inch long, each of which four Wings ſeem'd to conſiſt of two ſmall long Feathers, very curiouſly tufted, or haired on each ſide, with purely white, and exceedingly fine and ſmall Haires, proportion'd to the ſtalks or ſtems, out of which they grew, much like the tufts of a long wing-feather of ſome Bird, and their ſtalks or ſtems were, like thoſe, bended backwards and downwards, as may be plainly ſeen by the draughts of them in the Figure.

Obſerving one of theſe in my *Microſcope*, I found, in the firſt place, that all the Body, Legs, Horns and the Stalks of the Wings, were covered over with various kinds of curious white Feathers, which did, with handling or touching, eaſily rubb off and fly about, in ſo much that looking on my Fingers, with which I had handled this Moth, and perceiving on them little white ſpecks, I found by my *Microſcope*, that they were ſeveral of the ſmall Feathers of this little creature, that ſtuck up and down in the *rugoſities* of my Skin.

Next, I found that underneath theſe Feathers, the pretty Inſect was covered all over with a cruſted Shell, like other of thoſe Animals, but with one much thinner and tenderer.

Thirdly, I found, as in Birds alſo is notable, it had differing and appropriate kinds of Feathers, that covered ſeveral parts of its body.

Fourthly, ſurveying the parts of its body, with a more accurate and better Magnifying *Microſcope*, I found that the tufts or haires of its Wings were nothing elſe but a congeries, or thick ſet cluſter of ſmall *vimina* or twiggs, reſembling a ſmall twigg of Birch, ſtript or whitned, with which Bruſhes are uſually made, to beat out or bruſh off the duſt from Cloth and Hangings. Every one of the twiggs or branches that compoſed the Bruſh of the Feathers, appeared in this bigger Magnifying Glaſs (of which E F which repreſents $\frac{1}{24}$ part of an Inch, is the ſcale, as G is of the leſſer, which is only $\frac{1}{3}$) like the figure D. The Feathers alſo that covered a part of his Body, and were interſperſed among the bruſh of his Wings, I found, in the bigger Magnifying Glaſs, of the ſhape A, confiſting of a ſtalk or ſtem in the middle, and a ſeeming tuftedneſs or bruſhy part on each ſide. The Feathers that cover'd moſt part of his Body and the ſtalk of his wings, were, in the ſame *Microſcope*, much of the figure B, appearing of the ſhape of a ſmall Feather, and ſeemed tufted : thoſe which covered the Horns and ſmall parts of the Leggs, through the ſame *Microſcope*, appear'd of the ſhape C. Whether the tufts of any or all of theſe ſmall Feathers, confiſted of ſuch component particles as the Feathers of Birds, I much doubt, becauſe I find that Nature does not alwaies keep, or operate after the ſame method, in ſmaller and bigger creatures. And of this, we have particular Inſtances in the Wings of ſeveral creatures. For whereas, in Birds of all kinds, it compoſes each of the Feathers of which its Wing conſiſts, of ſuch an exceeding curious and moſt admirable and ſtupendious texture, as I elſe where ſhew, in the Obſervations on a Feather; we find it to alter its method quite, in the fabrick of the Wings of theſe minute creatures, compoſing ſome of thin extended membranes

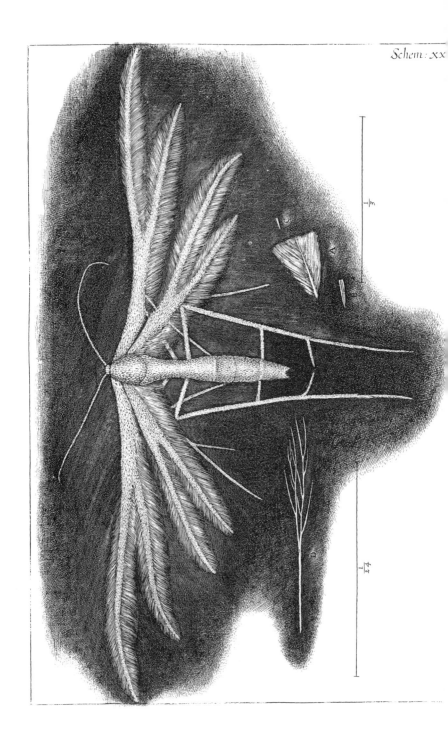

or fkins, fuch as the Wings of Dragon-flys; in others, thofe fkins are all over-grown, or pretty thick beftuck, with fhort brifles, as in Flefh-flies; in others, thofe filmes are covered, both on the upper and under fide, with fmall Feathers, plac'd almoft like the tyles on a Houfe, and are curioufly rang'd and adorn'd with moft lively colours, as is obfervable in Butter-flies, and feveral kinds of Moths; In others, inftead of their films, Nature has provided nothing, but a matter of half a fcore ftalks(if I well remember the number; for I have not lately met with any of thefe flys, and did not,when I firft obferv'd them, take fufficient notice of divers particulars) and each of thefe ftalks, with a few fingle branchings on each fide, refembling much the branched back-bone of a Herring or the like Fifh, or a thin hair'd Peacocks feather, the top or the eye being broken off. With a few of thefe on either fide(which it was able to fhut up or expand at pleafure, much like a Fann, or rather like the pofture of the feathers in a wing, which ly all one under another, when fhut, and by the fide of each other, when expanded) this pretty little grey Moth (for fuch was the creature I obferv'd, thus wing'd) could very nimbly, and as it feem'd very eafily move its *corpufcle*, through the Air,from place to place. Other Infects have their wings cas'd, or cover'd over,with certain hollow fhells, fhap'd almoft like thofe hollow Trayes, in which Butchers carry meat, whofe hollow fides being turn'd downwards, do not only fecure their folded wings from injury of the earth, in which moft of thofe creatures refide, but whilft they fly, ferves as a help to fuftain and bear them up. And thefe are obfervable in *Scarabees* and a multitude of other terreftrial *cruftaceous* Infects; in which we may yet further obferve a particular providence of Nature.

Now in all thefe kinds of wings, we obferve this particular, as a thing moft worthy remark; that where ever a wing confifts of difcontinued parts, the Pores or *interftitia* between thofe parts are very feldom, either much bigger,or much fmaller, then thefe which we here find between the particles of thefe brufhes, fo that it fhould feem to intimate, that the parts of the Air are fuch, that they will not eafily or readily, if at all, pafs through thefe Pores, fo that they feem to be ftrainers fine enough to hinder the particles of the Air (whether hinder'd by their bulk, or by their *agitation, circulation, rotation* or *undulation*, I fhall not here determine) from getting through them,and,by that means,ferve the Animal as well,if not better, then if they were little films. I fay, if not better, becaufe I have obferv'd that all thofe creatures, that have film'd wings, move them aboundantly quicker and more ftrongly,fuch as all kind of Flies and *Scarabees* and Batts, then fuch as have their wings covered with feathers, as Butter-flies and Birds, or twiggs, as Moths, which have each of them a much flower motion of their wings; That little ruggednefs perhaps of their wings helping them fomewhat,by taking better hold of the parts of the Air, or not fuffering them fo eafily to pafs by, any other way then one.

But what ever be the reafon of it, tis moft evident, that the fmooth wing'd Infects have the ftrongeft Mufcles or movent parts of their wings, and the other much weaker; and this very Infect, we are now defcribing,
had

had a very fmall *thorax* or middle part of his body, if compar'd to the length and number of his wings; which therefore, as he mov'd them very flowly, fo muft he move them very weakly. And this laft propriety do we find fomewhat obferv'd alfo in bigger kind of Flying creatures, Birds; fo that we fee that the Wifdom and Providence of the All-wife Creator, is not lefs fhewn in thefe fmall defpicable creatures, Flies and Moths, which we have branded with a name of ignominy, calling them Vermine, then in thofe greater and more remakable animate bodies, Birds.

I cannot here ftand to add any thing about the nature of flying, though, perhaps, on another occafion, I may fay fomething on that fubject, it being fuch as may deferve a much more accurate examination and fcrutiny then it has hitherto met with; For to me there feems nothing wanting to make a man able to fly, but what may be eafily enough fupply'd from the Mechanicks hitherto known, fave onely the want of ftrength, which the Mufcles of a man feem utterly uncapable of, by reafon of their fmalnefs and texture, but how even ftrength alfo may be mechanically made, an artificial Mufcle fo contriv'd, that thereby a man fhall be able to exert what ftrength he pleafes, and to regulate it alfo to his own mind, I may elfewhere endeavour to manifeft.

Obferv. XLVII. *Of the* Shepherd Spider, *or long legg'd* Spider.

THe Carter, Shepherd Spider, or long-legg'd Spider, has, for two particularities, very few fimilar creatures that I have met with; the firft, which is difcoverable onely by the *Microfcope*, and is in the firft and fecond *Figures* of the 31. *Scheme*, plainly defcrib'd, is the curious contrivance of his eyes, of which (differing from moft other Spiders) he has onely two, and thofe plac'd upon the top of a fmall pillar or hillock, rifing out of the middle of the top of its back, or rather the crown of its head, for they were fix'd on the very top of this pillar (which is about the heighth of one of the tranfverfe Diameters of the eye, and look'd on in another pofture, appear d much of the fhape, BCD) The two eyes, BB, were placed back to back, with the tranfparent parts, or the pupils, looking towards either fide, but fomewhat more forward then backwards. C was the column or neck on which they ftood, and D the crown of the head out of which that neck fprung.

Thefe eyes, to appearance, feem'd to be of the very fame ftructure with that of larger *binocular* creatures, feeming to have a very fmooth and very protuberant *Cornea*, and in the midft of it to have a very black pupil, incompaffed about with a kind of grey *Iris*, as appears by the *Figure*; whether it were able to move thefe eyes to and fro, I have not obferv'd, but 'tis not very likely he fhould, the pillar or neck C, feeming to be cover'd and ftiffen'd with a crufty fhell; but Nature, in probability, has fupply'd

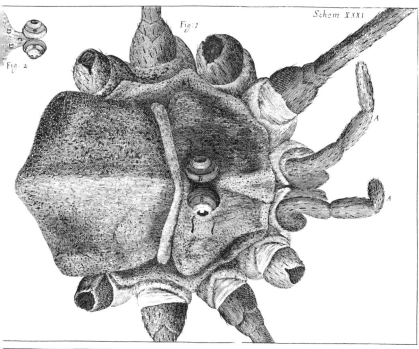

Fig: 1

Fig: 2

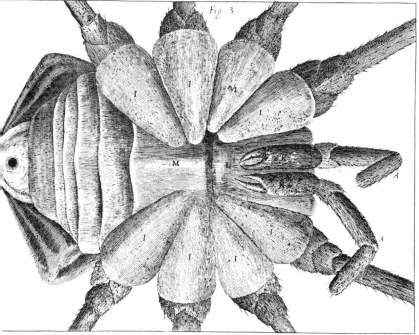

Fig: 3

ply'd that defect, by making the *Cornea* so very protuberant, and setting it so cleer above the shadowing or obstructing of its prospect by the body, that 'tis likely each eye may perceive,though not see distinctly, almost a *Hemisphere*, whence having so small and round a body plac'd upon such long leggs, it is quickly able so to wind, and turn it, as to see any thing distinct. This creature, as do all other Spiders I have yet examin'd, does very much differ from most other Insects in the Figure of its eyes; for I cannot, with my best *Microscope*, discover its eyes to be any ways knobb'd or pearl'd like those of other Insects.

The second Peculiarity which is obvious to the eye, is also very remarkable, and that is the prodigious length of its leggs, in proportion to its small round body, each legg of this I drew, being above sixteen times the length of its whole body, and there are some which have them yet longer, and others that seem of the same kind,that have them a great deal shorter; the eight leggs are each of them jointed, just like those of a Crab, but every of the parts are spun out prodigiously longer in proportion; each of these leggs are terminated in a small case or shell,shap'd almost like that of a Musle-shell, as is evident in the third *Figure* of the same *Scheme* (that represents the appearance of the under part or belly of the creature) by the shape of the protuberant *conical* body, I I I I, &c. These are as 'twere plac'd or fasten'd on to the protuberant body of the Insect,which is to be suppos'd very high at M,making a kind of blunt cone whereof M is to be suppos'd the *Apex*, about which greater cone of the body,the smaller cones of the leggs are plac'd,each of them almost reaching to the top in so admirable a manner, as does not a little manifest the wisdom of Nature in the contrivance; for these long Leavers (as I may so call them)of the legs,having not the advantage of a long end on the other side of the *hypomochlion* or centers on which the parts of the leggs move, must necessarily require a vast strength to move them, and keep the body ballanc'd and suspended, in so much, that if we should suppose a man's body suspended by such a contrivance, an hundred and fifty times the strength of a man would not keep the body from falling on the breast. To supply therefore each of these leggs with its proper strength, Nature has allow'd to each a large Chest or Cell, in which is included a very large and strong Muscle, and thereby this little Animal is not onely able to suspend its body upon less then these eight, but to move it very swiftly over the tops of grass and leaves.

Nor are these eight leggs so prodigiously long, but the ninth, and tenth, which are the two claws, K K, are as short, and serve in steed of a *proboscis*, for those seem'd very little longer then his mouth; each of them had three parts, but very short, the joints K K, which represented the third, being longer then both the other. This creature, seems (which I have several times with pleasure observ'd) to throw its body upon the prey, insteed of its hands, not unlike a hunting Spider, which leaps like a Cat at a Mouse. The whole Fabrick was a very pretty one, and could I have dissected it,I doubt not but I should have found as many singularities within it as without,perhaps, for the most part,not unlike the

the parts of a Crab, which this little creature does in many things, very much refemble; the curiofity of whofe contrivance, I have in another place examin'd. I omit the defcription of the horns, A A, of the mouth, L L, which feem'd like that of a Crab ; the fpecklednefs of his fhell, which proceeded from a kind of feathers or hairs, and the hairinefs of his leggs, his large *thorax* and little belly, and the like, they being manifefted by the Figure ; and fhall onely take notice that the three parts of the body, namely, the head, breaft, and belly, are in this creature ftrangely confus'd, fo that 'tis difficult to determine which is which, as they are alfo in a Crab ; and indeed, this feems to be nothing elfe, but an Air-crab, being made more light and nimble, proportionable to the *medium* wherin it refides ; and as Air feems to have but one thoufandth part of the body of Water, fo does this Spider feem not to be a thoufandth part of the bulk of a Crab.

Obferv. X L V I I I. *Of the hunting* Spider, *and feveral other forts of* Spiders.

THe hunting Spider is a fmall grey Spider, prettily befpeck'd with black fpots all over its body, which the *Microfcope* difcovers to be a kind of feathers like thofe on Butterflies wings, or the body of the white Moth I lately defcrib'd. Its gate is very nimble by fits, fometimes running, and fometimes leaping, like a Grafhopper almoft, then ftanding ftill, and fetting it felf on its hinder leggs, it will very nimbly turn its body, and look round it felf every way : It has fix very confpicuous eyes, two looking directly forwards, plac'd juft before ; two other, on either fide of thofe, looking forward and fide-ways; and two other about the middle of the top of its back or head, which look backwards and fide-wards ; thefe feem'd to be the biggeft. The furface of them all was very black, fphærical, purely polifh'd, reflecting a very cleer and diftinct Image of all the ambient objects, fuch as a window, a man's hand, a white Paper, or the like. Some other properties of this Spider, obferv'd by the moft accomplifh'd Mr. *Evelyn*, in his travels in *Italy*, are moft emphatically fet forth in the Hiftory hereunto annexed, which he was pleas'd upon my defire to fend me in writing.

Of all the forts of Infects, there is none has afforded me more divertifements then the *Venatores*, which are a fort of *Lupi*, that have their Denns in the rugged walls, and crevices of our houfes ; a fmall brown and delicately fpotted kind of Spiders, whofe hinder leggs are longer then the reft.

Such I did frequently obferve at *Rome*, which efpying a Fly at three or four yards diftance, upon the Balcony (where I ftood) would

would not make directly to her, but craul under the Rail, till being arriv'd to the *Antipodes*, it would steal up, seldom missing its aim; but if it chanced to want any thing of being perfectly oppofite, would at firft peep, immediatly flide down again, till taking better notice, it would come the next time exactly upon the Fly's back: But, if this hapn'd not to be within a competent leap, then would this Infect move fo foftly, as the very fhadow of the Gnomon feem'd not to be more imperceptible, unlefs the Fly mov'd; and then would the Spider move alfo in the fame proportion, keeping that juft time with her motion, as if the fame Soul had animated both thofe little bodies; and whether it were forwards, backwards, or to either fide, without at all turning her body, like a well mannag'd Horfe: But, if the capricious Fly took wing, and pitch'd upon another place behind our Huntrefs, then would the Spider whirle its body fo nimbly about, as nothing could be imagin'd more fwift; by which means, fhe always kept the head towards her prey, though to appearance, as immovable, as if it had been a Nail driven into the Wood, till by that indifcernable progrefs (being arriv'd within the fphere of her reach) fhe made a fatal leap (fwift as Lightning) upon the Fly, catching him in the pole, where fhe never quitted hold till her belly was full, and then carried the remainder home. I have beheld them inftructing their young ones, how to hunt, which they would fometimes difcipline for not well obferving; but, when any of the old ones did (as fometimes) mifs a leap, they would run out of the field, and hide them in their crannies, as afham'd, and haply not be feen abroad for four or five hours after; for fo long have I watched the nature of this ftrange Infect, the contemplation of whofe fo wonderfull fagacity and addrefs has amaz'd me; nor do I find in any chafe whatfoever, more cunning and Stratagem obferv'd: I have found fome of thefe Spiders in my Garden, when the weather (towards the Spring)

E e   is

is very hot, but they are nothing fo eager of hunting as they are in *Italy*.

There are multitudes of other forts of Spiders, whofe eyes, and moft other parts and properties, are fo exceedingly different both from thofe I have defcrib'd, and from one another, that it would be almoft endlefs, at leaft too long for my prefent Effay, to defcribe them, as fome with fix eyes, plac'd in quite another order; others with eight eyes; others with fewer, and fome with more. They all feem to be creatures of prey, and to feed on other fmall Infects, but their ways of catching them feem very differing : the Shepherd Spider by running on his prey; the Hunting Spider by leaping on it, other forts weave Nets, or Cobwebs, whereby they enfnare them, Nature having both fitted them with materials and tools, and taught them how to work and weave their Nets, and to lie perdue, and to watch diligently to run on any Fly, as foon as ever entangled.

Their thread or web feems to be fpun out of fome vifcous kind of excrement, lying in their belly, which, though foft when drawn out, is, prefently by reafon of its fmalnefs, hardned and dried by the ambient Air. Examining feveral of which with my *Microfcope*, I found them to appear much like white Horf-hair, or fome fuch tranfparent horny fubftance, and to be of very differing magnitudes; fome appearing as bigg as a Pigg's briffle, others equal to a Horfs-hair; other no bigger then a man's hair; others yet fmaller and finer. I obferv'd further, that the radiating chords of the web were much bigger, and fmoother then thofe that were woven round, which feem'd fmaller, and all over knotted or pearl'd, with fmall tranfparent Globules, not unlike fmall Cryftal Beads or feed Pearls, thin ftrung on a Clew of Silk; which, whether they were fo fpun by the Spider, or by the adventitious moifture of a fogg (which I have obferv'd to cover all thefe filaments with fuch Cryftalline Beads) I will not now difpute.

Thefe threads were fome of them fo fmall, that I could very plainly, with the *Microfcope*, difcover the fame confecutions of colours as in a *Prifme*, and they feem'd to proceed from the fame caufe with thofe colours which I have already defcrib'd in thin plated bodies.

Much refembling a Cobweb, or a confus'd lock of thefe Cylinders, is a certain white fubftance which, after a fogg, may be obferv'd to fly up and down the Air; catching feveral of thefe, and examining them with my *Microfcope*, I found them to be much of the fame form, looking moft like to a flake of Worfted prepar'd to be fpun, though by what means they fhould be generated, or produc'd, is not eafily imagined : they were of the fame weight, or very little heavier then the Air; and 'tis not unlikely, but that thofe great white clouds, that appear all the Summer time, may be of the fame fubftance.

Obferv.

Schem. XXX. II

Obſerv. XLIX. *Of an* Ant *or* Piſmire.

THis was a creature, more troubleſom to be drawn, then any of the reſt, for I could not, for a good while, think of a way to make it ſuffer its body to ly quiet in a natural poſture; but whil'ſt it was alive, if its feet were fetter'd in Wax or Glew , it would ſo twiſt and wind its body, that I could not any wayes get a good view of it ; and if I killed it, its body was ſo little, that I did often ſpoile the ſhape of it, before I could throughly view it: for this is the nature of theſe minute Bodies, that as ſoon, almoſt, as ever their life is deſtroy'd, their parts immediately ſhrivel, and loſe their beauty ; and ſo is it alſo with ſmall Plants, as I inſtanced before, in the deſcription of Moſs. And thence alſo is the rea-ſon of the variations in the beards of wild Oats, and in thoſe of Muſk-graſs ſeed, that their bodies, being exceeding ſmall, thoſe ſmall variations which are made in the ſurfaces of all bodies, almoſt upon every change of Air, eſpecially if the body be porous, do here become ſenſible, where the whole body is ſo ſmall, that it is almoſt nothing but ſurface ; for as in vegetable ſubſtances, I ſee no great reaſon to think, that the moiſture of the Aire(that, ſticking to a wreath'd beard, does make it untwiſt)ſhould evaporate, or exhale away, any faſter then the moiſture of other bodies, but rather that the avolation from, or acceſs of moiſture to, the ſurfaces of bodies being much the ſame , thoſe bodies become moſt ſenſible of it, which have the leaſt proportion of body to their ſurface. So is it alſo with Animal ſubſtances ; the dead body of an Ant, or ſuch little creature, does almoſt inſtantly ſhrivel and dry, and your objeƈt ſhall be quite an-other thing, before you can half delineate it, which proceeds not from the extraordinary exhalation, but from the ſmall proportion of body and jui-ces, to the uſual drying of bodies in the Air, eſpecially if warm. For which inconvenience, where I could not otherwiſe remove it, I thought of this expedient.

I took the creature, I had deſign'd to delineate, and put it into a drop of very well reƈtified ſpirit of Wine, this I found would preſently diſpatch, as it were, the Animal, and being taken out of it, and lay'd on a paper, the ſpirit of Wine would immediately fly away, and leave the Animal dry, in its natural poſture, or at leaſt, in a conſtitution, that it might eaſi-ly with a pin be plac'd, in what poſture you deſired to draw it, and the limbs would ſo remain, without either moving, or ſhriveling. And thus I dealt with this Ant, which I have here delineated, which was one of ma-ny, of a very large kind, that inhabited under the Roots of a Tree, from whence they would ſally out in great parties, and make moſt grievous havock of the Flowers and Fruits, in the ambient Garden, and return back again very expertly, by the ſame wayes and paths they went.

It was more then half the bigneſs of an Earwig, of a dark brown, or reddiſh colour, with long legs, on tne hinder of which it would ſtand up, and raife its head as high as it could above the ground, that it might ftare the further about it, juft after the fame manner as I have alfo obferv'd a hunting Spider to do: and putting my finger towards them, they have at firft all run towards it, till almoft at it; and then they would ftand round about it, at a certain diftance, and fmell, as it were, and confider whether they fhould any of them venture any further, till one more bold then the reft venturing to climb it, all the reft, if I would have fuffered them, would have immediately followed : many fuch other feemingly rational actions I have obferv'd in this little Vermine with much pleafure, which would be too long to be here related ; thofe that defire more of them may fatisfie their curiofity in *Ligons* Hiftory of the *Barbadoes*.

Having infnar'd feveral of thefe into a fmall Box, I made choice of the talleft grown among them, and feparating it from the reft, I gave it a Gill of Brandy, or Spirit of Wine, which after a while e'en knock'd him down dead drunk, fo that he became movelefs, though at firft putting in he ftruggled for a pretty while very much, till at laft, certain bubbles iffuing out of its mouth, it ceafed to move ; this (becaufe I had before found them quickly to recover again, if they were taken out prefently) I fuffered to lye above an hour in the Spirit; and after I had taken it out, and put its body and legs into a natural pofture, remained movelefs about an hour; but then, upon a fudden, as if it had been awaken out of a drunken fleep, it fuddenly reviv'd and ran away; being caught, and ferv'd as before, he for a while continued ftruggling and ftriving, till at laft there iffued feveral bubbles out of its mouth, and then, *tanquam animam expiraffet*, he remained movelefs for a good while ; but at length again recovering, it was again redipt, and fuffered to lye fome hours in the Spirit ; notwithftanding which, after it had layen dry fome three or four hours, it again recovered life and motion : Which kind of Experiments, if profecuted, which they highly deferve, feem to me of no inconfiderable ufe towards the invention of the *Latent Scheme*, (as the Noble *Verulam* calls it) or the hidden, unknown Texture of Bodies.

Of what Figure this Creature appear'd through the *Microfcope*, the 32. *Scheme* (though not fo carefully graven as it ought) will reprefent to the eye, namely, That it had a large head A A, at the upper end of which were two protuberant eyes, pearl'd like thofe of a Fly, but fmaller B B ; out of the Nofe, or foremoft part, iffued two horns C C, of a fhape fufficiently differing from thofe of a blew Fly, though indeed they feem to be both the fame kind of Organ, and to ferve for a kind of fmelling ; beyond thefe were two indented jaws D D, which he open'd fide-wayes, and was able to gape them afunder very wide ; and the ends of them being armed with teeth, which meeting went between each other, it was able to grafp and hold a heavy body, three or four times the bulk and weight of its own body : It had only fix legs, fhap'd like thofe of a Fly, which, as I fhewed before, is an Argument that it is a winged Infect, and though I could not perceive any fign of them in the middle part of its body (which feem'd to confift of three joints or pieces

ces E F G, out of which fprung two legs, yet 'tis known that there are of them that have long wings, and fly up and down in the air.

The third and laft part of its body I I I was bigger and larger then the other two, unto which it was joyn'd by a very fmall middle, and had a kind of loofe fhell, or another diftinct part of its body H, which feem'd to be interpos'd, and to keep the *thorax* and belly from touching.

The whole body was cas'd over with a very ftrong armour, and the belly I I I was covered likewife with multitudes of fmall white fhining brifles; the legs, horns, head, and middle parts of its body were beftuck with hairs alfo, but fmaller and darker.

Obferv. L. *Of the wandring* Mite.

IN *September* and *October*, 1661. I obferv'd in *Oxford* feveral of thefe little pretty Creatures to wander to and fro, and often to travel over the plains of my Window. And in *September* and *October*. 1663. I obferv'd likewife feveral of thefe very fame Creatures traverfing a window at *London*, and looking without the window upon the fubjacent wall, I found whole flocks of the fame kind running to and fro among the fmall groves and thickets of green mofs, and upon the curioufly fpreading vegetable blew or yellow mofs, which is a kind of a Mufhrome or Jews-ear.

Thefe Creatures to the naked eye feemed to be a kind of black Mite, but much nimbler and ftronger then the ordinary Cheefe-Mites; but examining them in a *Microfcope*, I found them to be a very fine crufted or fhell'd Infect, much like that reprefented in the firft Figure of the three and thirtieth *Scheme*, with a protuberant oval fhell A, indented or pitted with an abundance of fmall pits, all covered over with little white brifles, whofe points all directed backwards.

It had eight legs, each of them provided with a very fharp tallon, or claw at the end, which this little Animal, in its going, faftned into the pores of the body over which it went. Each of thefe legs were beftuck in every joynt of them with multitudes of fmall hairs, or (if we refpect the proportion they bore to the bignefs of the leg) turnpikes, all pointing towards the claws.

The *Thorax*, or middle parts of the body of this Creature, was exceeding fmall, in refpect both of the head and belly, it being nothing but that part which was covered by the two fhells B B, though it feem'd to grow thicker underneath: And indeed, if we confider the great variety Nature ufes in proportioning the three parts of the body, the *Head, Thorax,* and *Belly*) we fhall not wonder at the fmall proportion of this *Thorax*, nor at the vafter bulk of the belly, for could we exactly anatomife this little Creature, and obferve the particular defigns of each part, we fhould doubtlefs, as we do in all her more manageable

nageable and tractable fabricks, find much more reason to admire the excellency of her contrivance and workmanship, then to wonder, it was not made otherwise.

The head of this little Insect was shap'd somewhat like a Mite's, that is, it had a long snout, in the manner of a Hogs, with a knobbed ridge running along the middle of it, which was bestuck on either side with many small brisles, all pointing forward, and two very large pikes or horns, which rose from the top of the head, just over each eye, and pointed forward also. It had two pretty large black eyes on either side of the head E E, from one of which I could see a very bright reflection of the window, which made me ghess, that the *Cornea* of it was smooth, like those of bigger Insects. Its motion was pretty quick and strong, it being able very easily to tumble a stone or clod four times as big as its whole body.

At the same time and place, and divers times since, I have observed with my *Microscope*, another little Insect, which, though I have not annexed the picture of, may be worth noting, for its exceeding nimbleness as well as smalness; it was as small as a Mite, with a body deep and ridged, almost like a Flea; it had eight blood-red legs, not very long, but slender; and two horns or feelers before. Its motion was so exceeding quick, that I have often lost sight of one I have observed with my naked eye; and though, when it was not frighted, I was able to follow the motions of some with my *Microscope*; yet if it vvere never so little startled, it posted avvay vvith such speed, and turn'd and vvinded it self so quick, that I should presently lose sight of it.

When I first observ'd the former of these Insects, or Mites, I began to conjecture, that certainly I had found out the vagabond Parents of those Mites we find in Cheeses, Meal, Corn, Seeds, musty Barrels, musty Leather, &c. these little Creatures, vvandring to and fro every vvhither, might perhaps, as they vvere invited hither and thither by the musty steams of several putrifying bodies, make their invasions upon those new and pleasing territories, and there spending the remainder of their life, which might be perhaps a day, or thereabouts, in very plentiful and riotous living, might leave their off-spring behind them, which by the change of the soil and Country they now inhabite, might be quite alter'd from the hew of their *primogenitors*, and, like *Mores* tranflated into Northern *European* Climates, after a little time, change both their skin and shape. And this seems yet more probable in these Insects, because that the soil or body they inhabit, seems to be almost half their parent, for it not only hatches and brings those little eggs, or seminal principles, to perfection, but seems to augment and nourish them also before they are hatch'd or shaped; for it is obvious enough to be observ'd, that the eggs of many other Insects, and particularly of Mites, are increas'd in bulk after they are laid out of the bodies of the Insects, and plump'd sometimes into many times their former bigness, so that the bodies they are laid in being, as it were, half their mothers, we shall not wonder that it should have such an active power to change their forms. We find by

relations,

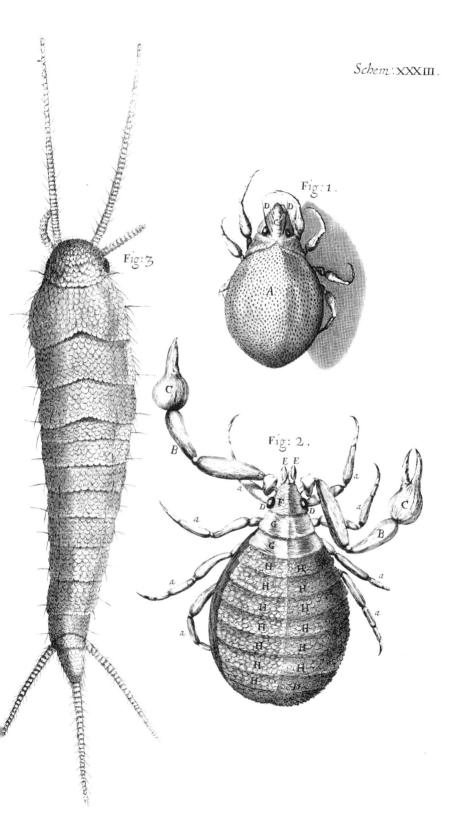

Fig: 3

Fig: 1.

Fig: 2.

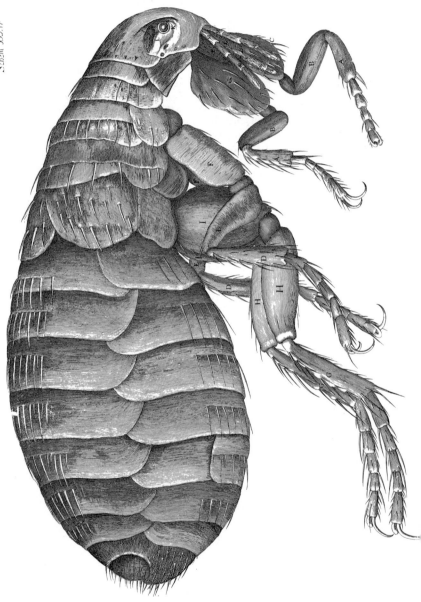

relations how much the *Negro* Women do befmeer the of-fpring of the *Spaniard*, bringing forth neither white-fkinn'd nor black, but tawny hided *Mulattos*.

Now, though I propound this as probable, I have not yet been fo farr certify'd by Obfervations as to conclude any thing, either pofitively or negatively, concerning it. Perhaps, fome more lucky diligence may pleafe the curious Inquirer with the difcovery of this, to be a truth, which I now conjecture, and may thereby give him a fatisfactory account of the caufe of thofe creatures, whofe original feems yet fo obfcure, and may give him caufe to believe, that many other animate beings, that feem alfo to be the mere product of putrifaction, may be innobled with a Pedigree as ancient as the firft creation, and farr exceed the greateft beings in their numerous Genealogies. But on the other fide, if it fhould be found that thefe, or any other animate body, have no immediate fimilar Parent, I have in another place fet down a conjectural *Hypothefis* whereby thofe *Phænomena* may likely enough be folv'd, wherein the infinite wifdom and providence of the Creator is no lefs rare and wonderfull.

Obferv. L I. *Of the* Crab-like *Infect.*

REading one day in *Septemb.* I chanced to obferve a very fmal creature creep over the Book I was reading, very flowly; having a *Microfcope* by me, I obferv'd it to be a creature of a very unufual form, and that not lefs notable; fuch as is defcrib'd in the fecond *Figure* of the 33. *Scheme*. It was about the bignefs of a large Mite, or fomewhat longer, it had ten legs, eight of which, A A A A, were topt with veryfharp claws, and were thofe upon which he walk'd, feeming fhap'd much like thofe of a Crab, which in many other things alfo this little creature refembled; for the two other claws, B B, which were the formoft of all the ten, and feem'd to grow out of his head, like the horns of other Animals, were exactly form'd in the manner of Crabs or Lobfters claws, for they were fhap'd and jointed much like thofe reprefened in the *Scheme* and the ends of them were furnifh'd with a pair of claws or pincers, C C, which this little animal did open and fhut at pleafure : It feem'd to make ufe of thofe two horns or claws both for feelers and holders; for in its motion it carried thefe aloft extended before, moving them to and fro, juft as a man blindfolded would do his hands when he is fearfull of running againft a wall, and if I put a hair to it, it would readily take hold of it with thefe claws, and feem to hold it faft. Now, though thefe horns feem'd to ferve him for two ufes, namely, for feeling and holding; yet he feem'd neither blind, having two fmall black fpots, D D, which by the make of them, and the bright reflection from them feem'd to be his eyes; nor did it want other hands, having another pair of claws, E E, very neer plac'd to its mouth, and feem'd adjoining to it.

The whole body was cafed over with armour-fhells, as is ufuall in all

thofe

thofe kinds of *cruftaceous* creatures, efpecially about their bellies, and feem'd of three kinds; the head F feem'd cover'd with a kind of fcaly fhell, the *thorax* with two fmooth fhells, or Rings, G G, and the belly with eight knobb'd ones. I could not certainly find whether it had under thefe laft fhells any wings, but I fufpect the contrary; for I have not found any wing'd Infect with eight leggs, two of thofe leggs being always converted into wings, and, for the moft part, thofe that have but fix, have wings.

This creature, though I could never meet with more then one of them, and fo could not make fo many examinations of it as otherwife I would, I did notwithftanding, by reafon of the great curiofity that appear'd to me in its fhape, delineate it, to fhew that, in all likelihood, Nature had crouded together into this very minute Infect, as many, and as excellent contrivances, as into the body of a very large Crab, which exceeds it in bulk, perhaps, fome Millions of times; for as to all the apparent parts, there is a greater rather then a lefs multiplicity of parts, each legg has as many parts, and as many joints as a Crabs, nay, and as many hairs or brifles; and the like may be in all the other vifible parts; and 'tis very likely, that the internal curiofities are not lefs excellent: It being a general rule in Nature's proceedings, that where fhe begins to difplay any excellency, if the fubject be further fearch'd into, it will manifeft, that there is not lefs curiofity in thofe parts which our fingle eye cannot reach, then in thofe which are more obvious.

Obferv. L I I. *Of the fmall Silver-colour'd* Book-worm.

AS among greater Animals there are many that are fcaled, both for ornament and defence, fo are there not wanting fuch alfo among the leffer bodies of Infects, whereof this little creature gives us an Inftance. It is a fmall white Silver-fhining Worm or Moth, which I found much converfant among Books and Papers, and is fuppos'd to be that which corrodes and eats holes through the leaves and covers; it appears to the naked eye, a fmall gliftering Pearl-colour'd Moth, which upon the removing of Books and Papers in the Summer, is often obferv'd very nimbly to fcud, and pack away to fome lurking cranney, where it may the better protect it felf from any appearing dangers. Its head appears bigg and blunt, and its body tapers from it towads the tail, fmaller and fmaller, being fhap'd almoft like a Carret.

This the *Microfcopical* appearance will more plainly manifeft, which exhibits, in the third *Figure* of the 33. *Scheme,* a conical body, divided into fourteen feveral partitions, being the appearance of fo many feveral fhels, or fhields that cover the whole body, every of thefe fhells are again cover'd or tiled over with a multitude of thin tranfparent fcales, which, from the multiplicity of their reflecting furfaces, make the whole Animal appear of a perfect Pearl-colour.

Which

Which, by the way, may hint us the reason of that so much admired appearance of those so highly esteem'd bodies, as also of the like in mother of Pearl-shells, and in multitudes of other shelly Sea-substances; for they each of them consisting of an infinite number of very thin shells or laminated orbiculations, cause such multitudes of reflections, that the compositions of them together with the reflections of others that are so thin as to afford colours (of which I elsewhere give the reason) gives a very pleasant reflection of light. And that this is the true cause, seems likely, first, because all those so appearing bodies are compounded of multitudes of plated substances. And next that, by ordering any trasparent substance after this manner, the like *Phænomena* may be produc'd ; this will be made very obvious by the blowing of Glass into exceeding thin shells, and then breaking them into scales, which any lamp-worker will presently do; for a good quantity of these scales, laid in a heap together, have much the same resemblance of Pearls. Another way, not less instructive and pleasant, is a way which I have several times done, which is by working and tossing, as 'twere, a parcel of pure crystalline glass whilst it is kept glowing hot in the blown flame of a Lamp for, by that means, that purely transparent body will be so divided into an infinite number of plates, or small strings, with interpos'd aerial plates and *fibres*, that from the multiplicity of the reflections from each of those internal surfaces, it may be drawn out into curious Pearl-like or Silver wire, which though small, will yet be opacous; the same thing I have done with a composition of red *Colophon* and *Turpentine*, and a little Bee's Wax, and may be done likewise with Birdlime, and such like glutinous and transparent bodies : But to return to our description.

The small blunt head of this Insect was furnish'd on either side of it with a cluster of eyes, each of which seem'd to contain but a very few, in comparison of what I had observ'd the clusters of other Insects to abound with; 'each of these clusters were beset with a row of small brisles, much like the *cilia* or hairs on the eye-lids, and, perhaps, they serv'd for the same purpose. It had two long horns before, which were streight, and tapering towards the top, curiously ring d or knobb'd, and brisled much like the Marsh Weed, call'd Horse-tail, or Cats-tail, having at each knot a fring'd Girdle, as I may so call it, of smaller hairs, and several bigger and larger brisles, here and there dispers'd among them : besides these, it had two shorter horns, or feelers, which were knotted and fring'd, just as the former, but wanted brisles, and were blunt at the ends; the hinder part of the creature was terminated with three tails, in every particular resembling the two longer horns that grew out of the head : The leggs of it were scal'd and hair'd much like the rest, but are not exprefs'd in this *Figure*, the Moth being intangled all in Glew, and so the leggs of this appear'd not through the Glass which looked perpendicularly upon the back.

This Animal probably feeds upon the Paper and covers of Books, and perforates in them several small round holes, finding, perhaps, a convenient nourishment in those husks of Hemp and Flax, which have pass'd through ſo many ſcourings, waſhings, dreſſings and dryings,' as the parts of old Paper muſt neceſſarily have ſuffer'd; the digeſtive faculty, it ſeems, of theſe little creatures being able yet further to work upon thoſe ſtubborn parts, and reduce them into another form.

And indeed, when I conſider what a heap of Saw-duſt or chips this little creature (which is one of the teeth of Time) conveys into its intrals. I cannot chuſe but remember and admire the excellent contrivance of Nature, in placing in Animals ſuch a fire, as is continually nouriſhed and ſupply'd by the materials convey'd into the ſtomach, and *fomented* by the bellows of the lungs; and in ſo contriving the moſt admirable fabrick of Animals, as to make the very ſpending and waſting of that fire, to be inſtrumental to the procuring and collecting more materials to augment and cheriſh it ſelf, which indeed ſeems to be the principal end of all the contrivances obſervable in bruit Animals.

Obſerv. LIII. *Of a* Flea.

THe ſtrength and beauty of this ſmall creature, had it no other relation at all to man, would deſerve a deſcription.

For its ſtrength, the *Microſcope* is able to make no greater diſcoveries of it then the naked eye, but onely the curious contrivance of its leggs and joints, for the exerting that ſtrength, is very plainly manifeſted, ſuch as no other creature, I have yet obſerv'd, has any thing like it; for the joints of it are ſo adapted, that he can, as 'twere, fold them ſhort one within another, and ſuddenly ſtretch, or ſpring them out to their whole length, that is, of the fore-leggs, the part A, of the 34. *Scheme*, lies within B, and B within C, parallel to, or ſide by ſide each other; but the parts of the two next, lie quite contrary, that is, D without E, and E without F, but parallel alſo; but the parts of the hinder leggs, G, H and I, bend one within another, like the parts of a double jointed Ruler, or like the foot, legg and thigh of a man; theſe ſix leggs he clitches up altogether, and when he leaps, ſprings them all out, and thereby exerts his whole ſtrength at once.

But, as for the beauty of it, the *Microſcope* manifeſts it to be all over adorn'd with a curiouſly poliſh'd ſuit of *ſable* Armour, neatly jointed, and beſet with multitudes of ſharp pinns, ſhap'd almoſt like Porcupine's Quills, or bright conical Steel-bodkins; the head is on either ſide beautify'd with a quick and round black eye K, behind each of which alſo appears a ſmall cavity, L, in which he ſeems to move to and fro a certain thin film beſet with many ſmall tranſparent hairs, which probably may be his ears; in the forepart of his head, between the two fore-leggs, he has two ſmall long jointed feelers, or rather ſmellers, M M, which have four joints, and are hairy, like thoſe of ſeveral other creatures; between theſe, it has a ſmall *proboſcis*, or *probe*, N N O, that ſeems to conſiſt of a tube,

tube N N, and a tongue or fucker O, which I have perceiv'd him to flip in and out. Befides thefe, it has alfo two chaps or biters P P, which are fomewhat like thofe of an Ant, but I could not perceive them tooth'd; thefe were fhap'd very like the blades of a pair of round top'd Scizers, and were opened and fhut juft after the fame manner; with thefe Inftruments does this little bufie Creature bite and pierce the fkin, and fuck out the blood of an Animal, leaving the fkin inflamed with a fmall round red fpot. Thefe parts are very difficult to be difcovered, becaufe, for the moft part, they lye covered between the fore-legs. There are many other particulars, which, being more obvious, and affording no great matter of information, I fhall pafs by, and refer the Reader to the Figure.

Obferv. LIV. *Of a Loufe.*

THis is a Creature fo officious, that 'twill be known to every one at one time or other, fo bufie, and fo impudent, that it will be intruding it felf in every ones company, and fo proud and afpiring withall, that it fears not to trample on the beft, and affects nothing fo much as a Crown; feeds and lives very high, and that makes it fo faucy, as to pull any one by the ears that comes in its way, and will never be quiet till it has drawn blood: it is troubled at nothing fo much as at a man that fcratches his head, as knowing that man is plotting and contriving fome mifchief againft it, and that makes it oftentime fculk into fome meaner and lower place, and run behind a mans back, though it go very much againft the hair; which ill conditions of it having made it better known then trufted, would exempt me from making any further defcription of it, did not my faithful *Mercury*, my *Microfcope*, bring me other information of it. For this has difcovered to me, by means of a very bright light caft on it, that it is a Creature of a very odd fhape; it has a head fhap'd like that expreft in 35. *Scheme* marked with A, which feems almoft Conical, but is a little flatted on the upper and under fides, at the biggeft part of which, on either fide behind the head (as it were, being the place where other Creatures ears ftand) are placed its two black fhining goggle eyes B B, looking backwards, and fenced round with feveral fmall *cilia* or hairs that incompafs it, fo that it feems this Creature has no very good forefight: It does not feem to have any eye-lids, and therefore perhaps its eyes were fo placed, that it might the better cleanfe them with its fore-legs; and perhaps this may be the reafon, why they fo much avoid and run from the light behind them, for being made to live in the fhady and dark recefles of the hair, and thence probably their eye having a great aperture, the open and clear light, efpecially that of the Sun, muft needs very much offend them; to fecure thefe eyes from receiving any injury from the hairs through which it pafles, it has

two

two horns that grow before it, in the place where one would have thought the eyes fhould be ; each of thefe C C hath four joynts, which are fringed, as 'twere, with fmall briftles, from which to the tip of its fnout D, the head feems very round and tapering, ending in a very fharp nofe D, which feems to have a fmall hole, and to be the paffage through which he fucks the blood.   Now whereas if it be plac'd on its back, with its belly upwards, as it is in the 35. *Scheme,* it feems in feveral Pofitions to have a refemblance of chaps, or jaws, as is reprefented in the Figure by E E, yet in other poftures thofe dark ftrokes difappear ; and having kept feveral of them in a box for two or three dayes, fo that for all that time they had nothing to feed on, I found, upon letting one creep on my hand, that it immediately fell to fucking, and did neither feem to thruft its nofe very deep into the fkin, nor to open any kind of mouth, but I could plainly perceive a fmall current of blood, which came directly from its fnout, and paft into its belly ; and about A there feem'd a contrivance, fomewhat refembling a Pump, pair of Bellows, or Heart, for by a very fwift *fyftole* and *diaftole* the blood feem'd drawn from the nofe, and forced into the body.   It did not feem at all, though I viewed it a good while as it was fucking, to thruft more of its nofe into the fkin then the very fnout D, nor did it caufe the leaft difcernable pain, and yet the blood feem'd to run through its head very quick and freely, fo that it feems there is no part of the fkin but the blood is difpers'd into, nay, even into the *cuticula* ; for had it thruft its whole nofe in from D to C C, it would not have amounted to the fuppofed thicknefs of that *tegument,* the length of the nofe being not more then a three hundredth part of an inch.   It has fix legs, covered with a very tranfparent fhell, and joynted exactly like a Crab's, or Lobfter's ; each leg is divided into fix parts by thefe joynts, and thofe have here and there feveral fmall hairs ; and at the end of each leg it has two claws, very properly adapted for its peculiar ufe, being thereby inabled to walk very fecurely both on the fkin and hair ; and indeed this contrivance of the feet is very curious, and could not be made more commodioufly and compendioufly, for performing both thefe requifite motions, of walking and climbing up the hair of a mans head, then it is : for, by having the leffer claw (a) fet fo much fhort of the bigger (b) when it walks on the fkin the fhorter touches not, and then the feet are the fame with thofe of a Mite, and feveral other fmall Infects, but by means of the fmall joynts of the longer claw it can bend it round, and fo with both claws take hold of a hair, in the manner reprefented in the Figure, the long tranfparent Cylinder F F F, being a Man's hair held by it.

The *Thorax* feem'd cas'd with another kind of fubftance then the belly, namely, with a thin tranfparent horny fubftance, which upon the fafting of the Creature did not grow flaccid ; through this I could plainly fee the blood, fuck'd from my hand, to be varioufly diftributed, and mov'd to and fro ; and about G there feem'd a pretty big white fubftance, which feem'd to be moved within its *thorax* ; befides, there appear'd very many fmall milk-white veffels, which croft over the breaft

between

between the legs, out of which, on either fide, were many fmall branchings, thefe feem'd to be the veins and arteries, for that which is analogus to blood in all Infects is milk-white.

The belly is covered with a tranfparent fubftance likewife, but more refembling a fkin then a fhell, for 'tis grain'd all over the belly juft like the skin in the palms of a man's hand, and when the belly is empty, grows very flaccid and wrinkled; at the upper end of this is placed the ftomach H H, and perhaps alfo the white fpot I I may be the liver or *pancreas*, which by the *periftaltick* motion of the guts, is a little mov'd to and fro, not with a *fyftole* and *diaftole*, but rather with a thronging or juftling motion. Viewing one of thefe Creatures, after it had fafted two dayes, all the hinder part was lank and flaccid, and the white fpot I I hardly mov'd, moft of the white branchings difappear'd, and moft alfo of the rednefs or fucked blood in the guts, the *periftaltick* motion of which was fcarce difcernable; but upon the fuffering it to fuck, it prefently fill'd the fkin of the belly, and of the fix fcolop'd embofments on either fide, as full as it could be ftuft; the ftomach and guts were as full as they could hold; the *periftaltick* motion of the gut grew quick, and the juftling motion of I I accordingly; multitudes of milk-white veffels feem'd quickly filled, and turgid, which were perhaps the veins and arteries, and the Creature was fo greedy, that though it could not contain more, yet it continued fucking as faft as ever, and as faft emptying it felf behind: the digeftion of this Creature muft needs be very quick, for though I perceiv'd the blood thicker and blacker when fuck'd, yet, when in the guts, it was of a very lovely ruby colour, and that part of it, which was digefted into the veins, feemed white; whence it appears, that a further digeftion of blood may make it milk, at leaft of a refembling colour: What is elfe obfervable in the figure of this Creature, may be feen by the 35. *Scheme*.

Obferv. LV. *Of* Mites.

THe leaft of *Reptiles* I have hitherto met with, is a Mite, a Creature whereof there are fome fo very fmall, that the fharpeft fight, unaffifted with Glaffes, is not able to difcern them, though, being white of themfelves, they move on a black and fmooth furface; and the Eggs, out of which thefe Creatures feem to be hatch'd, are yet fmaller, thofe being ufually not above a four or five hundredth part of a well grown Mite, and thofe well grown Mites not much above one hundredth of an inch in thicknefs; fo that according to this reckoning there may be no lefs then a million of well grown Mites contain'd in a cubick inch, and five hundred times as many Eggs.

Notwithftanding which minutenefs a good *Microfcope* difcovers thofe fmall movable fpecks to be very prettily fhap'd Infects, each of them furnifh'd nifh'd with eight well fhap'd and proportion'd legs, which are each of them joynted or bendable in eight feveral places, or joynts, each of which is covered, for the moft part, with a very tranfparent fhell, and the lower end of the fhell of each joynt is fringed with feveral fmall hairs; the contrivance of the joynts feems the very fame with that of Crabs and Lobfters legs, and like thofe alfo, they are each of them terminated with a very fharp claw or point; four of thefe legs are fo placed, that they feem to draw forwards, the other four are placed in a quite contrary pofition, thereby to keep the body backwards when there is occafion.

Fig. 1.
Schem. 36. The body, as in other larger Infects, confifts of three regions or parts; the hinder or belly A, feems covered with one intire fhell, the middle, or cheft, feems divided into two fhells B C. which running one within the other, the Mite is able to fhrink in and thruft out as it finds occafion, as it can alfo the fnout D. The whole body is pretty tranfparent, fo that being look'd on againft the light, divers motions within its body may be perceived; as alfo all the parts are much more plainly delineable, then in other poftures, to the light. The fhell, efpecially that which covers the back, is curioufly polifht, fo that 'tis eafie to fee, as in a *convex* Looking-glafs, or *foliated* Glafs-ball, the picture of all the objects round about; up and down, in feveral parts of its body, it has feveral fmall long white hairs growing out of its fhell, which are often longer then the whole body, and are reprefented too fhort in the firft and fecond Figures; they feem all pretty ftraight and plyable, fave only two upon the fore-part of its body, which feem to be the horns, as may be feen in the Figures; the firft whereof is a profpect of a fmaller fort of Mites (which are ufually more plump) as it was *paſſant* to and fro; the fecond is the profpect of one fixt on its tail (by means of a little mouthglew rub'd on the object plate) exhibiting the manner of the growing of the legs, together with their feveral joynts.

This Creature is very much diverfify'd in fhape, colour, and divers other properties, according to the nature of the fubftance out of which it feems to be ingendred and nourifhed, being in one fubftance more long, in another more round, in fome more hairy, in others more fmooth, in this nimble, in that flow, here pale and whiter, there browner, blacker, more tranfparent, *&c.* I have obferved it to be refident almoft on all kinds of fubftances that are mouldy, or putrifying, and have feen it very nimbly mefhing through the thickets of mould, and fometimes to lye *dormant* underneath them; and 'tis not unlikely, but that it may feed on that vegetating fubftance, *ſpontaneous Vegetables* feeming a food proper enough for *ſpontaneous Animals,*

But whether indeed this Creature, or any other, be fuch or not, I cannot pofitively, from any Experiment, or Obfervation, I have yet made, determine. But, as I formerly hinted, it feems probable, that fome kind of wandring Mite may fow, as 'twere, the firft feeds, or lay the firft eggs, in thofe places, which Nature has inftructed them to know convenient for the hatching and nourifhing their young; and though perhaps the

prime

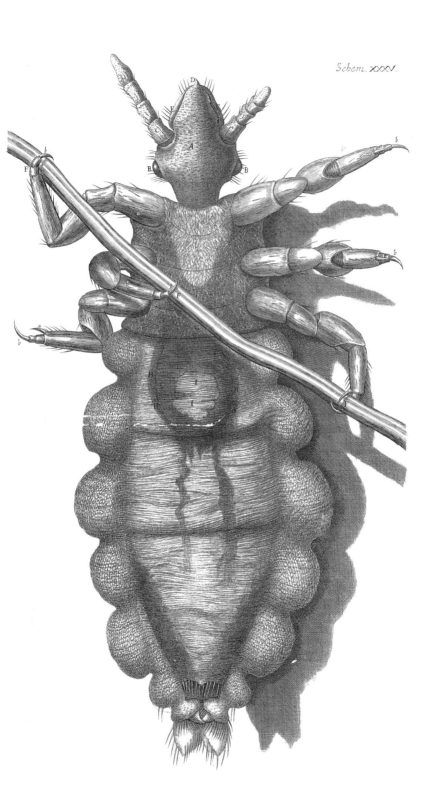

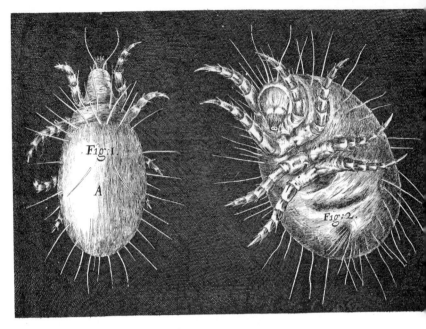

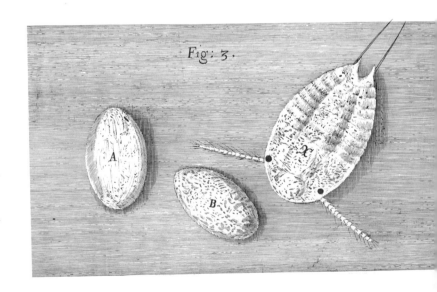

prime Parent might be of a shape very differing from what the off-spring, after a little while, by reason of the substance they feed on, or the Region (as 'twere) they inhabite; yet perhaps even one of these alter'd progeny, wandering again from its native soil, and lighting on by chance the same place from whence its prime Parent came, and there set-tling, and planting, may produce a generation of Mites of the same shapes and properties with the first wandring Mite : And from some such accidents as these, I am very apt to think, the most sorts of Animals, ge-nerally accounted *spontaneous* , have their *origination* , and all those va-rious sorts of Mites, that are to be met with up and down in divers pu-trifying substances, may perhaps be all of the same kind, and have sprung from one and the same sort of Mites at the first.

Observ. LVI. *Of a small Creature hatch'd on a Vine.*

THere is, almost all the Spring and Summer time, a certain small, round, white Cobweb, as 'twere, about the bigness of a Pea, which sticks very close and fast to the stocks of Vines nayl'd against a warm wall : being attentively viewed, they seem cover'd, upon the upper side of them, with a small husk, not unlike the scale, or shell of a Wood-louse, or Hog-louse, a small Insect usually found about rotten wood, which upon touching presently rouls it self into the form of a pepper-corn: Separating several of these from the stock , I found them, with my *Microscope*, to consist of a shell, which now seemed more likely to be the husk of one of these Insects : And the fur seem'd a kind of cobweb, consisting of abundance of small filaments, or sleaves of cobwebs. In the midst of this, if they were not hatch'd, and run away before, the time of which hatching was usually about the latter end of *June*, or begin-ning of *July*, I have often found abundance of small brown Eggs, such as A and B in the second Figure of the 36. *Scheme*, much about the big-ness of Mites Eggs; and at other times, multitudes of small Insects, sha-ped exactly like that in the third Figure marked with X. Its head large, almost half the bigness of its body, which is usual in the *fœtus* of most Creatures. It had two small black eyes *a a*, and two small long joynted and brisled horns *b b*. The hinder part of its body seem'd to consist of nine scales, and the last ended in a forked tayl, much like that of a *Cu-tio*, or Wood-louse, out of which grew two long hairs; they ran to and fro very swiftly, and were much of the bigness of a common Mite, but some of them less : The longest of them seem'd not the hundredth part of an inch, and the Eggs usually not above half as much. They seemed to have six legs, which were not visible in this I have here deline-ated, by reason they were drawn under its body.

If these Minute creatures were *Wood-lice*(as indeed from their own shape and frame, the skin, or shell, that grows on them, one may with great probability ghefs) it affords us an Inftance, whereof perhaps there are not many like in Nature, and that is, of the prodigious increafe of thefe Creatures, after they are hatch'd and run about : for a common Wood-loufe, of about half an inch long, is no lefs then a hundred and twenty five thoufand times bigger then one of thefe, which though indeed it feems very ftrange, yet I have obferved the young ones of fome Spiders have almoft kept the fame proportion to their Dam.

This, methinks, if it be fo, does in the next place hint a Quæry, which may perhaps deferve a little further examination : And that is, Whether there be not many of thofe minute Creatures, fuch as Mites, and the like, which, though they are commonly thought of otherwife, are only the *pully*, or young ones, of much bigger Infects, and not the generating, or parent Infect, that has layd thofe Eggs ; for having many times obferv'd thofe Eggs, which ufually are found in great abundance where Mites are found, it feems fomething ftrange, that fo fmall an Animal fhould have an Egg fo big in proportion to its body. Though on the other fide, I muft confefs, that having kept divers of thofe Mites inclofed in a box for a good while, I did not find them very much augmented beyond their ufual bignefs.

What the husk and cobweb of this little white fubftance fhould be, I cannot imagine, unlefs it be, that the old one, when impregnated with Eggs, fhould there ftay, and fix it felf on the Vine, and dye, and all the body by degrees fhould rot, fave only the husk, and the Eggs in the body : And the heat, or fire, as it were, of the approaching Sun-beams fhould vivifie thofe Relicts of the corrupted Parent, and out of the afhes, as 'twere, (as it is fabled of the *Phœnix*) fhould raife a new offfpring for the perpetuation of the *fpecies*. Nor will the cobweb, as it were, in which thefe Eggs are inclos'd, make much againft this Conjecture ; for we may, by thofe cobwebs that are carried up and down the Air after a Fog (which with my *Microfcope* I have difcovered to be made up of an infinite company of fmall filaments or threads) learn, that fuch a texture of body may be otherwife made then by the fpinning of a Worm.

Obferv. LVII. *Of the* Eels *in Vinegar.*

OF thefe fmall Eels, which are to be found in divers forts of Vinegar, I have little to add befides their Picture, which you may find drawn in the third Figure of the 25. *Scheme*: That is, they were fhaped much like an Eel, fave only that their nofe A, (which was a little more opacous then the reft of their body) was a little fharper, and longer, in proportion to their body, and the wrigling motion of their body feem'd to be onely upwards and downwards, whereas that of Eels is onely fide wayes : They feem'd to have a more opacous part
about

about B, which might, perhaps,be their Gills ; it seeming always the same proportionate distant from their nose, from which, to the tip of their tail, C, their body seem'd to taper.

Taking several of these out of their Pond of Vinegar, by the net of a small piece of filtring Paper, and laying them on a black smooth Glass plate, I found that they could wriggle and winde their body, as much almost as a Snake, which made me doubt, whether they were a kind of Eal or Leech.

I shall add no other observations made on this minute Animal, being prevented herein by many excellent ones already publish'd by the ingenious, Doctor *Power*, among his *Microscopical* Observations, save onely that a quantity of Vinegar repleat with them being included in a small Viol, and stop'd very close from the ambient air, all the included Worms in a very short time died, as if they had been stifled.

And that their motion seems (contrary to what we may observe in the motion of all other Insects) exceeding slow. But the reason of it seems plain, for being to move to and fro after that manner which they do, by waving onely, or wrigling their body ; the tenacity, or glutinousness, and the density or resistance of the fluid *medium* becomes so exceeding sensible to their extremely minute bodies,that it is to me indeed a greater wonder that they move them so fast as they do,then that they move them no faster.. For what a vastly greater proportion have they of their superficies to their bulk, then Eels or other larger Fishes, and next, the tenacity and density of the liquor being much the same to be moved,both by the one and the other, the resistance or impediment thence arising to the motions made through it, must be almost infinitely greater to the small one then to the great. This we find experimentally verify'd in the Air, which though a *medium* a thousand times more rarify'd then the water,the resistance of it to motions made through it,is yet so sensible to very minute bodies,that a Down-feather(the least of whose parts seem yet bigger then these Eels, and many of them almost incomparably bigger, such as the quill and stalk) is suspended by it, and carried to and fro as if it had no weight.

Observ. LVIII. *Of a new Property in the* Air, *and several other transparent* Mediums *nam'd* Inflection, *whereby very many considerable* Phænomena *are attempted to be solv'd, and divers other uses are hinted.*

SInce the Invention (and perfecting in some measure) of *Telescopes*, it has been observ'd by several, that the Sun and Moon neer the Horizon, are disfigur'd (losing that exactly-smooth terminating circular limb, which they are observ'd to have when situated neerer the Zenith) and are bounded with an edge every way (especially upon the right and left

sides)

fides) ragged and indented like a Saw : which inequality of their limbs, I have further obferv'd, not to remain always the fame, but to be continually chang'd by a kind of fluctuating motion, not unlike that of the waves of the Sea ; fo as that part of the limb, which was but even now nick'd or indented in, is now protuberant, and will prefently be finking again ; neither is this all but the whole body of the Luminaries, do in the *Telefcope*, feem to be deprefs'd and flatted, the upper, and more efpecially the under fide appearing neerer to the middle then really they are, and the right and left appearing more remote: whence the whole *Area* feems to be terminated by a kind of Oval. It is further obferv'd, that the body, for the moft part, appears red, or of fome colour approaching neer unto it, as fome kind of yellow ; and this I have always mark'd, that the more the limb is flatted or ovalled, the more red does the body appear, though not always the contrary.   It is further obfervable, that both fix'd Stars and Planets, the neerer they appear to the Horizon, the more red and dull they look, and the more they are obferv'd to twinkle ; in fo much, that I have feen the Dog-ftarr to vibrate fo ftrong and bright a radiation of light, as almoft to dazle my eyes, and prefently, almoft to difappear. It is alfo obfervable, that thofe bright fcintillations neer the Horizon, are not by much fo quick and fudden in their confecutions of one another, as the nimbler twinklings of Stars neerer the Zenith. This is alfo notable, that the Starrs neer the Horizon, are twinkled with feveral colours ; fo as fometimes to appear red, fometimes more yellow, and fometimes blue, and this when the Starr is a pretty way elevated above the Horizon.   I have further, very often feen fome of the fmall Starrs of the fifth or fixth magnitude, at certain times to difappear for a fmall moment of time, and again appear more confpicuous, and with a greater lufter. I have feveral times, with my naked eye, feen many fmaller Starrs, fuch as may be call'd of the feventh or eighth magnitude to appear for a fhort fpace, and then vanifh, which, by directing a fmall *Telefcope* towards that part they appear'd and difappeard in ; I could prefently find to be indeed fmall Starrs fo fituate, as I had feen them with my naked eye, and to appear twinkling like the ordinary vifible Stars ; nay, in examining fome very notable parts of the Heaven, with a three foot Tube , me thought I now and then, in feveral parts of the conftellation, could perceive little twinklings of Starrs, making a very fhort kind of apparition, and prefently vanifhing, but noting diligently the places where they thus feem'd to play at boe-peep, I made ufe of a very good twelve foot Tube, and with that it was not uneafie to fee thofe, and feveral other degrees of fmaller Starrs, and fome fmaller yet, that feem'd again to appear and difappear, and thefe alfo by giving the fame Object-glafs a much bigger aperture, I could plainly and conftantly fee appear in their former places ; fo that I have obferv'd fome twelve feveral magnitudes of Starrs lefs then thofe of the fix magnitudes commonly recounted in the Globes.

    It has been obferv'd and confirm'd by the accurateft Obfervations of the beft of our modern Aftronomers, that all the Luminous bodies appear above the Horizon, when they really are below it. So that the
                                                                        Sun

Sun and Moon have both been seen above the Horizon, whil'st the Moon has been in an Eclipse. I shall not here instance in the great refractions, that the tops of high mountains, seen at a distance, have been found to have; all which seem to argue the Horizontal refraction, much greater then it is hitherto generally believ'd.

I have further taken notice, that not onely the Sun, Moon and Starrs, and high tops of mountains have suffer'd these kinds of refraction, but Trees, and several bright Objects on the ground: I have often taken notice of the twinkling of the reflections of the Sun from a Glass-window at a good distance, and of a Candle in the night, but that is not so conspicuous and in observing the setting Sun, I have often taken notice of the tremulation of the Trees and Bushes, as well as of the edges of the Sun. Divers of these *Phænomena* have been taken notice of by several, who have given several reasons of them, but I have not yet met with any altogether satisfactory, though some of their conjectures have been partly true, but parly also false. Setting my self therfore upon the inquiry of these *Phænomena*, I first endeavour'd to be very diligent in taking notice of the several particulars and circumstances observable in them; and next, in making divers particular Experiments, that might cleer some doubts, and serve to determine, confirm, and illustrate the true and adæquate cause of each; and upon the whole, I find much reason to think, that the true cause of all these *Phænomena* is from the *inflection*, or *multiplicate refraction* of those Rays of light within the body of the *Atmosphere*, and that it does not proceed from a *refraction* caus'd by any terminating *superficies* of the Air above, nor from any such exactly defin d *superficies* within the body of the *Atmosphere*.

This Conclusion is grounded upon these two Propositions:

First, that a *medium*, whose parts are unequally *dense*, and mov'd by various motions and transpositions as to one another, will produce all these visible effects upon the Rays of light, without any other *coefficient* cause.

Secondly, that there is in the Air or *Atmosphere*, such a variety in the constituent parts of it, both as to their *density* and *rarity*, and as to their divers mutations and positions one to another.

By *Density* and *Rarity*, I understand a property of a transparent body, that does either more or less refract a Ray of light (coming obliquely upon its superficies out of a third *medium*) toward its perpendicular: As I call Glass a more dense body then Water, and Water a more rare body then Glass, because of the refractions (more or less deflecting towards the perpendicular) that are made in them, of a Ray of light out of the Air that has the same inclination upon either of their superficies.

So as to the business of Refraction, spirit of Wine is a more *dense* body then Water, it having been found by an accurate Instrument that measures the angles of Refractions to Minutes that for the same refracted angle of 30:00′ in both those *Mediums*, the angle of incidence in Water was but 41°. 3′5. but the angle of the incidence in the trial with spirit of Wine was 42°: 45′. But as to gravity, Water is a more *dense* body then

spirit

ſpirit of Wine, for the proportion of the ſame Water, to the ſame very well rectify'd ſpirit of Wine was, as 21. to 19.

So as to Refraction, Water is more Denſe then Ice; for I have found by a moſt certain Experiment, which I exhibited before divers illuſtrious Perſons of the *Royal Society*, that the Refraction of Water was greater then that of Ice, though ſome conſiderable Authors have affirm'd the contrary, and though the Ice be a very hard, and the Water a very fluid body.

That the former of the two preceding Propoſitions is true, may be manifeſted by ſeveral Experiments: As firſt, if you take any two liquors differing from one another in denſity, but yet ſuch as will readily mix: as Salt Water, or Brine, & Freſh; almoſt any kind of Salt diſſolv'd in Water, and filtrated, ſo that it be cleer, ſpirit of Wine and Water; nay, ſpirit of Wine, and ſpirit of Wine, one more highly rectify'd then the other, and very many other liquors; if (I ſay) you take any two of theſe liquors, and mixing them in a Glaſs Viol, againſt one ſide of which you have fix'd or glued a ſmall round piece of Paper, and ſhaking them well together (ſo that the parts of them may be ſomewhat diſturb'd and move up and down) you endeavour to ſee that round piece of Paper through the body of the liquors; you ſhall plainly perceive the Figure to wave, and to be indented much after the ſame manner as the limb of the Sun through a *Teleſcope* ſeems to be, ſave onely that the mutations here, are much quicker. And if, in ſteed of this bigger Circle, you take a very ſmall ſpot, and faſten and view it as the former, you will find it to appear much like the twinkling of the Starrs, though much quicker: which two *Phænomena* (for I ſhall take notice of no more at preſent, though I could inſtance in multitudes of others) muſt neceſſarily be caus d by an *inflection* of the Rays within the terminating ſuperficies of the compounded *medium*, ſince the ſurfaces of the tranſparent body through which the Rays paſs to the eye, are not at all altered or chang'd.

This *inflection* (if I may ſo call it) I imagine to be nothing elſe, but a *multiplicate refraction*, cauſed by the unequal *denſity* of the conſtituent parts of the *medium*, whereby the motion, action or progreſs of the Ray of light is hindred from proceeding in a ſtreight line, and *inflected* or *deflected* by a *curve*. Now, that it is a *curve* line is manifeſt by this Experiment: I took a Box, ſuch as A D G E, in the firſt *Figure* of the 37. *Scheme*, whoſe ſides A B C D, and E F G H, were made of two ſmooth flat plates of Glaſs, then filling it half full with a very ſtrong ſolution of Salt, I filled the other half with very fair freſh water, then expoſing the opacous ſide, D H G C, to the Sun, I obſerv'd both the *refraction* and *inflection* of the Sun beams, I D & K H, and marking as exactly as I could, the points, P, N, O, M, by which the Ray, K H, paſſed through the compounded *medium*, I found them to be in a *curve* line; for the parts of the *medium* being continually more denſe the neerer they were to the bottom, the Ray *p f* was continually more and more deflected downwards from the ſtreight line.

This Inflection may be mechanically explained, either by Monſieur

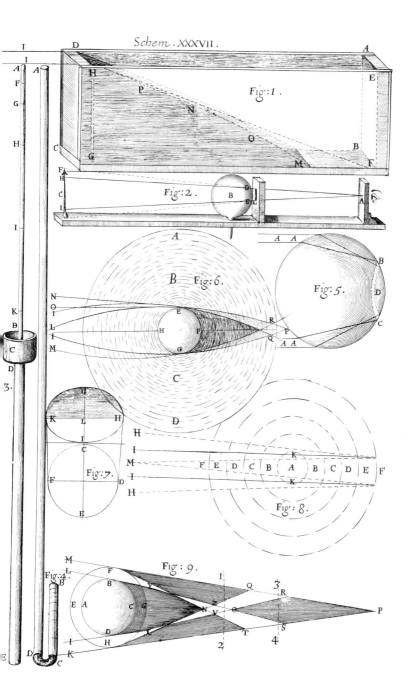

Schem. XXXVII.

Fig: 1.

Fig: 2.

Fig: 5.

Fig: 6.

Fig: 7.

Fig: 8.

Fig: 9.

Fig: 4.

Des Cartes principles, by conceiving the Globuls of the third Element to find lefs and lefs refiftance againft that fide of them which is downwards, or by a way, which I have further explicated in the Inquifition about Colours, to be from an obliquation of the pulfe of light, whence the ruder part is continually promoted, and confequently refracted towards the perpendicular, which cuts the Orbs at right angles. What the particular Figure of the *Curve line*, defcrib'd by this way of light, is, I fhall not now ftand to examine, efpecially fince there may be fo many forts of it as there may be varieties of the Pofitions of the *intermediat* degrees of *denfity* and *rarity* between the bottom and the top of the inflecting Medium.

I could produce many more Examples and Experiments, to illuftrate and prove this firft Propofition, *viz.* that there is fuch a conftitution of fome bodies as will caufe inflection. As not to mention thofe I have obferv'd in *Horn, Tortoife-fhell, tranfparent Gums,* and *refinous Subftances :* The *veins* of Glafs, nay, of melted *Cryftal,* found, and much complained of by Glafs-grinders, and others, might fufficiently demonftrate the truth of it to any diligent Obfervator.

But that, I prefume, I have by this Example given proof fufficient (*viz. ocular demonftration*) to evince, that there is fuch a modulation, or bending of the rayes of light, as I have call'd *inflection*, differing both from *reflection,* and *refraction* (fince they are both made in the fuperficies, this only in the middle) ; and likewife, that this is able or fufficient to produce the effects I have afcribed to it.

It remains therefore to fhew, that there is fuch a property in the Air, and that it is fufficient to produce all the above mentioned *Phænomena,* and therefore may be the principal, if not the only caufe of them.

Firft, That there is fuch a property, may be proved from this, that the parts of the Air are fome of them more condens'd, others more rarified, either by the differing heat, or differing preffure it fuftains, or by the fomewhat heterogeneous vapours interfpers'd through it. For as the Air is more or lefs rarified, fo does it more or lefs refract a ray of light (that comes out of a denfer medium) from the perpendicular. This you may find true, if you make tryal of this Experiment.

Take a fmall Glafs-bubble, made in the form of that in the fecond Figure of the 37. *Scheme,* and by heating the Glafs very hot, and thereby very much rarifying the included Air, or, which is better, by rarifying a fmall quantity of water, included in it, into vapours, which will expel the moft part, if not all the Air, and then fealing up the fmall neck of it, and letting it cool, you may find, if you place it in a convenient Inftrument, that there will be a manifeft difference, as to the refraction.

As if in this fecond Figure you fuppofe A to reprefent a fmall fight or hole, through which the eye looks upon an object, as C, through the Glafs-bubble B, and the fecond fight L ; all which remain exactly fixt in their feveral places, the object C being fo cized and placed, that it may juft feem to touch the upper and under edge of the hole L : and fo all of it be feen through the fmall Glafs-ball of rarified Air ; then by

breaking

breaking off the fmall feal'd neck of the Bubble (without at all ftirring the fights, objeƈt, or glafs) and admitting the external Air, you will find your felf unable to fee the utmoft ends of the objeƈt; but the terminating rayes A E and A D (which were before refraƈted to G and F by the rarified Air) will proceed almoft direƈtly to I and H; which alteration of the rayes (feeing there is no other alteration made in the Organ by which the Experiment is tryed, fave only the admiffion, or exclufion of the condens'd Air) muft neceffarily be caufed by the variation of the *medium* contain'd in the Glafs B; the greateft difficulty in the making of which Experiment, is from the uneven furfaces of the bubble, which will reprefent an uneven image of the objeƈt.

Now, that there is fuch a difference of the upper and under parts of the Air, is clear enough evinc'd from the late improvement of the *Torricellian* Experiment, which has been tryed at the tops and feet of Mountains; and may be further illuftrated , and inquired into, by a means, which fome whiles fince I thought of, and us'd, for the finding by what degrees the Air paffes from fuch a degree of Denfity to fuch a degree of Rarity. And another, for the finding what preffure was requifite to make it pafs from fuch a degree of Rarefaƈtion to a determinate Denfity: Which Experiments, becaufe they may be ufeful to illuftrate the prefent Inquiry, I fhall briefly defcribe.

Fig. 3. I took then a fmall Glafs-pipe A B, about the bignefs of a Swans quill, and about four foot long, which was very equally drawn, fo that, as far as I could perceive, no one part was bigger then another: This Tube (being open at both ends) I fitted into another fmall Tube D E, that had a fmall bore juft big enough to contain the fmall Pipe, and this was feal'd up at one, and open at the other, end; about which open end I faftned a fmall wooden box C with cement, fo that filling the bigger Tube, and part of the box, with Quickfilver, I could thruft the fmaller Tube into it, till it were all covered with the Quickfilver: Having thus done, I faftned my bigger Tube againft the fide of a wall , that it might ftand the fteadier , and plunging the fmall Tube cleer under the *Mercury* in the box, I ftopt the upper end of it very faft with cement, then lifting up the fmall Tube, I drew it up by a fmall pully, and a ftring that I had faftned to the top of the Room, and found the height of the *Mercurial Cylinder* to be about twenty nine inches.

Then letting down the Tube again, I opened the top, and then thruft down the fmall Tube, till I perceived the Quickfilver to rife within it to a mark that I had plac'd juft an inch from the top; and immediately clapping on a fmall peice of cement that I had kept warm, I with a hot Iron feal'd up the top very faft, then letting it cool (that both the cement might grow hard , and more efpecially , that the Air might come to its temper, natural for the Day I try'd the Experiment in) I obferv'd diligently, and found the included Air to be exaƈtly an Inch.

Here you are to take notice, that after the Air is feal'd up, the top of the Tube is not to be elevated above the fuperficies of the Quickfilver

in

in the box, till the furface of that within the Tube be equal to it, for the Quickfilver (as I have elfewhere prov'd) being more heterogeneous to the Glafs then the Air, will not naturally rife up fo high within the fmall Pipe, as the fuperficies of the *Mercury* in the box ; and therefore you are to obferve, how much below the outward fuperficies of the *Mercury* in the box, that of the fame in the Tube does ftand, when the top being open, free ingrefs is admitted to the outward Air.

Having thus done, I permitted the *Cylinder*, or fmall Pipe, to rife out of the box, till I found the furface of the Quickfilver in the Pipe to be two inches above that in the box, and found the Air to have expanded it felf but one fixteenth part of an inch ; then drawing up the fmall pipe, till I found the height of the Quickfilver within to be four inches above that without, I obferved the Air to be expanded only $\frac{1}{7}$ of an inch more then it was at firft, and to take up the room of $1\frac{1}{7}$ inch : then I raifed the Tube till the Cylinder was fix inches high, and found the Air to take up $1\frac{2}{9}$ inches of room in the Pipe ; then to 8, 10, 12. *&c.* the expanfion of the Air that I found to each of which Cylinders are fet down in the following Table ; where the firft row fignifies the height of the *Mercurial Cylinder* ; the next, the expanfion of the Air ; the third, the preffure of the *Atmofphere*, or the higheft *Cylinder* of *Mercury*, which was then neer thirty inches : The laft fignifies the force of the Air fo expanded, which is found by fubftracting the firft row of numbers out of the third ; for having found, that the outward Air would then keep up the Quickfilver to thirty inches, look whatever of that height is wanting muft be attributed to the Elater of the Air depreffing. And therefore having the Expanfion in the fecond row, and the height of the fubjacent *Cylinder* of *Mercury* in the firft, and the greateft height of the *Cylinder* of *Mercury*, which of it felf counterballances the whole preffure of the *Atmofphere* ; by fubftracting the numbers of the firft row out of the numbers of the third, you will have the meafure of the *Cylinders* fo depreft, and confequently the force of the Air, in the feveral Expanfions, regiftred.

The

The height of the Cylinder of Mercury, that, together with the Elater of the included Air, ballanced the pressure of the Atmosphere.	The Expansion of the Air.	The height of the Mercury that counter-ballanc'd the Atmosphere	The strength of the Elater of the expanded Air.
00	01	30	30
02	$01\frac{1}{16}$	30	28
04	$01\frac{1}{7}$	30	26
06	$01\frac{2}{9}$	30	24
08	$01\frac{1}{3}$	30	22
10	$01\frac{1}{12}$	30	20
12	$01\frac{2}{3}$	30	18
14	$01\frac{5}{6}$	30	16
16	$02\frac{2}{27}$	30	14
18	$02\frac{4}{9}$	30	12
20	03	30	10
22	$03\frac{7}{9}$	30	8
24	$05\frac{7}{13}$	30	6
25	$06\frac{2}{3}$	30	5
26	$08\frac{1}{2}$	30	4
$26\frac{1}{4}$	$09\frac{1}{2}$	30	$3\frac{3}{4}$
$26\frac{1}{2}$	$10\frac{3}{4}$	30	$3\frac{1}{2}$
$26\frac{3}{4}$	13	30	$3\frac{1}{4}$
27	$15\frac{1}{2}$	30	3

I had

I had feveral other Tables of my Obfervations, and Calculations, which I then made; but it being above a twelve month fince I made them; and by that means having forgot many circumftances and particulars, I was refolved to make them over once again, which I did *Auguft* the fecond 1661. with the very fame Tube which I ufed the year before, when I firft made the Experiment (for it being a very good one, I had carefully preferv'd it:) And after having tryed it over and over again; and being not well fatisfied of fome particulars, I, at laft, having put all things in very good order, and being as attentive, and obfervant, as poffibly I could, of every circumftance requifite to be taken notice of, did regifter my feveral Obfervations in this following Table. In the making of which, I did not exactly follow the method that I had ufed at firft; but, having lately heard of Mr. *Townly's* Hypothefis, I fhap'd my courfe in fuch fort, as would be moft convenient for the examination of that *Hypothefis*; the event of which you have in the latter part of the laft Table.

The other Experiment was, to find what degrees of force were requifite to comprefs, or condenfe, the Air into fuch or fuch a bulk.

The manner of proceeding therein was this : I took a Tube about five foot long, one of whofe ends was fealed up, and bended in the form of a *Syphon*, much like that reprefented in the fourth Figure of the 37. *Scheme*, one fide whereof A D, that was open at A, was about fifty inches long, the other fide B C, fhut at B, was not much above feven inches long; then placing it exactly perpendicular, I pour'd in a little Quickfilver, and found that the Air B C was $6\frac{7}{8}$ inches, or very near to feven; then pouring in Quickfilver at the longer Tube, I continued filling of it till the Air in the fhorter part of it was contracted into half the former dimenfions, and found the height exactly nine and twenty inches; and by making feveral other tryals, in feveral other degrees of condenfation of the Air, I found them exactly anfwer the former *Hypothefis*.

But having (by reafon it was a good while fince I firft made) forgotten many particulars, and being much unfatisfied in others, I made the Experiment over again, and, from the feveral tryals, collected the former part of the following Table : Where in the row next the left hand 24. fignifies the dimenfions of the Air, fuftaining only the preffure of the *Atmofphere*, which at that time was equal to a *Cylinder* of *Mercury* of nine and twenty inches: The next Figure above it (20) was the dimenfions of the Air induring the firft compreffion, made by a *Cylinder* of *Mercury* $5\frac{4}{16}$ high, to which the preffure of the *Atmofphere* nine and twenty inches being added, the elaftick ftrength of the Air fo compreft will be found $34\frac{4}{16}$, &c.

H h   A

A Table of the Elaſtick power of the Air,
both *Experimentally and Hypothetically calculated,*
according to its various *Dimenſions.*

The dimenſions of the included Air.	The height of the *Mercurial Cylinder* counterpois'd by the *Atmoſphere.*	The *Mercurial Cylinder* added, or taken from the former.	The ſum or difference of theſe two *Cylinders.*	What they ought to be according to the *Hypotheſis.*
12	29 †	29 $=$	58	58
13	29 †	$24\frac{11}{6}=$	$53\frac{11}{16}$	$53\frac{7}{13}$
14	29 †	$20\frac{3}{16}=$	$49\frac{3}{16}$	$49\frac{5}{7}$
16	29 †	$14=$	43	$43\frac{1}{2}$
18	29 †	$9\frac{1}{8}=$	$38\frac{1}{8}$	$38\frac{2}{3}$
20	29 †	$5\frac{3}{16}=$	$34\frac{3}{16}$	$34\frac{4}{5}$
24	29	$0=$	29	29
48	29—	$14\frac{5}{8}=$	$14\frac{3}{8}$	$14\frac{1}{2}$
96	29—	$22\frac{1}{8}=$	$6\frac{7}{8}$	$7\frac{1}{8}$
192	20—	$25\frac{5}{8}=$	$3\frac{3}{8}$	$3\frac{5}{8}$
384	29—	$27\frac{2}{8}=$	$1\frac{6}{8}$	$1\frac{7}{16}$
576	29—	$27\frac{7}{8}=$	$1\frac{1}{8}$	$1\frac{5}{24}$
768	29—	$28\frac{1}{8}=$	$0\frac{7}{8}$	$0\frac{7}{4}{8}$
960	29—	$28\frac{3}{8}=$	$0\frac{5}{8}$	$0\frac{4}{5}{8}$
1152	29—	$28\frac{7}{16}=$	$0\frac{9}{16}$	$0\frac{10}{16}$

From

From which Experiments, I think, we may fafely conclude, that the Elater of the Air is reciprocal to its extenfion, or at leaft very neer. So that to apply it to our prefent purpofe (which was indeed the chief caufe of inventing thefe wayes of tryal) we will fuppofe a *Cylinder* indefinitely extended upwards, [I fay a *Cylinder*, not a piece of a *Cone*, becaufe, as I may elfewhere fhew in the Explication of Gravity, that *triplicate* proportion of the fhels of a Sphere, to their refpective diameters, I fuppofe to be removed in this cafe by the decreafe of the power of Gravity] and the preffure of the Air at the bottom of this *Cylinder* to be ftrong enough to keep up a *Cylinder* of *Mercury* of thirty inches: Now becaufe by the moft accurate tryals of the moft illuftrious and incomparable Mr. *Boyle*, publifhed in his defervedly famous Pneumatick Book, the weight of Quickfilver, to that of the Air here below, is found neer about as fourteen thoufand to one: If we fuppofe the parts of the *Cylinder* of the *Atmofphere* to be every where of an equal denfity, we fhall (as he there deduces) find it extended to the height of thirty five thoufand feet, or feven miles: But becaufe by thefe Experiments we have fomewhat confirm'd the hypothefis of the reciprocal proportion of the Elaters to the Extenfions we fhall find, that by fuppofing this *Cylinder* of the *Atmofphere* divided into a thoufand parts, each of which being equivalent to thirty five feet, or feven geometrical paces, that is, each of thefe divifions containing as much Air as is fuppos'd in a *Cylinder* neer the earth of equal diameter, and thirty five foot high, we fhall find the lowermoft to prefs againft the furface of the Earth with the whole weight of the above mentioned thoufand parts; the preffure of the bottom of the fecond againft the top of the firft to be $1000 - 1 = 999$. of the third againft the fecond to be $1000 - 2 = 998$. of the fourth againft the third to be $1000 - 3 = 997$. of the uppermoft againft the 999. or that next below it, to be $1000 - 999 = 1$. fo that the extenfion of the lowermoft next the Earth, will be to the extenfion of the next below the uppermoft, as 1. to 999. for as the preffure fuftained by the 999. is to the preffure fuftain'd by the firft, fo is the extenfion of the firft to the extenfion of the 999. fo that, from this hypothetical calculation, we fhall find the Air to be indefinitely extended: For if we fuppofe the whole thicknefs of the Air to be divided, as I juft now inftanced, into a thoufand parts, and each of thofe under differing Dimenfions, or Altitudes, to contain an equall quantity of Air, we fhall find, that the firft *Cylinder*, whofe Bafe is fuppofed to lean on the Earth, will be found to be extended $35\frac{35}{999}$ foot; the fecond equal Divifion, or *Cylinder*, whofe *bafis* is fuppofed to lean on the top of the firft, fhall have its top extended higher by $35\frac{70}{998}$; the third $35\frac{105}{997}$; the fourth $35\frac{140}{996}$; and fo onward, each equal quantity of Air having its dimenfions meafured by 35. and fome additional number expreft alwayes in the manner of a fraction, whofe numerator is alway the number of the place multipli'd by 35. and whofe denominator is alwayes the preffure of the *Atmofphere* fuftain'd by that part, fo that by this means we may eafily calculate the height of 999. divifions of thofe 1000. divifions, I fuppos'd; whereas the uppermoft

may

may extend it felf more then as high again, nay, perhaps indefinitely, or beyond the Moon ; for the Elaters and Expanfions being in reciprocal proportions, fince we cannot yet find the *plus ultra*, beyond which the Air will not expand it felf, we cannot determine the height of the Air : for fince, as we have fhewn, the proportion will be alway as the preffure fuftain'd by any part is to 35. fo 1000. to the expanfion of that part; the multiplication or product therefore of the preffure, and expanfion, that is, of the two extream proportionals, being alwayes equal to the product of the means, or 35000. it follows, fince that Rectangle or Product may be made up of the multiplication of infinite diverfities of numbers, that the height of the Air is alfo indefinite ; for fince (as far as I have yet been able to try) the Air feems capable of an indefinite Expanfion, the preffure may be decreafed in *infinitum*, and confequently its expanfion upwards indefinite alfo.

There being therefore fuch a difference of denfity, and no Experiment yet known to prove a *Saltus*, or fkipping from one degree of rarity to another much differing from it, that is, that an upper part of the Air fhould fo much differ from that immediately *fubjacent* to it, as to make a diftinct fuperficies, fuch as we obferve between the Air and Water, &c. But it being more likely, that there is a continual increafe of rarity in the parts of the Air, the further they are removed from the furface of the Earth : It will hence neceffarily follow, that (as in the Experiment of the falt and frefh Water) the ray of Light paffing obliquely through the Air alfo, which is of very different denfity, will be continually, and infinitely inflected, or bended, from a ftreight, or direct motion.

This granted, the reafon of all the above recited *Phænomena*, concerning the appearance of the Celeftial Bodies, will very eafily be deduced. As,

Firft, The rednefs of the Sun, Moon, and Stars, will be found to be caufed by the inflection of the rays within the *Atmofphere*. That it is not really in or near the luminous bodies, will, I fuppofe, be very eafily granted, feeing that this rednefs is obfervable in feveral places differing in Longitude, to be at the fame time different, the fetting and rifing Sun of all parts being for the moft part red :

And fecondly, That it is not meerly the colour of the Air interpos'd, will, I fuppofe, without much more difficulty be yielded, feeing that we may obferve a very great *interftitium* of Air betwixt the Object and the Eye, makes it appear of a dead blew, far enough differing from a red, or yellow.

But thirdly, That it proceeds from the refraction, or inflection, of the rays by the *Atmofphere*, this following Experiment will, I fuppofe, fufficiently manifeft.

Take a fphærical Cryftalline Viol, fuch as is defcrib'd in the fifth Figure A B C D, and, having fill'd it with pure clear Water, expofe it to the Sun beams; then taking a piece of very fine *Venice* Paper, apply it againft that fide of the Globe that is oppofite to the Sun, as againft the fide

fide B C, and you fhall perceive a bright red Ring to appear, caus'd by the refraction of the Rays, A A A A, which is made by the Globe; in which Experiment, if the Glafs and Water be very cleer, fo that there be no Sands nor bubbles in the Glafs, nor dirt in the Water, you fhall not perceive any appearance of any other colour. To apply which Experiment, we may imagine the *Atmofphere* to be a great tranfparent Globe, which being of a fubftance more denfe then the other, or (which comes to the fame) that has its parts more denfe towards the middle, the Sun beams that are tangents, or next within the tangents of this Globe, will be refracted or inflected from their direct paffage towards the center of the Globe, whence, according to the laws of refractions made in a triangular *Prifm*, and the generation of colour fet down in the defcription of Mufcovi-glafs, there muft neceffarily appear a red colour in the *tranfitus* or paffage of thofe tangent Rays. To make this more plain, we will fuppofe (in the fixth *Figure*) A B C D, to reprefent the Globe of the *Atmofphere*, E F G H to reprefent the opacous Globe of the Earth, lying in the midft of it, neer to which, the parts of the Air, fuftaining a very great preffure, are thereby very much condens'd, from whence thofe Rays that are by inflection made tangents to the Globe of the Earth, and thofe without them, that pafs through the more condens'd part of the *Atmofphere*, as fuppofe between A and E, are by reafon of the inequality of the *medium*, inflected towards the center, whereby there muft neceffarily be generated a red colour, as is more plainly fhewn in the former cited place; hence whatfoever opacous bodies (as vapours, or the like) fhall chance to be elevated into thofe parts, will reflect a red towards the eye; and therefore thofe evenings and mornings appear reddeft, that have the moft ftore of vapours and halituous fubftances exhaled to a convenient diftance from the Earth; for thereby the inflection is made the greater, and thereby the colour alfo the more intenfe; and feveral of thofe exhalations being opacous, reflect feveral of thofe Rays, which, through an *Homogeneous* tranfparent *medium* would pafs unfeen; and therefore we fee, that when there chances to be any clouds fituated in thofe Regions they reflect a ftrong and vivid red. Now, though one great caufe of the rednefs may be this inflection, yet I cannot wholly exclude the colour of the vapours themfelves, which may have fomething of rednefs in them, they being partly nitrous, and partly fuliginous; both which fteams tinge the Rays that pafs through them, as is made evident by looking at bodies through the fumes of *Aqua fortis*, or fpirit of *Nitre* [as the newly mentioned Illuftrious Perfon has demonftrated] and alfo through the fmoak of a Fire or Chimney.

Having therefore made it probable at leaft, that the morning and evening rednefs may partly proceed from this inflection or refraction of the Rays, we fhall next fhew, how the Oval Figure will be likewife eafily deduced.

Suppofe we therefore, E F G H in the fixth *Figure* of the 37. *Scheme*, to reprefent the Earth; A B C D, the *Atmofpere*; E I, and E L, two Rays coming from the Sun, the one from the upper, the other from the neather Limb.

Limb, thefe Rays, being by the *Atmofphere* inflected, appear to the eye at E, as if they had come from the points, N and O; and becaufe the Ray L has a greater inclination upon the inequality of the *Atmofphere* then I, therefore muft it fuffer a greater inflection, and confequently be further elevated above its true place, then the Ray I, which has a lefs inclination, will be elevated above its true place; whence it will follow, that the lower fide appearing neerer the upper then really it is, and the two *lateral* fides, viz. the right and left fide, fuffering no fenfible alteration from the inflection, at leaft what it does fuffer, does rather increafe the vifible Diameter then diminifh it, as I fhall fhew by and by, the Figure of the luminous body muft neceffarily appear fomewhat *Elliptical.*

This will be more plain, if in the feventh *Figure* of th 37. *Scheme* we fuppofe A B to reprefent the fenfible Horizon; C D E F, the body of the Sun really below it; G H I K, the fame appearing above it, elevated by the inflection of the *Atmofphere* : For if, according to the beft obfervation, we make the vifible Diameter of the Sun to be about three or four and thirty minutes, and the Horizontal refraction according to *Ticho* be thereabout, or fomewhat more, the lower limb of the Sun E, will be elevated to I; but becaufe, by his account, the point C will be elevated but 29. minutes, as having not fo great an inclination upon the inequality of the Air, therefore I G, which will be the apparent refracted perpendicular Diameter of the Sun, will be lefs then C G, which is but 29. minutes, and confequently fix or feven minutes fhorter then the unrefracted apparent Diameter. The parts, D and F, will be likewife elevated to H and K, whofe refraction, by reafon of its inclination, will be bigger then that of the point C, though lefs then that of E; therefore will the femidiameter I L, be fhorter then L G, and confequently the under fide of the appearing Sun more flat then the upper.

Now, becaufe the Rays from the right and left fides of the Sun, &c. have been obferv'd by *Ricciolo* and *Grimaldus*, to appear more diftant one from another then really they are, though (by very many Obfervations that I have made for that purpofe, with a very good *Telefcope*, fitted with a divided Ruler) I could never perceive any great alteration, yet there being really fome, it will not be amifs, to fhew that this alfo proceeds from the refraction or inflection of the *Atmofphere*; and this will be manifeft, if we confider the *Atmofphere* as a tranfparent Globe, or at leaft a tranfparent fhell, encompaffing an opacous Globe, which, being more denfe then the *medium* encompaffing it, refracts or inflects all the entring parallel Rays into a point or focus, fo that wherefoever the Obfervator is plac'd within the *Atmofphere*, between the focus and the luminous body, the *lateral* Rays muft neceffarily be more converg'd towards his eye by the refraction or inflection, then they would have been without it; and therefore the Horizontal Diameter of the luminous body muft neceffarily be augmented.

This might be more plainly manifeft to the eye by the fixth *Figure*; but becaufe it would be fomwhat tedious, and the thing being obvious enough

enough to be imagin'd by any one that attentively confiders it, I fhall rather omit it, and proceed to fhew, that the mafs of Air neer the furface of
theEarth,confifts,or is made up,of parcels,which do very much differ from
one another in point of denfity and rarity ; and confequently the Rays of
light that pafs through them will be varioufly inflected,here one way,and
there another,according as they pafs fo or fo through thofe differing parts;
and thofe parts being always in motion,either upwards or downwards,or
to the right or left, or in fome way compounded of thefe, they do by this
their motion inflect the Rays, now this way, and prefently that way.

This irregular,unequal and unconftant inflection of the Rays of light,
is the reafon why the limb of the *Sun, Moon, Jupiter, Saturn, Mars,* and
Venus, appear to wave or dance; and why the body of the Starrs appear
to tremulate or twinkle, their bodies,by this means, being fometimes magnify'd,and fometimes diminifhed; fometimes elevated, otherwhiles deprefs'd; now thrown to the right hand, and then to the left.

And that there is fuch a property or unequal diftribution of parts, is
manifeft from the various degrees of heat and cold that are found in the
Air; from whence will follow a differing denfity and rarity, both as to
quantity and refraction; and likewife from the vapours that are interpos'd, (which,by the way, I imagine,as to refraction or inflection, to do
the fame thing, as if they were rarify'd Air; and that thofe vapours that
afcend,are both lighter, and lefs denfe, then the ambient Air which boys
them up; and that thofe which defcend, are heavier and more denfe)
The firft of thefe may be found true, if you take a good thick piece of
Glafs,and heating it pretty hot in the fire, lay it upon fuch another piece
of Glafs, or hang it in the open Air by a piece of Wire, then looking
upon fome far diftant Object (fuch as a Steeple or Tree) fo as the Rays
from that Object pafs directly over the Glafs before they enter your eye,
you fhall find fuch a tremulation and wavering of the remote Object, as
will very much offend your eye: The like tremulous motion you may
obferve to be caus'd by the afcending fteams of Water, and the like.
Now, from the firft of thefe it is manifeft, that from the rarifaction of the
parts of the Air,by heat,there is caus'd a differing refraction,and from the
afcenfion of the more rarify'd parts of the Air, which are thruft up by the
colder, and therefore more condens'd and heavie, is caus'd an undulation or wavering of the Object; for I think, that there are very few
will grant, that Glafs, by as gentle a heat as may be endur'd by ones
hand, fhould fend forth any of its parts in fteams or vapours, which does
not feem to be much wafted by that violent fire of the green Glafs-houfe;
but, if yet it be doubted, let Experiment be further made with that body that is accounted, by Chymifts and others, the moft ponderous and
fix'd in the world; for by heating of a piece of Gold, and proceeding in
the fame manner, you may find the fame effects.

This trembling and fhaking of the Rays, is more fenfibly caus'd by an
actual flame, or quick fire, or any thing elfe heated glowing hot; as by
a Candle, live Coal, red-hot Iron, or a piece of Silver, and the like: the
fame alfo appears very confpicuous, if you look at an Object betwixt
which

which and your eye, the rifing fmoak of fome Chimney is interpos'd; which brings into my mind what I had once the opportunity to obferve, which was, the Sun rifing to my eye juft over a Chimney that fent forth a copious fteam of fmoak; and taking a fhort *Telefcope*, which I had then by me, I obferv'd the body of the Sun, though it was but juft peep'd above the Horizon, to have its underfide, not onely flatted, and prefs'd inward, as it ufually is when neer the Earth; but to appear more protuberant downwards then if it had fuffered no refraction at all; and befides all this, the whole body of the Sun appear'd to tremble or dance, and the edges or limb to be very ragged or indented, undulating or waving, much in the manner of a flag in the Wind.

This I have likewife often obferv'd in a hot Sunfhiny Summer's day, that looking on an Object over a hot ftone, or dry hot earth, I have found the Object to be undulated or fhaken, much after the fame manner. And if you look upon any remote Object through a *Telefcope* (in a hot Summer's day efpecially) you fhall find it likewife to appear tremulous. And further, if there chance to blow any wind, or that the air between you and the Object be in a motion or current, whereby the parts of it, both rarify'd and condens'd, are fwiftly remov'd towards the right or left, if then you obferve the Horizontal ridge of a Hill far diftant, through a very good *Telefcope*, you fhall find it to wave much like the Sea, and thofe waves will appear to pafs the fame way with the wind.

From which, and many other Experiments, tis cleer that the lower Region of the Air, efpecially that part of it which lieth neereft to the Earth, has, for the moft part, its conftituent parcels varioufly agitated, either by heat or winds, by the firft of which, fome of them are made more rare, and fo fuffer a lefs refraction; others are interwoven, either with afcending or defcending vapours; the former of which being more light, and fo more rarify'd, have likewife a lefs refraction; the latter being more heavie, and confequently more denfe, have a greater.

Now, becaufe that heat and cold are equally diffus'd every way; and that the further it is fpread, the weaker it grows; hence it will follow, that the moft part of the under Region of the Air will be made up of feveral kinds of *lentes*, fome whereof will have the properties of *Convex*, others of *Concave glaffes*; which, that I may the more intelligibly make out, we will fuppofe in the eighth *Figure* of the 37. *Scheme*, that A reprefents an afcending vapour, which, by reafon of its being fomewhat *Heterogeneous* to the ambient Air, is thereby thruft into a kind of Globular form, not any where terminated, but gradually finifhed, that is, it is moft rarity'd in the middle about A, fomewhat more condens'd about B B, more then that about C C; yet further, about D D, almoft of the fame denfity with the ambient Air about E E; and laftly, inclofed with the more denfe Air F F, fo that from A, to F F, there is a continual increafe of denfity. The reafon of which will be manifeft, if we confider the rifing vapour to be much warmer then the ambient heavie Air; for by the coldnefs of the ambient Air, the fhell E E will be more refrigerated then D D, and that then C C, which will be yet more then B B, and that

more

more then A ; so that from F to A, there is a continual increase of heat, and consequently of rarity ; from whence it will necessarily follow, that the Rays of light will be inflected or refracted in it, in the same manner as they would be in a *Concave-glass* ; for the Rays GKI, GKI will be inflected by GKH, GKH, which will easily follow from what I before explained concerning the inflection of the *Atmosphere*.

On the other side, a descending vapour, or any part of the air included by an ascending vapour, will exhibit the same effects with a *Convex lens* ; for, if we suppose, in the former Figure, the quite contrary constitution to that last describ'd ; that is, the ambient Air F F being hotter then any part of that matter within any circle, therefore the coldest part must necessarily be A, as being farthest remov'd from the heat, all the intermediate spaces will be gradually discriminated by the continuall mixture of heat and cold, so that it will be hotter at E E, then D D, in D D then C C, in C C then B B, and in B B then A. From which, a like refraction and condensation will follow ; and consequently a lesser or greater refraction, so that every included part will refract more then the including, by which means the Rays, GKI, GKI, coming from a Starr, or some remote Object, are so inflected, that they will again concurr and meet, in the point M. By the interposition therefore of this descending vapour the visible body of the Star, or other Object, is very much augmented, as by the former it was diminished.

From the quick consecutions of these two, one after another, between the Object and your eye, caused by their motion upwards or downwards, proceeding from their levity or gravity, or to the right or left, proceeding from the wind, a Starr may appear, now bigger, now less, then really it would otherwise without them ; and this is that property of a Starr, which is commonly call'd twinkling, or scintillation.

The reason why a Star will now appear of one colour, now of another, which for the most part happens when 'tis neer the Horizon, may very easily be deduc'd from its appearing now in the middle of the vapour, other whiles neer the edge ; for if you look against the body of a Starr with a *Telescope* that has a pretty deep *Convex* Eye-glass, and so order it, that the Star may appear sometimes in one place, and sometimes in another of it ; you may perceive this or that particular colour to be predominant in the apparent Figure of the Starr, according as it is more or less remote from the middle of the *Lens*. This I had here further explain'd, but that it does more properly belong to another place.

I shall therefore onely add some few Quæries, which the consideration of these particulars hinted, and so finish this Section.

And the first I shall propound is, Whether there may not be made an artificial transparent body of an exact Globular Figure that shall so inflect or refract all the Rays, that, coming from one point, fall upon any *Hemisphere* of it ; that every one of them may meet on the opposite side, and cross one another exactly in a point ; and that it may do the like also with all the Rays that, coming from a *lateral* point, fall upon any other *Hemisphere* ; for if so, there were to be hoped a perfection of *Dioptricks*, and

and a tranſmigration into heaven, even whil'ſt we remain here upon earth in the fleſh, and a deſcending or penetrating into the center and inner-moſt receſſes of the earth, and all earthly bodies; nay, it would open not onely a cranney, but a large window (as I may ſo ſpeak) into the Shop of Nature, whereby we might be enabled to ſee both the tools and opera-tors, and the very manner of the operation it ſelf of Nature; this, could it be effected, would as farr ſurpaſs all other kind of perſpectives as the vaſt extent of Heaven does the ſmall point of the Earth, which diſtance it would immediately remove, and unite them, as 'twere, into one, at leaſt, that there ſhould appear no more diſtance between them then the length of the Tube, into the ends of which theſe Glaſſes ſhould be n ſerted: Now, whether this may not be effected with parcels of Glaſs of ſeveral denſities, I have ſometimes proceeded ſo farr as to doubt (though in truth, as to the general, I have wholly deſpair'd of it) for I have often obſerv'd in Optical Glaſſes a very great variety of the parts, which are commonly called Veins; nay, ſome of them round enough (for they are for the moſt part, drawn out into ſtrings) to conſtitute a kind of *lens*.

This I ſhould further proceed to ope, had any one been ſo in-quiſitive as to have found out the way of making any tranſparent body, either more denſe or more rare; for then it might be poſſible to compoſe a Globule that ſhould be more denſe in the middle of it, then in any other part, and to compoſe the whole bulk, ſo as that there ſhould be a continual gradual tranſition from one degree of denſity to another; ſuch as ſhould be found requiſite for the deſired inflection of the *tranſmigra-ting* Rays; but of this enough at preſent, becauſe I may ſay more of it when I ſet down my own Trials concerning the melioration of *Dioptricks*, where I ſhall enumerate with how many ſeveral ſubſtances I have made both *Microſcopes*, and *Teleſcopes*, and by what and how many, ways: Let ſuch as have leiſure and opportunity farther conſider it.

The next Quæry ſhall be, whether by the ſame collection of a more denſe body then the other, or at leaſt, of the denſer part of the other, there might not be imagin'd a reaſon of the apparition of ſome new fix'd Stars, as thoſe in the Swan, *Caſſiope's Charr, Serpentarius, Piſcis, Ce-tus*, &c.

Thirdly, Whether it be poſſible to define the height of the *Atmoſphere* from this inflection of the Rays, or from the Quickſilver Experiment of the rarifaction or extenſion of the Air.

Fourthly, Whether the diſparity between the upper and under Air be not ſometimes ſo great, as to make a reflecting ſuperficies; I have had ſe-veral Obſervations which ſeem to have proceeded from ſome ſuch cauſe, but it would be too long to relate and examine them. An Experiment, alſo ſomewhat analogous to this, I have made with Salt-water and Freſh, which two liquors, in moſt Poſitions, ſeem'd the ſame, and not to be ſepa-rated by any determinate ſuperficies, which ſeparating ſurface yet in ſome other Poſitions did plainly appear.

And if ſo, Whether the reaſon of the equal bounding or *terminus* of the under parts of the clouds may not proceed from this cauſe ; whether,

ſecondly,

secondly, the Reason of the apparition of many Suns may not be found out, by confidering how the Rays of the Sun may so be reflected, as to defcribe a pretty true Image of the body, as we find them from any regular Superficies. Whether alfo this may not be found to caufe the apparition of fome of thofe *Parelii*, or counterfeit Suns, which appear coloured, by refracting the Rays fo, as to make the body of the Sun appear in quite another place then really it is. But of this more elfewhere.

5. Whether the *Phænomena* of the Clouds may not be made out by this diverfity of denfity in the upper and under parts of the Air, by fuppofing the Air above them to be much lighter then they themfelves are, and they themfelves to be yet lighter then that which is fubjacent to them, many of them feeming to be the fame fubftance with the Cobwebs that fly in the Air after a Fog.

Now that fuch a conftitution of the Air and Clouds, if fuch there be, may be fufficient to perform this effect, may be confirm'd by this Experiment.

Make as ftrong a Solution of Salt as you are able, then filling a Glafs of fome depth half full with it, fill the other half with frefh Water, and poyfe a little Glafs-bubble, fo as that it may fink pretty quick in frefh Water, which take and put into the aforefaid Glafs, and you fhall find it to fink till it comes towards the middle, where it will remain fixt, without moving either upwards or downwards. And by a fecond Experiment, of poifing fuch a bubble in water, whofe upper part is warmer, and confequently lighter, then the under, which is colder and heavier; the manner of which follows in this next Quæry, which is,

6. Whether the rarifaction and condenfation of Water be not made after the fame manner, as thofe effects are produc'd in the Air by heat; for I once pois'd a feal'd up Glafs-bubble fo exactly, that never fo fmall an addition would make it fink, and as fmall a detraction make it fwim, which fuffering to reft in that Veffel of Water for fome time, I alwayes found it about noon to be at the bottom of the Water, and at night, and in the morning, at the top : Imagining this to proceed from the Rarifaction of the Water, caus'd by the heat, I made tryal, and found moft true ; for I was able at any time, either to deprefs, or raife it, by heat and cold ; for if I let the Pipe ftand for fome time in cold water, I could eafily raife the Bubble from the bottom, whither I had a little afore detruded it, by putting the fame Pipe into warm Water. And this way I have been able, for a very confiderable time, to keep a Bubble fo poys'd in the Water, as that it fhould remain in the middle, and neither fink, nor fwim : For gently heating the upper part of the Pipe with a Candle, Coal, or hot Iron, till I perceived the Bubble begin to defcend, then forbearing, I have obferved it to defcend to fuch or fuch a ftation, and there to remain fufpended for fome hours, till the heat by degrees were quite vanifhed , when it would again afcend to its former place. This I have alfo often obferved naturally performed by the heat of the Air, which being able to rarifie the upper parts of the Water fooner then the lower, by reafon of its immediate contact, the heat of the Air

has fometimes fo flowly increafed, that I have obferved the Bubble to be fome hours in paffing between the top and bottom.

7. Whether the appearance of the *Pike* of *Tenerif*, and feveral other high Mountains, at fo much greater a diftance then feems to agree with their refpective heights , be not to be attributed to the *Curvature* of the vifual Ray, that is made by its paffing obliquely through fo differingly *Denfe* a Medium from the top to the eye very far diftant in the Horizon : For fince we have already, I hope, made it very probable , that there is fuch an *inflection* of the Rays by the differing denfity of the parts of the Air ; and fince I have found , by feveral Experiments made on places comparatively not very high, and have yet found the preffure fuftain'd by thofe parts of the Air at the top and bottom, and alfo their differing Expanfions very confiderable : Infomuch that I have found the preffure of the *Atmofphere* lighter at the top of St. *Paul's* Steeple in *London* (which is about two hundred foot high) then at the bottom by a fixtieth or fiftieth part, and the expanfion at the top greater then that at the bottom by neer about fo much alfo ; for the *Mercurial Cylinder* at the bottom was about 39. inches, and at the top half an inch lower ; the Air alfo included in the Weather-glafs,that at the bottom fill d only 155. fpaces, at the top fill'd 158. though the heat at the top and bottom was found exactly the fame with a fcal'd *Thermometer*: I think it very rational to fuppofe , that the greateft Curvature of the Rays is made neareft the Earth,and that the inflection of the Rays, above 3. or 4. miles upwards, is very inconfiderable, and therefore that by this means fuch calculations of the height of Mountains, as are made from the diftance they are vifible in the Horizon,from the fuppofal that that Ray is a ftraight Line (that from the top of the Mountain is, as'twere, a Tangent to the Horizon whence it is feen) which really is a *Curve*, is very erroneous. Whence, I fuppofe,proceeds the reafon of the exceedingly differing Opinions and Affertions of feveral Authors, about the height of feveral very high Hills.

8. Whether this Inflection of the Air will not very much alter the fuppofed diftances of the Planets, which feem to have a very great dependence upon the Hypothetical refraction or inflection of the Air, and that refraction upon the hypothetical height and denfity of the Air: For fince (as I hope) I have here fhewn the Air to be quite otherwife then has been hitherto fuppos'd, by manifefting it to be, both of a vaft , at leaft an uncertain, height, and of an unconftant and irregular denfity ; It muft neceffarily follow, that its inflection muft be varied accordingly : And therefore we may hence learn, upon what fure grounds all the Aftronomers hitherto have built, who have calculated the diftance of the Planets from their Horizontal *Parallax* ; for fince the Refraction and *Parallax* are fo nearly ally'd, that the one cannot be known without the other, efpecially by any wayes that have been yet attempted, how uncertain muft the *Parallax* be, when the Refraction is unknown? And how eafie is it for Aftronomers to affign what diftance they pleafe to the Planets,and defend them,when they have fuch a curious *fubterfuge* as that of Refraction,wherein a very little variation will allow them liberty enough to place the Celeftial Bodies at what diftance they pleafe.

If therefore we would come to any certainty in this point, we muſt go other wayes to work ; and as I have here examined the height and refractive property of the Air by other wayes then are uſual, ſo muſt we find the Parallax of the Planets by wayes not yet practiſed ; and to this end, I cannot imagine any better way, then the Obſervations of them by two perſons at very far diſtant parts of the Earth, that lye as neer as may be under the ſame Meridian, or Degree of longitude, but differing as much in latitude, as there can be places conveniently found : Theſe two perſons, at certain appointed times, ſhould (as near as could be) both at the ſame time, obſerve the way of the *Moon, Mars, Venus, Jupiter*, and *Saturn*, amongſt the fixt Stars, with a good large *Teleſcope*, and making little Iconiſmes, or pictures, of the ſmall fixed Stars, that appear to each of them to lye in or near the way of the Center of the Planet, and the exact meaſure of the apparent Diameter ; from the comparing of ſuch Obſervations together, we might certainly know the true diſtance, or Parallax, of the Planet. And having any one true Parallax of theſe Planets, we might very eaſily have the other by their apparent Diameters, which the *Teleſcope* likewiſe affords us very accurately. And thence their motions might be much better known, and their Theories more exactly regulated. And for this purpoſe I know not any one place more convenient for ſuch an Obſervation to be made in, then in the Iſland of St. *Helena*, upon the Coaſt of *Africk*, which lyes about ſixteen degrees to the Southwards of the Line, and is very near, according to the lateſt Geographical Maps, in the ſame Meridian with *London* ; for though they may not perhaps lye exactly in the ſame, yet their Obſervations, being ordered according to what I ſhall anon ſhew, it will not be difficult to find the true diſtance of the Planet. But were they both under the ſame Meridian, it would be much better.

And becauſe Obſervations may be much eaſier, and more accurately made with good *Teleſcopes*, then with any other Inſtruments, it will not, I ſuppoſe, ſeem impertinent to explain a little what wayes I judge moſt fit and convenient for that particular. Such therefore as ſhall be the Obſervators for this purpoſe, ſhould be furniſhed with the beſt *Teleſcopes* that can be had, the longer the better and more exact will their Obſervations be, though they are ſomewhat the more difficultly manag'd. Theſe ſhould be fitted with a *Rete*, or divided Scale, plac'd at ſuch a diſtance within the Eye-glaſs,that they may be diſtinctly ſeen,which ſhould be the meaſures of minutes and ſeconds ; by this Inſtrument each Obſervator ſhould, at certain prefixt times, obſerve the Moon, or other Planet, in, or very near, the Meridian ; and becauſe it may be very difficult to find two convenient ſtations that will happen to be juſt under the ſame Meridian, they ſhall, each of them, obſerve the way of the Planet, both for an hour before, and an hour after, it arrive at the Meridian ; and by a line, or ſtroke, amongſt the ſmall fixed Stars, they ſhall denote out the way that each of them obſerv'd the Center of the Planet to be mov'd in for thoſe two hours : Theſe Obſervations each of them ſhall repeat for many dayes together, that both it may happen, that both of
them

them may sometimes make their Observations together, and that from divers Experiments we may be the better assured of what certainty and exactness such kind of Observations are like to prove. And because many of the Stars which may happen to come within the compass of such an *Iconism*, or Map, may be such as are only visible through a good *Telescope*, whose Positions perhaps have not been noted, nor their longitudes, or latitudes, any where remarked; therefore each Observator should indeavour to insert some fixt Star, whose longitude, and latitude, is known; or with his *Telescope* he shall find the Position of some notable *telescopical* Star, inserted in his Map, to some known fixt Star, whose place in the *Zodiack* is well defin'd.

Having by this means found the true distance of the Moon, and having observed well the *apparent Diameter* of it at that time with a good *Telescope*, it is easie enough, by one single Observation of the apparent Diameter of the Moon with a good Glass, to determine her distances in any other part of her *Orbit*, or *Dragon*, and consequently, some few Observations will tell us, whether she be mov'd in an *Ellipsis*, (which, by the way, may also be found, even now, though I think we are yet ignorant of her true distance) and next (which without such Observations, I think, we shall not be sure of) we may know exactly the bigness of that *Ellipsis*, or Circle, and her true velocity in each part, and thereby be much the better inabled to find out the true cause of all her Motions. And though, even now also, we may, by such Observations in one station, as here at *London*, observe the *apparent Diameter* and motion of the Moon in her *Dragon*, and consequently be inabled to make a better ghess at the *Species* or kind of Curve, in which she is mov'd, that is, whether it be sphærical, or *elliptical*, or neither, and with what proportional velocities she is carried in that Curve; yet till her true *Parallax* be known, we cannot determine either.

Next, for the true distance of the Sun, the best way will be, by accurate Observations, made in both these forementioned stations, of some convenient Eclipse of the Sun, many of which may so happen, as to be seen by both; for the *Penumbra* of the Moon may, if she be sixty Semidiameters distant from the Earth, and the Sun above seven thousand, extend to about seventy degrees on the Earth, and consequently be seen by Observators as far distant as *London*, and St. *Helena*, which are not full sixty nine degrees distant. And this would much more accurately, then any way that has been yet used, determine the Parallax, and distance, of the Sun; for as for the Horizontal Parallax I have already shewn it sufficiently uncertain; nor is the way of finding it by the Eclipse of the Moon any other then hypothetical; and that by the difference of the true and apparent quadrature of the Moon is less not uncertain, witness their Deductions from it, who have made use of it; for *Vendeline* puts that difference to be but 4′. 30″. whence he deduces a vast distance of the Sun, as I have before shewn. *Ricciolo* makes it full 30′. 00. but *Reinoldus*, and *Kircher*, no less then three degrees. And no wonder, for if we examine the *Theory*, we shall find it so complicated with uncertainties.

Firſt,

First, From the irregular surface of the Moon, and from several Parallaxes, that unless the *Dichotomy* happen in the *Nonagesimus* of the *Eclip-tick*, and that in the Meridian, *&c.* all which happen so very seldom, that it is almost impossible to make them otherwise then uncertainly. Besides, we are not yet certain, but that there may be somewhat about the Moon *analogus* to the Air about the Earth, which may cause a refraction of the light of the Sun, and consequently make a great difference in the apparent *dichotomy* of the Moon. Their way indeed is very rational and ingenious ; and such as is much to be preferr'd before the way by the Horizontal Parallax, could all the uncertainties be remov'd, and were the true distance of the Moon known.

But because we find by the Experiments of *Vendiline*, *Reinoldus*, &c. that Observations of this kind are very uncertain also : It were to be wisht, that such kind of Observations, made at two very distant stations, were promoted. And it is so much the more desirable, because, from what I have now shewn of the nature of the Air, it is evident, that the refraction may be very much greater then all the Astronomers hitherto have imagined it : And consequently, that the distance of the Moon, and other Planets, may be much lesse then what they have hitherto made it

For first, this Inflection, I have here propounded, will allow the shadow of the Earth to be much shorter then it can be made by the other *Hypothesis* of refraction, and consequently, the Moon will not suffer an Eclipse, unless it comes very much nearer the Earth then the Astronomers hitherto have suppofed it.

Secondly, There will not in this *Hypothesis* be any other shadow of the Earth, such as *Kepler* supposes, and calls the *Penumbra*, which is the shadow of the refracting *Atmosphere* ; for the bending of the Rays being altogether caus'd by *Inflection*, as I have already shewn , all that part which is ascribed by *Kepler*, and others after him, to the *Penumbra*, or dark part, which is without the *umbra terræ*, does clear vanish ; for in this *Hypothesis* there is no refracting surface of the Air, and consequently there çan be no shadows, such as appear in the ninth Figure of the 37. *Scheme*, where let A B C D represent the Earth, and E F G H the *Atmosphere*, which according to *Keplers* supposition, is like a Sphære of Water terminated with an exact surface E F G H, let the lines M F, L B, I D, K H, represent the Rays of the Sun ; 'tis manifest, that all the Rayes between L B, and I D, will be reflected by the surface of the Earth B A D, and consequently, the conical space B O D would be dark and obscure ; but, say the followers of *Kepler*, the Rays between M F, and L B, and between I D, and K H, falling on the *Atmosphere* , are refracted, both at their ingress and egress out of the *Atmosphere*, nearer towards the Axis of the spærical shadow C O, and consequently, inlighten a great part of that former dark Cone, and shorten, and contract, its top to N. And because of this Reflection of these Rays, say they, there is superinduc'd another shell of a dark Cone F P H, whose Apex P is yet further distant from the Earth : By this *Penumbra*, say they, the Moon is Eclipſed, for it alwayes paſſes between the lines 1 2, and 3 4.

To which I ſay, That if the Air be ſuch, as I have newly ſhewn it to be, and conſequently cauſe ſuch an inflection of the Rays that fall into it, thoſe dark *Penumbra's* F Y Z Q, H X V T, and O R P S, will all vaniſh. For if we ſuppoſe the Air indefinitely extended, and to be no where bounded with a determinate refracting ſurface, as I have ſhewn it uncapable of having, from the nature of it ; it will follow, that the Moon will no where be totally obſcured, but when it is below the Apex N, of the dark blunt Cone of the Earth's ſhadow:Now,from the ſuppoſition, that the Sun is diſtant about ſeven thouſand Diameters, the point N, according to calculation, being not above twenty five terreſtrial Semidiameters from the Center of the Earth : It follows, that whenſoever the Moon eclipſed is totally darkned, without afferding any kind of light, it muſt be within twenty five Semidiameters of the Earth, and conſequently much lower then any Aſtronomers have hitherto put it.

This will ſeem much more conſonant to the reſt of the ſecundary Planets ; for the higheſt of *Jupiter's* Moons is between twenty and thirty *Jovial Semidiameters* diſtant from the Center of *Jupiter* ; and the Moons of *Saturn* much about the ſame number of *Saturnial Semidiameters* from the Center of that Planet.

But theſe are but conjectures alſo,and muſt be determin'd by ſuch kind of Obſervations as I have newly mention'd.

Nor will it be difficult, by this *Hypotheſis*, to ſalve all the appearances of Eclipſes of the Moon, for in this *Hypotheſis* alſo, there will be,on each ſide of the ſhadow of the Earth, a *Penumbra*, not caus'd by the Refraction of the Air, as in the *Hypotheſis* of *Kepler* ; but by the faint inlightning of it by the Sun : For if, in the ſixth Figure,we ſuppoſe E S Q, and G S R,to be the Rays that terminate the ſhadow from either ſide of the Earth ; E S Q coming from the upper limb of the Sun, and G S R from the under ; it will follow, that the ſhadow of the Earth, within thoſe Rays, that is, the Cone G S E, will be totally dark. But the Sun being not a point, but a large *area* of light, there will be a ſecondary dark Cone of ſhadow E P G, which will be caus'd by the earth's hindring part of the Rays of the Sun from falling on the parts G P R, and E P Q. of which halved ſhadow, or *Penumbra*, that part will appear brighteſt which lyes neareſt the terminating Rayes G P, and E P, and thoſe darker that lye neareſt to G S, and E S : when therefore the Moon appears quite dark in the middle of the Eclipſe,ſhe muſt be below S,that is, between S and F ; when ſhe appears lighter near the middle of the Eclipſe, ſhe muſt paſs ſome where between R Q and S ; and when ſhe is alike light through the whole Eclypſe, ſhe muſt paſs between R Q, and P.

Obſerv.

Obferv. LIX. *Of multitudes of fmall* Stars *difcoverable by the* Telefcope.

HAving, in the laft Obfervation, premis'd fome particulars obfervable in the *medium*, through which we muft look upon *Cæleftial* Objects, I fhall here add one Obfervation of the Bodies themfelves ; and for a *fpecimen* I have made choice of the *Pleiades*, or feven Stars, commonly fo called (though in our time and Climate there appear no more then fix to the naked eye) and this I did the rather, becaufe the defervedly famous *Galileo*, having publifht a Picture of this *Afterifme*, was able, it feems, with his Glafs to difcover no more then thirty fix, whereas with a pretty good twelve foot *Telefcope*, by which I drew this 38 *Iconifm*, I could very plainly difcover feventy eight, placed in the order they are ranged in the Figure, and of as many differing Magnitudes as the *Afterisks*, wherewith they are Marked, do fpecifie ; there being no lefs then fourteen feveral Magnitudes of thofe Stars, which are compris'd within the draught, the biggeft whereof is not accounted greater then one of the third Magnitude ; and indeed that account is much too big, if it be compared with other Stars of the third Magnitude, efpecially by the help of a *Telefcope* ; for then by it may be perceiv'd, that its fplendor, to the naked eye, may be fomewhat augmented by the three little Stars immediately above it, which are near adjoyning to it. The *Telefcope* alfo difcovers a great variety, even in the bignefs of thofe, commonly reckon'd, of the firft, fecond, third, fourth, fifth, and fixth Magnitude ; fo that fhould they be diftinguifh'd thereby, thofe fix Magnitudes would, at leaft, afford no lefs then thrice that number of Magnitudes, plainly enough diftinguifhable by their Magnitude, and brightnefs ; fo that a good twelve foot Glafs would afford us no lefs then twenty five feveral Magnitudes. Nor are thefe all, but a longer Glafs does yet further, both more nicely diftinguifh the Magnitudes of thofe already noted, and alfo difcover feveral other of fmaller Magnitudes, not difcernable by the twelve foot Glafs : Thus have I been able, with a good thirty fix foot Glafs, to difcover many more Stars in the *Pleiades* then are here delineated, and thofe of three or four diftinct Magnitudes lefs then any of thofe fpots of the fourteenth Magnitude. And by the twinkling of divers other places of this *Afterifme*, when the Sky was very clear, I am apt to think, that with longer Glaffes, or fuch as would bear a bigger *aperture*, there might be difcovered multitudes of other fmall Stars, yet inconfpicuous. And indeed, for the difcovery of fmall Stars, the bigger the *aperture* be, the better adapted is the Glafs ; for though perhaps it does make the feveral fpecks more radiant, and glaring, yet by that means, uniting more Rays very near to one point, it does make many of thofe radiant points confpi-

cuous,

cuous, which, by putting on a lefs *aperture*, may be found to vanifh; and therefore, both for the difcovery of the fixt Star, and for finding the *Satellites* of *Jupiter*, before it be out of the day, or twilight, I alwayes leave the Object-glafs as clear without any *aperture* as I can, and have thereby been able to difcover the *Satellites* a long while before; I was able to difcern them, when the fmaller *apertures* were put on; and at other times, to fee multitudes of other fmaller Stars, which a fmaller *aperture* makes to difappear.

In that notable *Afterifm* alfo of the Sword of *Orion*, where the ingenious Monfieur *Hugens van Zulichem* has difcovered only three little Stars in a clufter, I have with a thirty fix foot Glafs, without any *aperture* (the breadth of the Glafs being about fome three inches and a half) difcover'd five, and the twinkling of divers others up and down in divers parts of that fmall milky Cloud.

So that 'tis not unlikely, but that the meliorating of *Telefcopes* will afford as great a variety of new Difcoveries in the Heavens, as better *Microfcopes* would among fmall terreftrial Bodies, and both would give us infinite caufe, more and more to admire the omnipotence of the Creator.

Obferv. L X. *Of the* Moon.

HAving a pretty large corner of the Plate for the feven Starrs, void, for the filling it up, I have added one fmall *Specimen* of the appearance of the parts of the Moon, by defcribing a fmall fpot of it, which, though taken notice of, both by the Excellent *Hevelius*, and called *Mons Olympus* (though I think fomewhat improperly, being rather a vale) and reprefented by the Figure X, of the 38. *scheme*, and alfo by the Learn'd *Ricciolus*, who calls it *Hipparchus*, and defcribes it by the Figure Y, yet how far fhort both of them come of the truth, may be fomewhat perceiv'd by the draught, which I have here added of it, in the Figure Z, (which I drew by a thirty foot Glafs, in *October* 1664. juft before the Moon was half inlightned) but much better by the Reader's diligently obferving it himfelf, at a convenient time, with a Glafs of that length, and much better yet with one of threefcore foot long; for through thefe it appears a very fpacious Vale, incompaffed with a ridge of Hills, not very high in comparifon of many other in the Moon, nor yet very fteep. The Vale it felf A B C D, is much of the figure of a Pear, and from feveral appearances of it, feems to be fome very fruitful place, that is, to have its furface all covered over with fome kinds of vegetable fubftances; for in all pofitions of the light on it, it feems to give a much fainter reflection then the more barren tops of the incompaffing Hills, and thofe a much fainter then divers other cragged, chalky, or rocky Mountains of the Moon. So that I am not unapt to think, that the Vale may have

Vegetables

Vegetables *analogus* to our Grafs, Shrubs, and Trees ; and moft of thefe incompaffing Hills may be covered with fo thin a vegetable Coat, as we may obferve the Hills with us to be, fuch as the fhort Sheep pafture which covers the Hills of *Salisbury* Plains.

Up and down in feveral parts of this place here defcrib'd (as there are multitudes in other places all over the furface of the Moon) may be perceived feveral kinds of pits, which are fhap'd almoft like a difh, fome bigger, fome lefs, fome fhallower, fome deeper, that is, they feem to be a hollow *Hemifphere*, incompaffed with a round rifing bank, as if the fubftance in the middle had been digg'd up, and thrown on either fide. Thefe feem to me to have been the effects of fome motions within the body of the Moon, *analogus* to our Earthquakes, by the eruption of which, as it has thrown up a brim, or ridge,round about, higher then the Ambient furface of the Moon, fo has it left a hole, or depreffion, in the middle, proportionably lower ; divers places refembling fome of thefe, I have obferv'd here in *England*, on the tops of fome Hills, which might have been caus'd by fome Earthquake in the younger dayes of the world. But that which does moft incline me to this belief, is, firft, the generality and diverfity of the Magnitude of thefe pits all over the body of the Moon. Next, the two experimental wayes, by which I have made a reprefentation of them.

The firft was with a very foft and well temper'd mixture of Tobacco-pipe clay and Water, into which, if I let fall any heavy body, as a Bullet, it would throw up the mixture round the place, which for a while would make a reprefentation, not unlike thefe of the Moon ; but confidering the ftate and condition of the Moon, there feems not any probability to imagine, that it fhould proceed from any caufe *analogus* to this ; for it would be difficult to imagine whence thofe bodies fhould come ; and next, how the fubftance of the Moon fhould be fo foft ; but if a Eubble be blown under the furface of it, and fuffer'd to rife, and break ; or if a Bullet, or other body, funk in it, be pull'd out from it, thefe departing bodies leave an impreffion on the furface of the mixture, exactly like thefe of the Moon, fave that thefe alfo quickly fubfide and vanifh. But the fecond, and moft notable, reprefentation was, what I obferv'd in a pot of boyling Alabafter, for there that powder being by the eruption of vapours reduc'd to a kind of fluid confiftence, if,whil'ft it boyls, it be gently remov'd befides the fire , the Alabafter prefently ceafing to boyl, the whole furface, efpecially that where fome of the laft Bubbles have rifen, will appear all over covered with fmall pits, exactly fhap'd like thefe of the Moon, and by holding a lighted Candle in a large dark Room, in divers pofitions to this furface, you may exactly reprefent all the *Phænomena* of thefe pits in the Moon, according as they are more or lefs inlightned by the Sun.

And that there may have been in the Moon fome fuch motion as this, which may have made thefe pits, will feem the more probable, if we fuppofe it like our Earth, for the Earthquakes here with us feem to proceed from fome fuch caufe, as the boyling of the pot of Ala-

bafter,

bafter, there feeming to be generated in the Earth from fome fubterraneous fires, or heat, great quantities of vapours, that is, of expanded aerial fubftances, which not prefently finding a paffage through the ambient parts of the Earth, do, as they are increafed by the fupplying and generating principles, and thereby (having not fufficient room to expand themfelves) extreamly condens'd, at laft overpower, with their *elaftick* properties, the refiftence of the incompaffing Earth, and lifting it up, or cleaving it, and fo fhattering of the parts of the Earth above it, do at length, where they find the parts of the Earth above them more loofe, make their way upwards, and carrying a great part of the Earth before them, not only raife a fmall brim round about the place, out of which they break, but for the moft part confiderable high Hills and Mountains, and when they break from under the Sea, divers times, mountainous Iflands; this feems confirm d by the *Vulcans* in feveral places of the Earth, the mouths of which, for the moft part, are incompaffed with a Hill of a confiderable height, and the tops of thofe Hills, or Mountains, are ufually fhap'd very much like thefe pits, or difhes, of the Moon : Inftances of this we have in the defcriptions of *Ætna* in *Sicily*, of *Hecla* in *Iceland*, of *Tenerif* in the *Canaries*, of the feveral *Vulcans* in *New-Spain*, defcrib'd by *Gage*, and more efpecially in the eruption of late years in one of the *Canary* Iflands. In all of which there is not only a confiderable high Hill raifed about the mouth of the *Vulcan*, but, like the fpots of the Moon, the top of thofe Hills are like a difh, or bafon. And indeed, if one attentively confider the nature of the thing, one may find fufficient reafon to judge, that it cannot be otherwife; for thefe eruptions, whether of fire, or fmoak, always rayfing great quantities of Earth before them, muft neceffarily, by the fall of thofe parts on either fide, raife very confiderable heaps.

Now, both from the figures of them, and from feveral other circumftances; thefe pits in the Moon feem to have been generated much after the fame manner that the holes in Alabafter, and the *Vulcans* of the Earth are made. For firft, it is not improbable, but that the fubftance of the Moon may be very much like that of our Earth, that is, may confift of an earthy, fandy, or rocky fubftance, in feveral of its fuperficial parts, which parts being agitated, undermin'd, or heav'd up, by eruptions of vapours, may naturally be thrown into the fame kind of figured holes, as the fmall duft, or powder of Alabafter. Next, it is not improbable, but that there may be generated, within the body of the Moon, divers fuch kind of internal fires and heats, as may produce fuch Exhalations; for fince we can plainly enough difcover with a *Telefcope*, that there are multitudes of fuch kind of eruptions in the body of the Sun it felf, which is accounted the moft noble Ætherial body, certainly we need not be much fcandaliz'd at fuch kind of alterations, or corruptions, in the body of this lower and lefs confiderable part of the univerfe, the Moon, which is only fecundary, or attendant, on the bigger, and more confiderable body of the Earth. Thirdly, 'tis not unlikely, but that fuppofing fuch a fandy or mouldring fubftance to

be

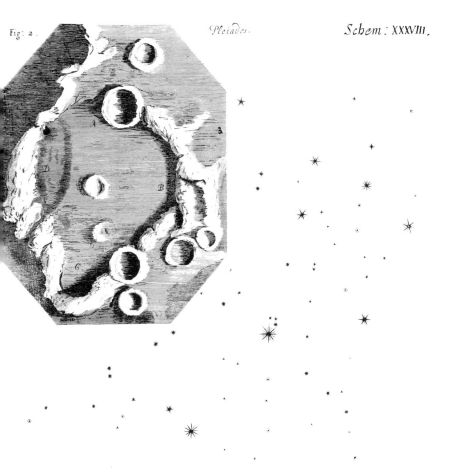

Fig: X.

Fig: Y.

Stellarum magnitudines

1   2   3   4   5   6   7   8   9 10 11 12 13 14

be there found, and suppofing alfo a poffibility of the generation of the internal *elaftical* body (whether you will call it air or vapours) 'tis not unlikely, I fay, but that there is in the Moon a principle of gravitation, fuch as in the Earth. And to make this probable, I think, we need no better Argument, then the roundnefs, or globular Figure of the body of the Moon it felf, which we may perceive very plainly by the *Telefcope*, to be (bating the fmall inequality of the Hills and Vales in it, which are all of them likewife fhap'd, or levelled, as it were, to anfwer to the center of the Moons body) perfectly of a Sphærical figure, that is, all the parts of it are fo rang'd (bating the comparitively fmall ruggednefs of the Hills and Dales) that the outmoft bounds of them are equally diftant from the Center of the Moon, and confequently, it is exceedingly probable alfo, that they are equidiftant from the Center of gravitation; and indeed, the figure of the fuperficial parts of the Moon are fo exactly fhap'd, according as they fhould be, fuppofing it had a gravitating principle as the Earth has, that even the figure of thofe parts themfelves is of fufficient efficacy to make the gravitation, and the other two fuppofitions probable: fo that the other fuppofitions may be rather prov'd by this confiderable Circumftance, or Obfervation, then this fuppos'd Explication can by them; for he that fhall attentively obferve with an excellent *Telefcope*, how all the Circumftances, notable in the fhape of the fuperficial parts, are, as it were, exactly adapted to fuit with fuch a principle, will, if he well confiders the ufual method of Nature in its other proceedings, find abundant argument to believe it to have really there alfo fuch a principle; for I could never obferve, among all the mountainous or prominent parts of the Moon (whereof there is a huge variety) that any one part of it was plac'd in fuch a manner, that if there fhould be a gravitating, or attracting principle in the body of the Moon, it would make that part to fall, or be mov'd out of its vifible pofture. Next, the fhape and pofition of the parts is fuch, that they all feem put into thofe very fhapes they are in by a gravitating power: For firft, there are but very few clifts, or very fteep declivities in the afcent of thefe Mountains; for befides thofe Mountains, which are by *Hevelius* call'd the *Apennine* Mountains, and fome other, which feem to border on the Seas of the Moon, and thofe only upon one fide, as is common alfo in thofe Hills that are here on the Earth; there are very few that feem to have very fteep afcents, but, for the moft part, they are made very round, and much refemble the make of the Hills and Mountains alfo of the Earth; this may be partly perceived by the Hills incompaffing this Vale, which I have here defcrib'd; and as on the Earth alfo, the middlemoft of thefe Hills feems the higheft, fo is it obvious alfo, through a good *Telefcope*, in thofe of the Moon; the Vales alfo in many are much fhap'd like thofe of the Earth, and I am apt to think, that could we look upon the Earth from the Moon, with a good *Telefcope*, we might eafily enough perceive its furface to be very much like that of the Moon.

Now whereas in this fmall draught, (as there would be multitudes if the whole Moon were drawn after this manner) there are feveral little
Ebullitions,

Ebullitions, or Dishes, even in the Vales themselves, and in the incompassing Hills also ; this will, from this suppossition, (which I have, I think, upon very good reason taken) be exceeding easily explicable ; for, as I have several times also obseri'd, in the surface of Alabaster so ordered, as I before describ'd, so may the later eruptions of vapours be even in the middle, or on the edges of the former ; and other succeeding these also in time may be in the middle or edges of these, &c. of which there are Instances enough in divers parts of the body of the Moon, and by a boyling pot of Alabaster will be sufficiently exemplifi'd.

To conclude therefore, it being very probable, that the Moon has a principle of gravitation, it affords an excellent distinguishing Instance in the search after the cause of gravitation, or attraction, to hint, that it does not depend upon the diurnal or turbinated motion of the Earth, as some have somewhat inconsiderately supposed and affirmed it to do ; for if the Moon has an attractive principle, whereby it is not only shap'd round, but does firmly contain and hold all its parts united, though many of them seem as loose as the sand on the Earth, and that the Moon is not mov'd about its Center ; then certainly the turbination cannot be the cause of the attraction of the Earth ; and therefore some other principle must be thought of, that will agree with all the secundary as well as primary Planets. But this, I confess, is but a probability, and not a demonstration, which (from any Observation yet made) it seems hardly capable of, though how successful future indeavours (promoted by the meliorating of Glasses, and observing particular circumstances) may be in this, or any other, kind, must be with patience expected.

F I N I S.

THE TABLE.

The TABLE.

L l

The Table.

The TABLE.

Observ.

The TABLE.

Obſerv.

The TABLE.

The TABLE.

of

The TABLE.

ERRATA.

E R R A T A.

IN the Preface, Page 7. line 18. read *feet :* line 24. read *Gilbert, Harvy.*
   Page 13. line ult. read *taste :* p. 34. l. 18. r. *small lens :* l. penult. r. *that proceeds from :* p. 40. l. 44. r. *when you:* p. 48. l. 34. r. *broadest:* p. 57. l. 39. dele *be:* p. 62. l. 36. r. *water-drop :* p. 64. l. 9. r. *duction of G A C H : l.* 35. r. *impressions :* p. 96. l. 33. r. *compose :* p. 100. l. 11. r. *Mersennus :* p. 106 l. 8. r. *extreamly :* p. 110 l. 8. r. *as :* l. 12. r. *those :* p. 112. l. 32. r. *Aldronandus, Wormius :* p. 121. l. 9. dele *of :* p. 128. l. 43, dele *from:* p. 129. l. 18. r. *fifth place :* p. 130. l. 29. r. *Aerial menstruum :* p. 136. l. 39. r. *knew how :* p. 144. l. 2. r. *parts of the :* p. 147. l. 36. r. *look'd on :* p. 161. l. 13. r. *body :* p. 162. l. 17. dele *only :* p. 166. l. 11. r. 22 : l. 12. dele the Semicolon: l. 17. r. *place :* p. 167. l. 40. r. 22 : p. 172. l. 18. r. *and first for the :* p. 198. l. 17. r. *and an artific.* p. 215, l. ult. r. *and from the :* p. 221. l. 4. r. *whence the under :* p. 234. l. 18. r. *to hope .* p. 238. l. 42. r. *is not less :* p. 240. l. 19. r. *Moon.*

INDEX

Reprinted from the 1780 abridgment.

F I N I S.

SUPPLEMENTARY INDEX

Micrographia *had no alphabetical Index, but was furnished with a very useful paginated Table of Contents printed on the preceding ten pages 247 to 256.*

An excellent analytical Index was added to Micrographia Restaurata *of 1745 and 1780, reprinted on pages 257 to 270. To this the present Index is supplementary.*